Cellular Adhesion
Molecular Definition to
Therapeutic Potential

NEW HORIZONS IN THERAPEUTICS

Smithkline Beecham Pharmaceuticals U.S. Research Symposia Series

Series Editors: Brian W. Metcalf and George Poste
SmithKline Beecham Pharmaceuticals, Philadelphia, Pennsylvania

---

Cellular Adhesion
Molecular Definition to Therapeutic Potential

Edited by

BRIAN W. METCALF
BARBARA J. DALTON *and*
GEORGE POSTE
SmithKline Beecham Pharmaceuticals
Philadelphia, Pennsylvania

Technical Editor:
JUDY SCHATZ

PLENUM PRESS • NEW YORK AND LONDON

Library of Congress Cataloging-in-Publication Data

Cellular adhesion : molecular definition to therapeutic potenial /
  edited by Brian W. Metcalf, Barbara J. Dalton, and George Poste.
     p.   cm. -- (New horizons in therapeutics)
  Includes bibliographical references and index.
  ISBN 0-306-44685-5
  1. Cell adhesion molecules--Physiological effect. 2. Cell
adhesion molecules--Therapeutic use. 3. Cell adhesion.
I. Metcalf, Brian W. II. Dalton, Barbara J. III. Poste, George.
IV. Series.
  [DNLM: 1. Cell Adhesion--physiology. 2. Cell Adhesion Molecules-
-physiology. 3. Integrins--physiology.   QH 623 C3936 1994]
QP552.C42C46 1994
591.87'6--dc20
DNLM/DLC
for Library of Congress   94-27164
                                                              CIP

ISBN 0-306-44685-5

©1994 Plenum Press, New York
A Division of Plenum Publishing Corporation
233 Spring Street, New York, N.Y. 10013

Printed in the United States of America

Contributors

Omid Abbassi, Department of Pediatrics, Baylor College of Medicine, Houston, Texas 77030

David H. Adams, The Experimental Immunology Branch, National Cancer Institute, National Institutes of Health, Bethesda, Maryland 20892

Fadia Ali, Department of Medicinal Chemistry, SmithKline Beecham Pharmaceuticals, King of Prussia, Pennsylvania 19406

John Bean, Department of Medicinal Chemistry, SmithKline Beecham Pharmaceuticals, King of Prussia, Pennsylvania 19406

Donald R. Bertolini, Department of Cellular Biochemistry, SmithKline Beecham Pharmaceuticals, King of Prussia, Pennsylvania 19406

James Callahan, Department of Medicinal Chemistry, SmithKline Beecham Pharmaceuticals, King of Prussia, Pennsylvania 19406

David A. Cheresh, Department of Immunology, The Scripps Research Institute, La Jolla, California 92037

J. W. Costerton, Department of Biological Sciences, University of Calgary, Calgary, Alberta, Canada T2N 1N4. Present address: Center for Biofilm Engineering, Montana State University, Bozeman, Montana 59717

Kathryn L. Crossin, Department of Neurobiology, The Scripps Research Institute, La Jolla, California 92037

Mark L. Entman, Section of Cardiovascular Sciences, Department of Medicine, Baylor College of Medicine, Houston, Texas 77030

David V. Erbe, Department of Immunology, Genentech, Inc., South San Francisco, California 94080

Lynne Flaherty, Department of Surgery, Harborview Medical Center, University of Washington, Seattle, Washington 98104

John M. Harlan, Department of Medicine, Harborview Medical Center, University of Washington, Seattle, Washington 98104

William Huffman, Department of Medicinal Chemistry, SmithKline Beecham Pharmaceuticals, King of Prussia, Pennsylvania 19406

David Jones, Department of Chemical Engineering, Rice University, Houston, Texas 77030

Kenneth Kopple, Department of Physical and Structural Chemistry, SmithKline Beecham Pharmaceuticals, King of Prussia, Pennsylvania 19406

Paul F. Koster, Department of Cardiovascular Pharmacology, SmithKline Beecham Pharmaceuticals, King of Prussia, Pennsylvania 19406

Gilbert L. Kukielka, Section of Cardiovascular Sciences, The Methodist Hospital, The DeBakey Heart Center, Department of Medicine, and Speros P. Martel Laboratory of Leukocyte Biology, Department of Pediatrics, Texas Children's Hospital, Baylor College of Medicine, Houston, Texas 77030

Laurence A. Lasky, Department of Immunology, Genentech, Inc., South San Francisco, California 94080

D. Euan MacIntyre, Department of Cellular and Molecular Immunology, Merck Research Laboratories, Rahway, New Jersey 07065

Rodger McEver, Department of Medicine and Biochemistry, University of Oklahoma Health Sciences Center, Oklahoma City, Oklahoma 73104

L. V. McIntire, Department of Chemical Engineering, Rice University, Houston, Texas 77030

Michele Mariscalco, Department of Pediatrics, Baylor College of Medicine, Houston, Texas 77030

George E. Mark, III, Department of Cellular and Molecular Biology, Merck Research Laboratories, Rahway, New Jersey 07065

Michael S. Mulligan, Department of Pathology, The University of Michigan, Ann Arbor, Michigan 48109

Andrew J. Nichols, Department of Cardiovascular Pharmacology, SmithKline Beecham Pharmaceuticals, King of Prussia, Pennsylvania 19406

Eduardo A. Padlan, National Institute of Diabetes and Digestive and Kidney Diseases, National Institutes of Health, Bethesda, Maryland 20892

Catherine Peishoff, Department of Physical and Structural Chemistry, SmithKline Beecham Pharmaceuticals, King of Prussia, Pennsylvania 19406

Leonard G. Presta, Department of Immunology, Genentech, Inc., South San Francisco, California 94080

Anne L. Prieto, Department of Neurobiology, The Scripps Research Institute, La Jolla, California 92037

Martin Ringwald, Max-Planck Institute for Immunobiology, Division of Molecular Embryology, W-7800 Freiburg, Germany. *Present address:* The Jackson Laboratory, Bar Harbor, Maine 04609

James M. Samanen, Department of Medicinal Chemistry, SmithKline Beecham Pharmaceuticals, King of Prussia, Pennsylvania 19406

Stephen Shaw, The Experimental Immunology Branch, National Cancer Institute, National Institutes of Health, Bethesda, Maryland 20892

Melvin Silberklang, Department of Cellular and Molecular Biology, Merck Research Laboratories, Rahway, New Jersey 07065

Irwin I. Singer, Department of Biochemical and Molecular Pathology, Merck Research Laboratories, Rahway, New Jersey 07065

C. Wayne Smith, Departments of Pediatrics and Microbiology and Immunology, Baylor College of Medicine, Houston, Texas 77030

K. B. Tan, Department of Cellular Biochemistry, SmithKline Beecham Pharmaceuticals, King of Prussia, Pennsylvania 19406

Yoshiya Tanaka, The Experimental Immunology Branch, National Cancer Institute, National Institutes of Health, Bethesda, Maryland 20892

Richard E. Valocik, Department of Cardiovascular Pharmacology, SmithKline Beecham Pharmaceuticals, King of Prussia, Pennsylvania 19406

Janice A. Vasko, Department of Cardiovascular Pharmacology, SmithKline Beecham Pharmaceuticals, King of Prussia, Pennsylvania 19406

Peter A. Ward, Department of Pathology, The University of Michigan, Ann Arbor, Michigan 48109

William I. Weis, Department of Biochemistry and Molecular Biophysics and Howard Hughes Medical Institute, Columbia University, New York, New York 10032. *Present address:* Department of Cell Biology, Stanford University School of Medicine, Stanford, California 94305

Robert K. Winn, Department of Surgery and Physiology–Biophysics, Harborview Medical Center, University of Washington, Seattle, Washington 98104

Samuel D. Wright, The Laboratory of Cellular Physiology and Immunology, The Rockefeller University, New York, New York 10021

Preface to the Series

The unprecedented scope and pace of discovery in modern biology and clinical medicine present remarkable opportunities for the development of new therapeutic modalities, many of which would have been unimaginable even a few years ago. This situation reflects the unprecedented progress being made not only in disciplines such as pharmacology, physiology, organic chemistry, and biochemistry that have traditionally made important contributions to drug discovery, but also in new disciplines such as molecular genetics, gene medicine, cell biology, and immunology that are now of sufficient maturity to further our understanding of the pathogenesis of disease and the development of novel therapies. Contemporary biomedical research, embracing the entire spectrum of biological organization from the molecular level to whole body function, is on the threshold of an era in which biological processes, including disease, can be analyzed in increasingly precise and mechanistic terms. The transformation of biology from a largely descriptive, phenomenological discipline to one in which the regulatory principles underlying biological organization can be understood and manipulated with ever-increasing predictability brings an entirely new dimension to the study of disease and the search for effective therapeutic modalities. In undergoing this transformation into an increasingly mechanistic discipline, biology and medicine are following the course already charted by the sister disciplines of chemistry and physics, albeit still far behind.

The consequences of these changes for biomedical research are profound: new concepts; new and increasingly powerful analytical techniques; new advances generated at a seemingly ever-rapid pace; an almost unmanageable glut of information dispersed in an increasing number of books and journals; and the task of integrating this information into a realistic experimental framework. Nowhere is the challenge more pronounced than in the pharmaceutical industry. Drug discovery and development have always required the successful coordination of multiple scientific disciplines. The need to assimilate more and more disciplines within the drug discovery process, the extraordinary pace of discovery in all disciplines, and the growing scientific and organizational complexity of coordinating increasingly ultra-specialized and resource-intensive scientific skills in an ever-enlarging framework of collaborative research activities represent formidable challenges for the pharmaceutical industry. These demands are balanced, however, by the excitement and the scale of the potential opportunities for achieving dramatic improvements in health care and the quality of human life over the next twenty years via the development of novel therapeutic modalities for effective treatment of major human and animal diseases.

It is against this background of change and opportunity that this symposium series, *New Horizons in Therapeutics*, was conceived as a forum for providing critical and up-to-date surveys of important topics in biomedical research in which significant advances are occurring and which offer new approaches to the therapy of disease. Each volume contains authoritative and topical articles written by investigators who have contributed significantly to their respective research fields. While individual articles discuss specialized topics, all papers in a single volume are related to a common theme. The level is advanced, directed primarily to the needs of the active research investigator and graduate students.

Editorial policy is to impose as few restrictions as possible on contributors. This is appropriate, since each volume is limited to the papers presented at the symposium and no attempt will be made to create a definitive monograph dealing with all aspects of the selected subject. Although each symposium volume provides a survey of recent research accomplishments, emphasis is also given to the examination of controversial and conflicting issues, to the presentation of new ideas and hypotheses, to the identification of important unsolved questions, and to future directions and possible approaches by which such questions might be answered.

The range of topics for the volumes in the symposium series is broad, and embraces the full repertoire of scientific disciplines that contribute to modern drug discovery and development. We thus look forward to the publication of another volume in what we hope is viewed as a worthy series of volumes that reflects the excitement and challenge of contemporary biomedical research in defining new horizons in therapeutics.

Brian W. Metcalf
George Poste

Philadelphia

Preface

This volume focuses on recent developments in our understanding of selected adhesion processes that may offer new approaches to developing therapeutics for a variety of diseases. The volume first introduces the molecules involved in key adhesive processes, then describes the biological consequences of several adhesive interactions, and closes with a description of the initial therapeutic approaches to antagonizing adhesion. These papers were originally presented at the SmithKline Beecham Pharmaceuticals Seventh U.S. Research Symposium held in Philadelphia in October of 1992.

In its simplest sense, cellular adhesion is the adherence of a cell to a surface. The cells of interest in the context of this volume are bacterial and mammalian, with an emphasis on leukocytes; surfaces can be other cells and tissues (such as bone), matrix proteins, or inanimate objects such as in-dwelling medical devices and catheters. Interaction between adhesion molecules usually results in a specific biological response.

Adhesion is a form of cellular communication, and represents the way a cell senses its environment through contact. Like hormones and cytokines, the soluble mediators used by cells for communication, adhesion molecules are defined molecular entities that recognize specific receptor structures on the surface to which they adhere. Recent activity has focused on defining the structure of the individual molecules responsible for many types of cellular adhesion. New adhesion proteins are being cloned with the help of specific antibodies and precise functional assays. Their counter-receptors are also being described in great molecular detail, so that we may begin to understand the precise molecular interactions between adhesive structures. Carbohydrates, with their tremendous molecular diversity, have recently been described as active participants in several adhesive processes. All of this work has led to an increased understanding of the importance of adhesion molecules and their function in biological systems, and provides a whole new series of targets for therapeutic intervention.

The roles of adhesion molecules are numerous, as evidenced by the following list of adhesion functions: in the bacterial world, adherence is required for colonization; in animals, it is required for the development of multicellular organisms and specific tissues within those organisms. Platelet aggregation is a response to vessel injury; the generation of an immune response requires intimate and specific contact; leukocyte attachment to endothelium and extravascular cells and matrix stimulates cells on the path to sites of tissue injury, infection, and inflammation; and the continual remodeling of bone is dependent upon the formation of tight junctions between the calcified bone matrix and the cells responsible for its reshaping and development. These biological processes will be targeted to test the therapeutic utility of the first inhibitors/regulators of adhesion molecule function, and will pave the way for more advances in the future.

Brian W. Metcalf
Barbara J. Dalton
George Poste

Philadelphia

Contents

Chapter 2

Integrin Modulating Factor and the Regulation of Leukocyte Integrins

Samuel D. Wright

Chapter 3

Structure–Function Aspects of Selectin–Carbohydrate Interactions

Laurence A. Lasky, Leonard G. Presta, and David V. Erbe

Chapter 4

Recognition of Cell Surface Carbohydrates by C-Type Animal Lectins

William I. Weis

Chapter 5

The Uvomorulin–Catenin Complex: Insights into the
Structure and Function of Cadherins

Martin Ringwald

Chapter 6

Cytotactin: A Substrate Adhesion Molecule with
Amphitropic Functions in Morphogenesis

Kathryn L. Crossin and Anne L. Prieto

Chapter 7

Adhesion Molecules and Bone Remodeling

Donald R. Bertolini and K. B. Tan

Part II. BIOLOGICAL CONSEQUENCES OF CELLULAR ADHESION

Chapter 8

Effects of Shear Stress on Leukocyte Adhesion

Omid Abbassi, David Jones, Michele Mariscalco, Rodger McEver,
L. V. McIntire, and C. Wayne Smith

Chapter 9

Blockade of Leukocyte Adhesion in in Vivo *Models of Inflammation*

Lynne Flaherty, John M. Harlan, and Robert K. Winn

Chapter 10

*Proadhesive Cytokine Immobilized on Endothelial Proteoglycan:
A New Paradigm for Recruitment of T Cells*

Yoshiya Tanaka, David H. Adams, and Stephen Shaw

Chapter 11

Lung Inflammation and Adhesion Molecules

Michael S. Mulligan and Peter A. Ward

Chapter 12

Adhesion Molecule-Dependent Cardiovascular Injury

Gilbert L. Kukielka and Mark L. Entman

Chapter 13

GPIIb/IIIa Antagonists as Novel Antithrombotic Drugs:
Potential Therapeutic Applications

Andrew J. Nichols, Janice A. Vasko, Paul F. Koster,
Richard E. Valocik, and James M. Samanen

Part III. TARGETING ADHESION MOLECULES—THERAPEUTIC POTENTIAL

Chapter 14

The Pathological Consequences of Bacterial Adhesion to Medical Devices:
A Practical Solution to the Problem of Device-Related Infections

J. W. Costerton

Chapter 15

Peptide Mimetics as Adhesion Molecule Antagonists

James Samanen, Fadia Ali, John Bean, James Callahan, William Huffman, Kenneth Kopple, Catherine Peishoff, and Andrew Nichols

Chapter 16

Derivation of Therapeutically Active Humanized and Veneered Anti-CD18 Antibodies

George E. Mark III, Melvin Silberklang, D. Euan MacIntyre, Irwin I. Singer, and Eduardo A. Padlan

I

MOLECULAR DEFINITION— ADHESION MOLECULE STRUCTURE AND FUNCTION

The Biological Function of β3 Integrins and Other Vitronectin Receptors

DAVID A. CHERESH

1. Introduction

Integrins are a family of cell surface molecules mediating diverse biological events involving cell–matrix as well as cell–cell interactions. The term *integrin* denotes a functional linkage between the extracellular matrix and the cells' interior, thus providing cellular responsiveness to the extracellular environment (Tamkun *et al.*, 1986; Buck and Horowitz, 1987). Integrin–ligand specificity is conferred by the subunit composition of the integrin heterodimer (a) complex (Hynes, 1987; Ruoslahti and Pierschbacher, 1987). In some cases, a given integrin heterodimer may recognize multiple ligands, or multiple integrins may recognize common ligands. The biological function underlying this intrinsic multiplicity and redundancy in ligand–integrin interaction is not well understood.

The αv integrin subunit is unique in that it associates with more than two β subunits, including β1 (Bodary and McLean, 1990; Vogel *et al.*, 1990), β3 (Cheresh and Spiro, 1987; Cheresh *et al.*, 1989a), β5 (Cheresh *et al.*, 1989b), β6 (Sheppard *et al.*, 1990), and β8 (Moyle *et al.*, 1991). While the αvβ1 and αvβ5 heterodimers appear to be restricted in their ligand binding specificity, the αvβ3 heterodimer interacts with multiple ligands in an Arg-Gly-Asp (RGD)-dependent

DAVID A. CHERESH • Department of Immunology, The Scripps Research Institute, La Jolla, California 92037.

Cellular Adhesion: Molecular Definition to Therapeutic Potential, edited by Brian W. Metcalf, Barbara J. Dalton, and George Poste. Plenum Press, New York, 1994.

manner (Cheresh, 1987; Cheresh and Spiro, 1987; Charo *et al.*, 1987). We recently reported that while both $\alpha v \beta 3$ and $\alpha v \beta 5$ contribute to melanoma and carcinoma cell adhesion to vitronectin, only $\alpha v \beta 3$ was found to cluster into focal contacts. In contrast, $\alpha v \beta 5$ failed to localize to focal contacts, but rather clustered in small patches distributed over much of the cell surface (Wayner *et al.*, 1991). Focal contacts, or adhesion plaques, are structures that traverse the plasma membrane linking the extracellular substrate with components of the cytoskeleton (Burridge *et al.*, 1988; Otey and Burridge, 1990). The ability of a cell to form focal contacts correlates with its ability to assemble an organized cytoskeleton which, in turn, is prerequisite for cell spreading and migration (Dejana *et al.*, 1987). Thus, $\alpha v \beta 3$ and $\alpha v \beta 5$ exhibit different responses upon binding immobilized vitronectin, and thereby have the potential to mediate distinct signals within the cell.

To investigate the biological role of these integrins we identified a cell line, FG human carcinoma (Cheresh *et al.*, 1989a), which fails to express $\alpha v \beta 3$ and thus adheres to vitronectin in an $\alpha v \beta 5$-dependent manner. These cells are unable to spread or migrate on this ligand. Upon transfection of FG cells with a cDNA encoding the human $\beta 3$ integrin subunit, we could express $\alpha v \beta 3$ in a functional form resulting in the ability of these cells to spread and migrate on a vitronectin or fibrinogen matrix. Our results thus demonstrate that two homologous integrins, expressed on the same cell and recognizing the same ligand, can promote distinct signals potentiating diverse biological activities in response to a given extracellular matrix.

2. Materials and Methods

2.1. Cells and Cell Culture

FG is a human pancreatic carcinoma cell line that fails to express mRNA for the $\beta 3$ integrin subunit (Cheresh *et al.*, 1989a). FG-A and FG-B are two stably transfected sublines; FG-B is transfected with a full-length cDNA encoding human $\beta 3$ gene, while FG-A is a subline transfected with vector alone. (M21 human melanoma cells were a gift from Dr. Donald Morton, Department of Surgery, University of California, Los Angeles.) All cells were grown in RPMI 1640 with 10% fetal bovine serum and 50 μg/ml gentamicin, and tested free from mycoplasma during these studies.

2.2. Antibodies

Integrin-specific monoclonal antibodies LM609 ($\alpha v \beta 3$, Cheresh and Spiro, 1987), LM142 (αv, Cheresh and Harper, 1987), P4C10 ($\beta 1$, Carter *et al.*, 1990), LM534 ($\beta 1$, Wayner *et al.*, 1991), and P3G2 ($\alpha v \beta 5$, Wayner *et al.*, 1991) were affinity purified from ascites on protein A–Sepharose (Maps Kit, Bio-Rad, Richmond, CA). Monoclonal antibody AP3 was a generous gift from Dr. Peter Newman, Blood Center of Southwestern Wisconsin, Milwaukee.

2.3. Adhesive Ligands

Vitronectin was prepared as described by Yatohgo *et al.* (1988). Fibrinogen was purified according to the method of Felding-Habermann *et al.* (1992). Collagen type I was purchased from Collaborative Research (New Bedford, MA).

2.4. cDNA Transfection of FG Cells

Full-length cDNA encoding the human β3 integrin subunit, kindly provided by Dr. Larry Fitzgerald (University of Utah, Salt Lake City, UT) was ligated into the expression vector pcDNA1*neo* (Invitrogen, San Diego, CA) and transfected into FG carcinoma cells using the lipofectin protocol (Calbiochem, La Jolla, CA). In brief, subconfluent adherent cells were incubated with 2 μg cDNA and 40 μg lipofectin for 24 hr and were allowed to recover for 48 hr prior to selection in neomycin for 2 weeks at a concentration of 500 μg/ml. Neomycin-resistant cells were expanded and the cell surface expression of αvβ3 was analyzed by flow cytometry using the αvβ3 complex-specific antibody, LM609. Immunoreactive cells were enriched with four consecutive rounds of fluorescence-activated cell sorting, after which we observed stable surface expression of αvβ3 in approximately 40% of the transfected cell line. Positive (FG-B) and negative (FG-A) transfectant populations were recovered and maintained *in vitro*.

2.5. Cell Surface Labeling and Immunoprecipitation

Cell surface proteins were ^{125}I-labeled using lactoperoxidase as previously described (Cheresh *et al.*, 1989a). Radiolabeled cells were lysed in *RIPA* buffer [10 mM Tris, pH 7.2, 150 mM NaCl, containing 1% Triton X-100, 1% deoxycholate, 0.1% SDS with 1% aprotinin (Sigma No. A6279) and 1 mM phenylmethanesulfonylfloride] prior to immunoprecipitation with integrin-specific monoclonal antibodies coupled to Sepharose (Pharmacia, Uppsala, Sweden) as previously described (Cheresh *et al.*, 1989a). Immunoprecipitates were analyzed by SDS-PAGE (Laemmli, 1970) under nonreducing conditions on 7.5% polyacrylamide gels, and radiolabeled species were visualized with autoradiography as previously described (Cheresh, 1987).

2.6. Adhesion Assay

Cell adhesion assays were performed as previously described with modifications (Wayner *et al.*, 1991). In brief, sterile, untreated, bacteriological-grade polystyrene 48-well cluster plates (Costar) were coated at 4°C overnight with adhesive ligands (200 μl at 10 μg/ml in PBS) and immediately prior to use were blocked with 5% BSA in PBS, pH 7.4. Cells were harvested with trypsin/EDTA, washed once and suspended in Hanks' balanced salt solution (HBSS) supplemented with 1% BSA, 1 mM $CaCl_2$, 1 mM $MgCl_2$, and 0.5 mM $MnCl_2$, and added to appropriate wells. Cells were permitted to adhere for various times at 37°C in 5% CO_2 in the presence or absence of purified monoclonal antibodies specific for various integrins (50 μg/ml) or the synthetic peptides (50 mM) GRGDSPK and SPGDRGK. Nonadherent cells were removed with gentle washing, and cell attachment was enumerated from dye uptake. Adherent cells were fixed with 3% paraformaldehyde, stained with 0.1% crystal violet, air-dried, and extracted in 10% acetic acid. Dye uptake was determined spectrophotometrically at 600 nm. Cell spreading was evaluated prior to staining in the above adhesion assay and was enumerated from random quadruplicate fields containing at least 100 cells.

2.7. Indirect Immunofluorescence

Cells were allowed to attach and spread on vitronectin- or fibrinogen-coated glass coverslips, and were stained by indirect immunofluorescence, as previously described (Wayner *et al.*, 1991). In brief, cells were fixed with 3% paraformaldehyde and permeabilized with 10 mM Tris, pH 7.4, 150 mM NaCl, 1 mM $CaCl_2$ containing 0.1% Triton X-100 for 1 min. Permeabilized cells were stained with mAb AP3 or antisera to human vitronectin (60 min at room temperature), washed, and counterstained (60 min at room temperature) with either rhodamine-conjugated goat anti-mouse IgG or fluorescein-conjugated goat anti-rabbit IgG (Tago, Burlingame, CA). Focal adhesion plaques were visualized by the exclusion of vitronectin-specific antibody from the cell–substrate contacts, as previously described (Wayner *et al.*, 1991). Fluorescence was detected with a Zeiss photomicroscope fitted with epifluorescence.

2.8. Cell Migration Assay

Cell migration assays were performed using modified Boyden chambers with a diameter of 6.5 mm, thickness of 10 μm, with a porous (8.0 μm) polystyrene membrane separating the two chambers (Transwell, Costar, Cambridge, MA). Soluble ligands (20 μg/ml) were placed in lower, migration chambers in 600 μl of serum-free RPMI 1640 supplemented with 1% ITS+ (Collaborative Research, New Bedford, Mass.), 50 μg/ml gentamicin, and 2.5 μg/ml amphotericin B (Sigma, St. Louis, MO.), and incubated at 37°C for 30 min prior to addition of cells. Subconfluent 24-hr cultures were harvested with trypsin/EDTA, washed once, and resuspended in the serum-free RPMI without adhesive ligands. Cells (50,000) were added in 100 μl to the upper chamber in the presence or absence of the synthetic peptides (50 mM) GRGDSPK, SPGDR-GK, or mAbs (50 μg/ml) to various integrins. Migration was measured after 48 hr incubation at 37°C by counting the number of cells recovered from the floor of the lower chamber. Nonadherent cellular debris was removed with gentle washing, and cells were fixed with 3% paraformaldehyde. Cell migration was quantitated from photographs taken under phase contrast of quadruplicate, random fields from duplicate samples. Random (nonspecific) migration was determined using the control protein BSA (20 μg/ml). Specific cell migration toward each matrix protein was calculated by subtracting BSA-mediated migration from total cell migration. In each case, random migration was less than or equal to 7% of total matrix-dependent migration.

3. Results

3.1. Integrins αvβ5 and αvβ3 Are Both Involved in Adhesion of M21 Melanoma and H2981 Carcinoma Cells to Vitronectin

In order to determine whether αvβ5 and αvβ3 were both involved in cell adhesion to vitronectin, M21 melanoma and H2981 carcinoma cells were allowed to attach to vitronectin in either the presence or absence of MAb P5H9 and LM609. As shown in Fig. 1, neither MAb alone (25 μg/ml) completely blocked M21 or H2981 cell adhesion to vitronectin. It is apparent that MAb LM609 significantly reduced the adhesion of M21 cells, while MAb P5H9 significantly reduced the adhesion of H2981 cells. These results are consistent with the relative levels of αvβ3 and αvβ5 expressed by these cells (Wayner et al., 1991). However, when these cells were allowed to react with a mixture of MAbs P5H9 and LM609, virtually all M21 cell adhesion to vitronectin was abolished (Fig. 1). The combined effects of both of these MAbs reduced H2981 cell adhesion by 80% (Fig. 1). As a control, MAb W6/32 directed to HLA-class I antigens present on these cells failed to inhibit cell adhesion alone or when combined with either vitronectin receptor MAb (Fig. 1). In addition, MAb P5H9 or LM609, either separately or together, had negligible effects on the adhesion of these cells to fibronectin, collagen, or laminin (data not shown). These results indicate that M21 and H2981 cells utilize both αvβ3 and αvβ5 in adhesion to vitronectin.

3.2. Integrins αvβ3 and αvβ5 Exhibit a Distinct Distribution on the Surface of M21 or H2981 Cells Attached to Vitronectin

In order to determine the distribution of αvβ3 and αvβ5 on cells attached to vitronectin, immunofluorescence experiments were performed. M21, H2981, or UCLA-P3 cells were allowed to attach and spread on coverslips coated with

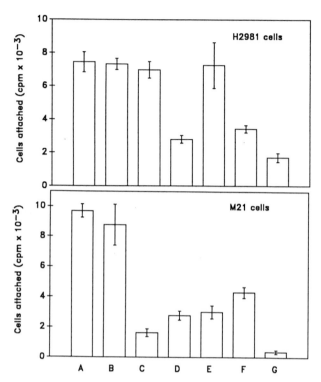

Figure 1. Integrins αvβ5 and αvβ3 are both involved in M21 and H2981 cell adhesion to vitronec-tin. Radiolabeled M21 cells (lower panel) or H2981 cells (upper panel) were allowed to adhere to wells coated with vitronectin as described in Materials and Methods. Cell adhesion was performed either in the absence of antibody (A) or in the presence (25 μg/ml) of the following: MAb W6/32 (B; anti-HLA, class I); MAb LM609 (C; anti-αvβ3); MAb P5H9 (D; anti-αvβ5); MAb W6/32 + MAb LM609 (E); MAb W6/32 + MAb P5H9 (F); or MAb LM609 + MAb P5H9 (G). Data are expressed as the cells attached (cpm) to the wells, which in the absence of antibody treatment, represents approximately 50% of the cells added to each well. Each bar represents the mean ± SD of triplicates.

vitronectin or various other ligands. As shown in Fig. 2, once M21 or H2981 cells attached to vitronectin, a focal distribution was detected upon staining with MAb AP3 directed to the β3 subunit (Fig. 2B,D). This staining pattern com-pletely codistributed with focal contacts as detected by these same cells stained with antivitronectin (Fig. 2A). In this case, antivitronectin was specifically ex-cluded from the regions of focal contact (Fig. 2 A,C,E, arrows), and these regions codistributed with the staining pattern of β3 on M21 cells or H2981 cells (Fig. 2B and D, respectively, arrows). In contrast, when H2981 cells were stained with anti-β5 (Fig. 2F), a punctate distribution was observed over the entire ventral surface (Fig. 2F), which showed little or minimal colocalization with focal contacts as depicted by antivitronectin exclusion (Fig. 2E). A composite staining pattern of β3 and β5 was observed (Fig. 2H) when H2981 was reacted with MAb LM142 directed to the αv subunit on these cells. Thus, it can clearly be seen that the immunolocalization of the αv subunit on these cells revealed both focal contact staining (identified in Fig. 2G, arrows) and the punctate nonfocal contact staining pattern.

When M21 (Fig. 3A,B) or H2981 cells (Fig. 3E,F) were attached to vitronectin and cells were stained with both anti-β3 (Fig. 3A,E) and anti-β5 (Fig. 3B,F), completely distinct patterns were observed. Thus, β3 was localized in a focal contact distribution (arrows), while β5 on the same cell was not. As expected, when M21 cells were stained with a MAb directed to vinculin (Fig. 3C), the focal contacts were detected. This staining pattern showed no colocalization with the same cells stained with the anti-β5 antibody (Fig. 3D). As shown in Fig. 3G and H, H2981 cells stained with antivinculin showed the expected focal contact distribution (arrows). When these cells were stained with an antibody to talin (Fig. 3I), there was significant colocalization with vinculin (Fig. 3H), but not β5 (Fig. 3D,F). Thus, when cells were attached to vitronectin, integrin αvβ3 was found in focal contacts colocalizing with vinculin and talin, while integrin αvβ5 was not. As a control, we examined the β3 distribution on H2981 and M21 cells attached to collagen. In this case, the cells did not organize β3 into focal contacts, but as expected, when MAb LM534 was directed to a nonfunctional epitope on β1, it located this subunit in focal contacts on collagen (data not shown). These results are consistent with cell adhesion experiments in which MAb P4C10, directed to a functional epitope on the β1 subunit, blocked collagen adhesion of all three cell lines (data not shown). Taken together, these results demonstrate that when cells are attached to vitronectin, αvβ3 and αvβ5 demonstrate distinct distributions on the cell surface even though both receptors are clearly involved in the attachment to this ligand.

UCLA-P3 cells readily attached to vitronectin, but failed to spread and form focal contacts. As shown in Fig. 4A, the β5 distribution on these cells showed a punctate staining pattern similar to those observed on M21 cells (Fig. 2) and H2981 cells (Figs. 2 and 3). MAb LM142 directed to αv, or a polyclonal antibody directed to αvβ3, which react with both receptors (Cheresh *et al.*, 1989a) on UCLA-P3 cells demonstrated this identical staining pattern, confirming the β5 staining pattern on these cells (data not shown). This lack of focal contact distribution is correlated with the inability of UCLA-P3 cells to organize actin during their adhesion to vitronectin (Fig. 4B). In contrast, M21 cells (Fig. 4D) or H2981 cells (Fig. 4F) attached to vitronectin demonstrated well-organized actin filaments as depicted on cells stained with rhodamine–phalloidin. As expected, the β3 distribution on M21 cells localized at the ends of actin filaments (Fig. 4C, arrows), whereas β5 on M21 cells (not shown) or H2981 cells did not (Fig. 4E). It is interesting to note that the actin filaments on vitronectin-attached

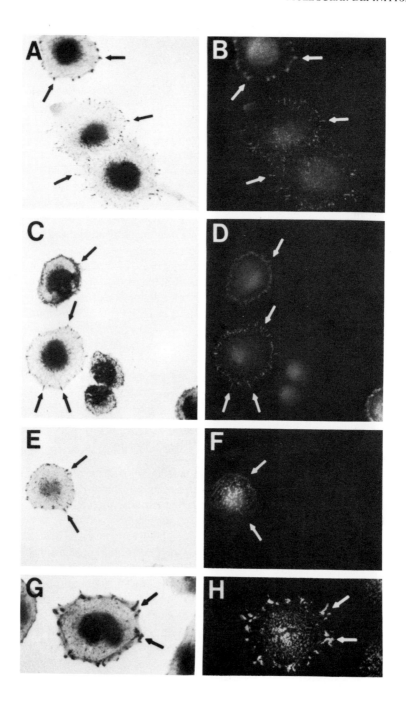

H2981 cells appeared to organize around the circumference of the ventral cell surface (Fig. 4F), whereas on M21 cells the actin filaments appeared to span the diameter of the cell (Fig. 4D).

The distinct localization of β3 and β5 on M21 and H2981 cells attached to vitronectin (as shown above) were based on experiments in which cells were allowed to attach and spread on vitronectin for 30 min. However, when M21 or H2981 cells were allowed to attach and spread for 120 min, we were still unable to observe focal contact distribution of αvβ5 (data not shown). These results indicate that integrin αvβ5, which is involved in cell attachment to vitronectin, is incapable of associating with the actin cytoskeleton, and therefore does not localize in focal contacts.

In order to compare the biological properties of the two vitronectin-binding integrins αvβ3 and αvβ5, we examined M21 cells, which primarily express αvβ3, and FG carcinoma cells, which express αvβ5 and not αvβ3 (Leavesley *et al.*, 1992). To investigate the biological consequences of αvβ3- or αvβ5-dependent cell adhesion, we examined the capacity of FG cells or M21 cells to spread on vitronectin. As shown in Fig. 5, FG cells readily attached to, but did not spread on, this matrix, even after 90 min. In contrast, these cells spread readily on a collagen matrix, indicating that the intrinsic ability of FG cells to spread is not deficient. M21 human melanoma cells, which utilize primarily αvβ3 to attach to vitronectin, spread rapidly on this matrix (Fig. 5). These data suggest that the homologous integrins αvβ3 and αvβ5 mediate differential biological functions in response to a common ligand.

3.3. Transfection of FG Cells with β3 cDNA Results in the Expression of Integrin αvβ3

In order to determine whether the inability of FG cells to spread on vitronectin is caused by the specific absence of integrin αvβ3, FG cells were transfected with an expression vector containing cDNA encoding the human β3 integrin subunit and a gene for neomycin resistance. After drug selection, two stable transfectant FG sublines were established from populations sorted for αvβ3

Figure 2. Immunofluorescence detection of αvβ3 and αvβ5 on the surface of M21 and H2981 cells. M21 (A, B) or H2981 (C–H) cells were allowed to attach and spread on vitronectin-coated coverslips for 30–45 min at 37°C, after which cells were fixed, permeabilized, and reacted with primary antibodies to vitronectin, either MAb 8E6 (A, C) or rabbit antivitronectin antibody (E, G). Localization of β3 (B, D) or β5 (F) on these cells was achieved using the primary antibodies MAb AP3 and rabbit anti-β5 peptide, respectively. The composite staining of both receptors was identified by MAb LM142 directed to αv (H). These antigens were detected with FITC-conjugated anti-mouse or rhodamine-conjugated anti-rabbit IgG. Arrows indicate focal contact staining as defined by anti-vitronectin exclusion in A, C, E, and G. Representative fields were photographed using a Zeiss microscope fitted with epifluorescence. Magnification 630×.

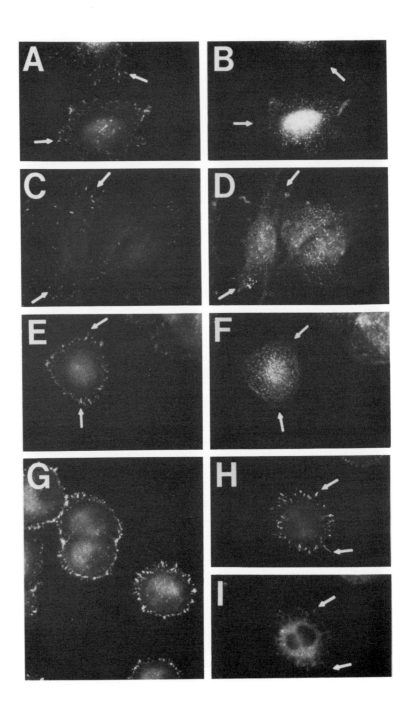

expression using FACS: FG-B, which expresses β3, and FG-A, which does not express β3. As shown in Fig. 6, immunoprecipitation analysis of ^{125}I-surface-labeled FG transfectants revealed that both cell types express equivalent levels of αvβ5 and β1 integrins. When MAb LM609 was directed to the αvβ3 complex, it immunoprecipitated this receptor from the β3-expressing FG-B subline and not from the FG-A (mock-transfected) subline. When MAb LM142 was directed to the αv subunit, it immunoprecipitated this subunit associated with multiple β subunits, including β5 and β6 (Busk *et al.*, 1992). These results demonstrate that transfecting FG cells with a cDNA encoding the β3 integrin subunit results in the stable expression of αvβ3 complexes at the cell surface. Moreover, these immunoprecipitation profiles indicate that the novel expression of β3 in the FG-B population does not alter the expression of other αv integrins, or β1 integrins, which remain at equivalent levels in both cell populations.

3.4. Expression of β3 in FG Cells Alters Their Biological Response to Vitronectin and Fibrinogen

In order to establish whether FG-B cells express the αvβ3 integrin in a functional form, cell adhesion assays were performed. Although both FG-A and FG-B cells attached to vitronectin, only FG-B cells could be inhibited with MAb LM609, directed to αvβ3 (55%, data not shown). This adhesion event resulted in FG-B cell spreading on both vitronectin (Fig. 7A–C) and fibrinogen (Fig. 7B, left panel) within 60 min. In contrast, the β3-negative, parental FG cells (Fig. 5) or mock-transfected FG-A cells not only failed to spread on vitronectin, but were unable to attach to fibrinogen (not shown). Moreover, FG-B cells that spread on both of these substrates expressed β3 in focal contacts (Fig. 7B,C), indicating that αvβ3 expression promoted the assembly of adhesion plaques and the organization of the microfilament cytoskeleton, leading to cell spreading on a vitronectin or fibrinogen matrix. Approximately 40% of the FG-B cell population expressed β3, and only these cells were found to spread on vitronectin or fibrinogen. In addition, MAb LM609 prevented cell spreading on vitronectin or fibrinogen, providing further evidence that αvβ3 plays a part in this process (data not shown).

In order to further investigate the biological role of integrin αvβ3, we examined the capacity of this receptor to potentiate FG-B cell migration. For these experiments we used modified Boyden chambers containing an 8.0-μm

---

Figure 3. αvβ5 and αvβ3 expressed on M21 and H2981 cells demonstrate distinct distributions. M21 (A–D) or H2981 (E–I) cells attached to vitronectin-coated coverslips were stained as in Fig. 3, except the primary antibodies used were MAb AP3 directed to β3 (A, E), rabbit anti-β5 (B, D, F), MAb directed to vinculin (C, G, H), or rabbit antitalin (I). The secondary antibodies used were anti mouse or anti rabbit labeled with FITC or rhodamine. Arrows refer to focal contacts as detected by β3 localization (A, E) or vinculin localization (C, G, H). Representative fields were photographed using a Zeiss microscope fitted with epifluorescence. Magnification 630×.

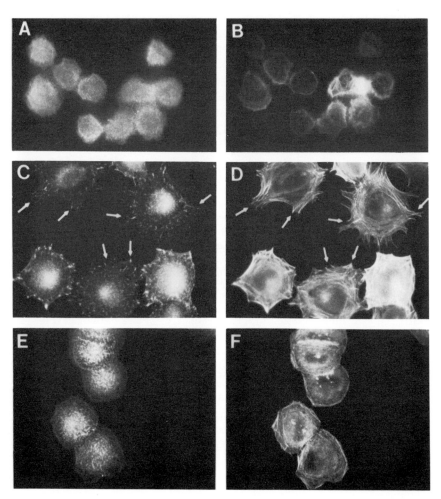

Figure 4. Integrin αvβ3 colocalizes with the ends of actin filaments, whereas αvβ5 does not. UCLA-P3 (A, B), M21 cells (C, D), or H2981 cells (E, F) were allowed to attach to vitronectin-coated coverslips for 30 min and were stained with anti-β5 (A, E) or the anti-β3 MAb AP3 (C), as described above. Actin was visualized with FITC (B, F) or rhodamine-conjugated phalloidin (D). Photos of representative fields were taken with a Zeiss microscope fitted with epifluorescence. When M21 or H2981 cells were allowed to spread for a longer period (1–2 hr), increased actin filament organization was observed, whereas UCLA-P3 cells did not spread, and had minimal actin organization. The 30 min time was chosen to correspond to the localization of β3 and β5. Magnification 1000× (A, B) or 630× (C–F).

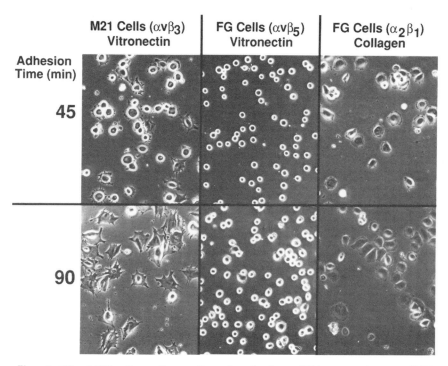

Figure 5. FG and M21 cell spreading on vitronectin and collagen. FG human carcinoma and M21 human melanoma cells were allowed to attach to vitronectin- or collagen-coated (10 μg/ml) polystyrene wells at 37°C. Nonadherent cells were removed with three gentle washes as described in Materials and Methods. Attached cells were photographed at 200× under phase-contrast using Kodak TMX100.

porous membrane. FG cells did not demonstrate significant or specific migration through the 8.0-μm pores separating these chambers in the presence of the control ligand BSA. However, both FG-A and FG-B cells readily migrated toward a collagen source, yet only FG-B cells, expressing αvβ3, were capable of migrating toward vitronectin or fibrinogen (Fig. 8). FG-B cell migration to both vitronectin and fibrinogen was completely inhibited by MAb LM609, demonstrating that αvβ3 was responsible for this event (Fig. 8). In contrast, MAb P3G2 directed to αvβ5 had no effect on FG-A or FG-B cell migration to all substrates tested (data not shown), indicating that this receptor failed to potentiate migration of these cells. In addition, an RGD-containing synthetic peptide that functionally inhibited αvβ3-mediated cell attachment (Cheresh *et al.*, 1989a), completely blocked αvβ3-dependent migration (data not shown). It is interesting that the migration of αvβ3-expressing FG-B cells toward collagen was not significantly inhibited by MAb P4C10 directed to β1 integrins, suggesting that there is an additional mechanism in these cells (Fig. 8). In fact, integrin αvβ3 has been

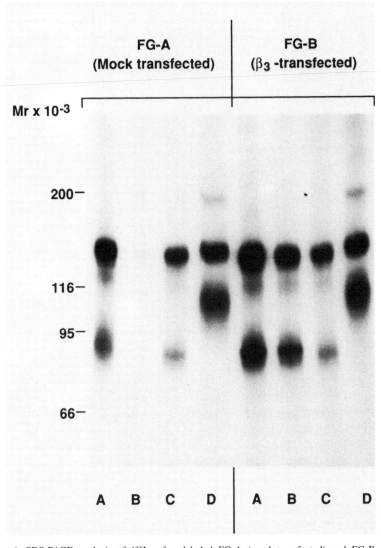

Figure 6. SDS-PAGE analysis of ¹²⁵I-surface-labeled FG-A (mock-transfected) and FG-B (β3-transfected) cells. FG carcinoma cells were transfected with a cDNA encoding full-length human β3 and neomycin resistance genes. After drug selection for 2 weeks, cells were sorted for αvβ3 expression by FACS and stable sublines were established. Cells were ¹²⁵I-surface-labeled and extracted with detergent as described in Materials and Methods. Detergent extracts were immunoprecipitated with integrin-specific MAbs coupled to Sepharose beads. MAbs LM142 (αv, lane A); LM609 (αvβ3, lane B); P3G2 (αvβ5, lane C); and LM534 (β1, lane D).

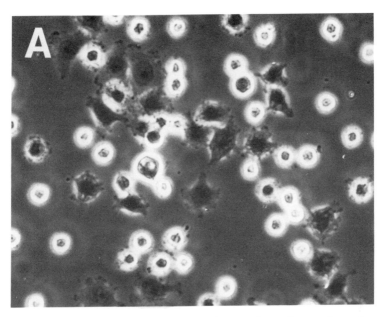

Figure 7. Expression of β3 promotes FG-B cell attachment and spreading on vitronectin and fibrinogen with β3 localization to focal contacts. β3-transfected FG-B cells were allowed to attach and spread on vitronectin-coated coverslips for 60 min. Nonadherent cells were removed with gentle washing and attached cells were evaluated for spreading with phase-contrast microscopy (A). For fluorescence experiments, FG-B cells were allowed to attach and spread on vitronectin- or fibrinogen-coated coverslips, fixed, permeabilized, and stained with MAb AP3 directed to the β3 integrin subunit, or antisera to vitronectin (B and C, respectively). β3 staining in cells spread on fibrinogen (B, left panel) and vitronectin (B, center and right panels) exactly codistributes with focal contacts, visualized by the exclusion of vitronectin-specific antibody from the cell–substrate contacts as described in Materials and Methods (C; β3: left panel, antivitronectin: right panel, arrows indicate colocalized staining). Representative cells were photographed with a Zeiss microscope fitted with epifluorescence. Magnification 100× (A) or 630× (B, C).

reported to recognize collagen in some cells (Kramer *et al.*, 1990), suggesting that, in addition to vitronectin and fibrinogen, the presence of αvβ3 may also contribute to collagen-dependent cell migration.

3.5. FG Cells Transfected with a β3/β5 Chimera Fail to Spread on Vitronectin

In order to determine whether the cytoplasmic domain of β5 was responsible for the lack of focal contact association of αvβ5, FG cells were transfected with a cDNA encoding a chimeric β3β5 protein. This cDNA was constructed to

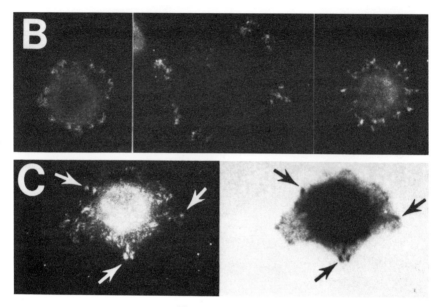

Figure 7. Continued.

include the ectodomain and transmembrane domain of β3 and the cytoplasmic tail of β5. FG cells were transfected and selected for expression of αvβ3 as detected by MAb LM609 immunofluorescent staining. Cells sorted for αvβ3 were propagated and examined for their ability to attach and spread on immobilized vitronectin. Cells allowed to attach showed no spreading capacity on vitronectin (not shown), as did the cell transfected with full-length β3 (see above). These data suggest that the cytoplasmic tail of β5 is structurally incompatible with cell spreading. Verification of this finding will be accomplished by generating a cDNA encoding for the β5 ectodomain and β3 cytoplasmic tail.

4. Discussion

This study was designed to characterize the biological properties of αvβ3 and αvβ5, two structurally and functionally related integrins that are often expressed on the same cell (Wayner *et al.*, 1991). These receptors bind vitronectin in an RGD-dependent manner and contain identical α subunits and structurally similar β subunits (Cheresh and Spiro, 1987; Cheresh *et al.*, 1989a; McLean *et al.*, 1990; Ramaswamy and Hemler, 1990; Suzuki *et al.*, 1990). However, the cytoplasmic tails of β3 and β5 are structurally distinct, thus raising the possi-

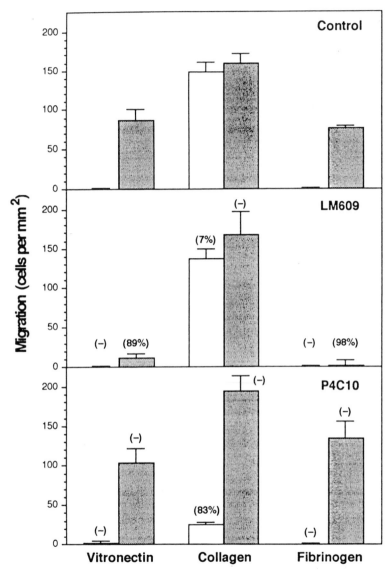

Figure 8. Antibody-specific inhibition of FG carcinoma cell migration. Vitronectin, collagen, and fibrinogen (20 μg/ml) in serum-free medium were placed into the lower chamber of a modified Boyden chamber separated by an 8.0-μm porous membrane from FG-A (open) or FG-B (shaded) cells placed into the upper chamber in the absence (control) or presence of MAbs LM609 (anti-αvβ3) or P4C10 (anti-β1). Migratory cells that had traversed the membrane into the lower chamber were counted (mean of four random fields ±SE) after 45 hr incubation at 37°C. Random, nonspecific migration (when BSA is placed into the lower chamber), which has been subtracted, was ≤7% of the total migration in response to any substrate.

bility that αvβ3 and αvβ5 mediate different biological signals in response to vitronectin.

This report provides several lines of evidence that these integrins mediate distinct biological responses to a vitronectin substrate. First, FG carcinoma cells utilize αvβ5 as their major vitronectin receptor, since they fail to express β3 mRNA or protein (Cheresh et al., 1989a), attach but fail to spread on vitronectin, and are incapable of attaching to fibrinogen. This is not the result of a general deficiency in FG cell spreading, since these cells readily attach and spread on collagen in an integrin β1-dependent manner. Second, M21 cells, which express αvβ3 as their major vitronectin receptor (Wayner et al., 1991), attach and spread on both vitronectin and fibrinogen with αvβ3 expressed in focal contacts at the end of actin filament bundles (Wayner et al., 1991). Third, transfection of β3-negative FG cells with a cDNA encoding β3 results in the surface expression of αvβ3, which enables these cells to spread on vitronectin and fibrinogen. Moreover, β3 localizes to focal contacts on these transfected FG cells, indicating that αvβ3 is directly involved in cell spreading on these matrix proteins. Finally, FG-B cells expressing the β3 gene product not only spread, but also acquire the ability to migrate in response to vitronectin and fibrinogen. The fact that this is the result of the presence of αvβ3 on these cells is demonstrated by the ability of MAb LM609 (anti-αvβ3) or an RGD-containing peptide to block this migration. These results provide a rationale for the expression of αvβ3 and αvβ5 on the same cell, where αvβ5 promotes simple adhesion while αvβ3 enables cells to modify their shape and mobility on vitronectin.

Although αvβ5 is the major vitronectin binding integrin on FG cells, we cannot completely exclude the possibility that additional vitronectin-binding integrins are expressed on these cells. For example, integrin αvβ1 apparently mediates vitronectin adhesion of certain cells (Bodary and McLean, 1990). It is conceivable that αvβ1 could play a minor role in FG cell and M21 cell attachment to vitronectin, since MAb P4C10 directed to β1 partially inhibited the attachment of both cell types to vitronectin (approximately 20%). However, immunoprecipitation analysis failed to detect αvβ1 on either cell type. In any event, the expression of αvβ1 or another vitronectin-binding integrin on FG cells does not account for measurable cell spreading and/or migration on vitronectin.

The cytoplasmic tail of the β5 subunit is structurally distinct from that of the β1 and β3 subunits (Fig. 9; Ramaswamy and Hemler, 1990; McLean et al., 1990; Suzuki et al., 1990), and thus may be responsible for a distinct signal transduction event. Alternatively, it is conceivable that this subunit is simply incapable of interacting with one or more cytoplasmic proteins thought to be involved in the assembly of focal contacts. For example, talin and α-actinin, two proteins found in focal contacts, have been shown to directly bind integrins (Horowitz et al., 1986). In fact, α-actinin directly binds to a peptide derived from the cytoplasmic tail of β1 (Otey et al., 1990). Based on a mutational

Figure 9. Amino acid sequence of integrin β1, β3, and β5 cytoplasmic tails. When β1, β3, and β5 are aligned for optimal sequence identity, an eight-amino-acid putative insert appears in β5 just prior to N(*), which is important for focal contact formation in β1 (Reszka and Horowitz, 1991). Shaded areas depict sequence identity.

analysis of the $\beta 1$ integrin, it appears that the structural basis of integrin focal contact formation depends on three domains with the cytoplasmic tail of $\beta 1$ (Reszka and Horowitz, 1991). It is noteworthy that $\beta 1$ and $\beta 3$ are extremely well conserved in each of these regions, while $\beta 5$ has little homology in the cytoplasmic domain. When these subunits are optimally aligned (Fig. 9), $\beta 5$ appears to contain an eight-amino-acid insert. This may be accounted for by alternative splicing, although this is currently speculative. The asparagine designated with * (N single-letter code) in Fig. 9 has been shown to be important for focal contact formation (Reszaka and Horowitz, 1991). Thus, this asparagine may be misaligned in $\beta 5$ because of this putative insert. Alternatively, it is conceivable that specific residues within this particular domain enable $\beta 5$ to associate with other molecules within the cytoplasm; it would thus not be available to interact with the actin cytoskeleton, as observed with $\beta 1$ and $\beta 3$ integrins. To test this hypothesis it will be necessary to examine the expression of truncated and/or chimeric integrin heterodimers in cells that normally fail to form focal contacts on a vitronectin substrate. It is also interesting to note that the cytoplasmic tail of $\beta 5$ contains five serine residues that are not found in $\beta 1$ or $\beta 3$ (Ramaswamy and Hemler, 1990; McLean et al., 1990; Suzuki et al., 1990). Perhaps phosphorylation of one or more of these prevents the localization of $\alpha v \beta 5$ to focal contacts. In fact, the βs subunit, which may be related to or identical to $\beta 5$, becomes phosphorylated on serine in response to activators of protein kinase C (Freed et al., 1989).

The differential ability of $\alpha v \beta 3$ and $\alpha v \beta 5$ to promote cell spreading and migration may have profound biological implications during events associated with development, wound healing, and neoplasia, where cell migration is known to take place. Thus, the genetic regulation of $\beta 3$ and $\beta 5$ may play a key role in determining the migratory status of a cell. Recent evidence supports this hypothesis. Nondifferentiated keratinocytes, which are known to be migratory, express both $\alpha v \beta 3$ and $\alpha v \beta 5$ integrins, while terminally differentiated keratinocytes, which do not migrate, no longer express $\alpha v \beta 3$ (Adams and Watt, 1991). Furthermore, the $\beta 3$ subunit expressed by subconfluent embryonic lung fibroblasts is downregulated when these cells reach confluence (Bates et al., 1991). The role of $\alpha v \beta 3$ in melanoma cell migration may be very relevant for the metastatic phenotype of these cells. In fact, $\alpha v \beta 3$ was found to be preferentially expressed on metastatic and vertically invasive primary lesions, whereas it was not detected on normal melanocytes, nevi, or horizontal primary melanoma (Albelda et al., 1990). The promiscuous ligand-binding capacity of $\alpha v \beta 3$ suggests that cells expressing this receptor can migrate on a wide variety of matrices and basement membranes. Our results demonstrate that fibrinogen, a known ligand for $\alpha v \beta 3$ (Cheresh, 1987), can promote the migration of $\beta 3$-transfected FG cells. The expression of $\alpha v \beta 3$ and the resulting phenotype may therefore play an important role in a wide range of migratory events involving multiple biological phenomena.

References

Adams, J. C., and Watt, F. M., 1991, Expression of β1, β3, β4 and β5 integrins by human epidermal keratinocytes and non-differentiating keratinocytes, *J. Cell Biol.* **115**:829–841.

Albelda, S. M., Mette, S. A., Elder, D. E., Stewart, R., Damjanovich, L., Herlyn, M., and Buck, C. A., 1990, Integrin distribution in malignant melanoma: Association of the β3 subunit with tumor progression, *Cancer Res.* **50**:6757–6764.

Bates, R. C., Rankin, L. M., Lucas, C. M., Scott, J. L., Krissansen, G. W., and Burns, G. F., 1991, Individual embryonic fibroblasts express multiple β chains in association with the αv integrin subunit, *J. Biol. Chem.* **266**:18593–18599.

Bodary, S. C., and McLean, J. W., 1990, The integrin β1 subunit associates with the vitronectin receptor αv to form a novel vitronectin receptor in a human embryonic kidney cell line, *J. Biol. Chem.* **265**:5938–5941.

Buck, C. A., and Horowitz, A. F., 1987, Integrin, a transmembrane glycoprotein complex mediating cell–substratum adhesion, *J. Cell Sci.* Suppl. **8**:231–250.

Burridge, K., Fath, K., Kelly, T., Nucholls, G., and Turner, C., 1988, Focal adhesions: Transmembrane junctions between the extracellular matrix and the cytoskeleton, *Annu. Rev. Cell Biol.* **4**:487–525.

Busk, M., Pytela, R., and Sheppard, D., 1992, Characterization of the integrin αvβ6 as fibronectin binding protein, *J. Biol. Chem.* in press.

Carter, W. G., Wayner, E. A., Bouchard, T. S., and Kaur, P., 1990, The role of integrins α2β1 and α3β1 in cell–cell and cell–substrate adhesion of human epidermal cells, *J. Cell Biol.* **110**:1387–1404.

Charo, I. F., Bekeart, L. S., and Phillips, D. R., 1987, Platelet glycoprotein Iib-IIIa-like proteins mediate endothelial cell attachment to adhesive proteins and the extracellular matrix, *J. Biol. Chem.* **262**:9935–9938.

Cheresh, D. A., 1987, Human endothelial cells synthesize and express an Arg-Gly-Asp-directed receptor involved in attachment to fibrinogen and von Willebrand factor, *Proc. Natl. Acad. Sci. USA* **84**:6471–6475.

Cheresh, D. A., and Harper, J. R., 1987, Arg-Gly-Asp recognition by a cell adhesion receptor requires its 130kDa α-subunit, *J. Biol. Chem.* **262**:1434–1437.

Cheresh, D. A., and Spiro, R. C., 1987, Biosynthetic and functional properties of an Arg-Gly-Asp-directed receptor involved in human melanoma cell attachment to vitronectin, fibrinogen and von Willebrand factor, *J. Biol. Chem.* **262**:17703–17711.

Cheresh, D. A., Smith, J. W., Cooper, H. M., and Quaranta, V., 1989a, A novel vitronectin receptor integrin (αvβx) is responsible for distinct properties of carcinoma cells, *Cell* **57**:59–69.

Dejana, E., Colella, S., Conforti, G., Abbadini, M., Gaboli, M., and Marchisio, P. C., 1987, Fibronectin and vitronectin regulate the organization of their respective Arg-Gly-Asp adhesion receptors in cultured human endothelial cells, *J. Cell Biol.* **107**:1215–1223.

Felding-Habermann, B., Ruggeri, Z. M., and Cheresh, D. A., 1992, Distinct biological consequences of integrin αvβ3-mediated melanoma cell adhesion to fibrinogen and its plasmic fragments, *J. Biol. Chem.* **267**:5070–5077.

Freed, E., Gailit, J., van der Geer, P., Ruoslahti, E., and Hunter, T., 1989, A novel integrin β subunit is associated with the vitronectin receptor α subunit (αv) in a human osteosarcoma cell line and is a substrate for protein kinase C, *EMBO J.* **8**:2955–2965.

Horowitz, A., Duggan, K., Buck, C., Beckerle, M. C., and Burridge, K., 1986, Interaction of plasma membrane fibronectin receptor with talin, a transmembrane linkage, *Nature (London)* **320**:531–533.

Hynes, R. O., 1987, Integrins: A family of cell surface receptors, *Cell* **48**:549–554.

Kramer, R. H., Cheng, Y.-F., and Clyman, R., 1990, Human microvascular endothelial cells use β1 and β3 integrin receptor complexes to attach to laminin, *J. Cell Biol.* **111**:1233–1243.

Laemmli, U. K., 1970, Cleavage of structural proteins during the assembly of the head of bacteriophage T4, *Nature (London)* **227**:680–685.

Leavesley, D. I., Ferguson, G. D., Wayner, E. A., and Cheresh, D. A., 1992, Requirement of the integrin β3 subunit for carcinoma cell spreading and migration on vitronectin and fibrinogen, *J. Cell Biol.* **117**:1101–1107.

McLean, J. W., Vestal, D. J., Cheresh, D. A., and Bodary, S. C., 1990, cDNA sequence of the human integrin β5 subunit, *J. Biol. Chem.* **265**:17126–17131.

Moyle, M., Napier, M. A., and McLean, J. W., 1991, Cloning and expression of a divergent integrin subunit β8, *J. Biol. Chem.* **266**:19650–19658.

Otey, C. A., and Burridge, K., 1990, Patterning of the membrane cytoskeleton by the extracellular matrix, *Semin. Cell Biol.* **1**:391–399.

Otey, C. A., Pavelko, F. M., and Burridge, K., 1990, An interaction between α-actinin and the β1 integrin subunit in vitro, *J. Cell Biol.* **111**:721–729.

Ramaswamy, H., and Hemler, M. E., 1990, Cloning, primary structure and properties of a novel human integrin β subunit, *EMBO J.* **9**:1561–1568.

Reszka, A. A., and Horowitz, A. F., 1991, Identification of avian integrin β1 subunit adhesion plaque localization sequences using site directed mutagenesis, *J. Cell Biol.* in press.

Ruoslahti, E., and Pierschbacher, M. D., 1987, New perspectives in cell adhesion: RGD and integrins, *Science* **238**:491–497.

Sheppard, D., Rozzo, C., Starr, L., Quaranta, V., Elre, D. J., and Pytela, R., 1990, Complete amino acid sequence of a novel integrin β subunit (β6) identified in epithelial cells using the polymerase chain reaction, *J. Biol. Chem.* **265**:11502–11507.

Suzuki, S., Huang, Z.-S., and Tanihara, H., 1990, Cloning of an integrin β subunit exhibiting high homology with integrin β3 subunit, *Proc. Natl. Acad. Sci. USA* **87**:5354–5358.

Tamkun, J. W., DeSimone, D. W., Fonda, D., Patel, R. S., Buck, C. A., Horowitz, A. F., and Hynes, R. O., 1986, Structure of integrin, a glycoprotein involved in the transmembrane linkage between fibronectin and actin, *Cell* **46**:271–282.

Vogel, B E., Tarone, G., Giancotti, F. G., Gailit, J., and Ruoslahti, E., 1990, A novel fibronectin receptor with an unexpected subunit composition (αvβ1), *J. Biol. Chem.* **265**:5934–5937.

Wayner, E. A., Orlando, R. A., and Cheresh, D. A., 1991, Integrins αvβ3 and αvβ5 contribute to cell attachment to vitronectin but differentially distribute on the cell surface, *J. Cell Biol.* **113**:919–929.

Yatohgo, T., Izumi, M., Kashiwagi, H., and Hiyashi, M., 1988, Novel purification of vitronectin from human plasma by heparin affinity chromatography, *Cell Struct. Funct.* **13**:281–292.

Integrin Modulating Factor and the Regulation of Leukocyte Integrins

SAMUEL D. WRIGHT

1. Introduction

This chapter reviews recent work describing rapid, reversible changes in the avidity of leukocyte integrins. The role of an allosteric effector, integrin modulating factor (IMF-1) in this type of regulation is discussed.

2. The Leukocyte Integrins

Recent work has identified a large family of proteins, the integrins, that are involved in cell adhesion. One group of integrins, the β2 integrins or CD18 antigens, are restricted in their cellular distribution to leukocytes. These comprise three glycoproteins, LFA-1 (CD11a/CD18), CR3 (CD11b/CD18, also known as MAC-1), and P150,95 (CD11c/CD18). These proteins participate in a wide variety of cell–cell and cell–substratum interactions by recognizing a variety of ligands (reviewed in Wright and Detmers, 1991). LFA-1 recognizes the broadly distributed protein ICAM-1, as well as ICAM-2 and -3, and participates in the interaction of cytotoxic T cells with targets, the interaction of NK cells with targets, and the interaction of monocytes, lymphocytes, and PMN with endothelium. CR3 was first recognized for its interaction with complement protein C3bi (Wright and Detmers, 1991; Wright et al., 1983), but it has been shown

SAMUEL D. WRIGHT • The Laboratory of Cellular Physiology and Immunology, The Rockefeller University, New York, New York 10021.
Cellular Adhesion: Molecular Definition to Therapeutic Potential, edited by Brian W. Metcalf, Barbara J. Dalton, and George Poste. Plenum Press, New York, 1994.

that it also recognizes fibrinogen (Wright *et al.*, 1988; Altieri *et al.*, 1988), factor X (Altieri and Edgington, 1988), an unidentified ligand on endothelial cells (Lo *et al.*, 1989b; Diamond *et al.*, 1990), and several microbial antigens (Wright, 1992). The ligand specificity of p150,95 is less clear, but it may participate in the interaction of cells with endothelium and surface-bound fibrinogen (Loike *et al.*, 1991; Postigo *et al.*, 1991). The importance of the CD18-dependent adhesion events is highlighted by studies on patients with a genetic deficiency in the β chain (CD18). Patients with the disease termed leukocyte adhesion deficiency (LAD) present with recurrent life-threatening infections (reviewed in Todd and Freyer, 1988). Phagocytes from these patients fail to bind C3bi-coated particles and unstimulated endothelial cells, and show defective cytocidal activity *in vitro*.

3. The Leukocyte Integrins Participate in the Movement of Cells out of the Vasculature

Several lines of evidence, described below, indicate that CD18 antigens play a crucial role, not only in phagocytosis and cell-mediated cytotoxicity, but also in diapedesis, the movement of leukocytes through an endothelial monolayer into tissues. (1) Anti-CD18 antibodies completely block the binding of stimulated PMN to unstimulated endothelial cells (EC) *in vitro* (Smith *et al.*, 1988; Lo *et al.*, 1989a; Harlan *et al.*, 1985). CR3 and LFA-1 appear to contribute equally to binding to endothelium, since the attachment of PMN is partially blocked with antibodies against either CR3 or LFA-1, and a combination of antibodies against both of these receptors completely eliminates binding (Lo *et al.*, 1989b). (2) PMN from LAD (CD18-deficient) patients cannot adhere to resting EC *in vitro* (Todd and Freyer, 1988). (3) Neither PMN nor monocytes accumulate in skin windows on LAD patients, and these patients fail to form pus (Todd and Freyer, 1988). (4) Blockade of CD18 with monoclonal antibodies in animals causes a complete inhibition of the movement of PMN into skin (Arfors *et al.*, 1987; Lundberg and Wright, 1990), bowel (Hernandez *et al.*, 1987), brain (Tuomanen *et al.*, 1989), peritoneal cavity (Rosen and Gordon, 1987), and muscle (Simpson *et al.*, 1988), which occurs in response to a variety of inflammatory stimuli. The chemotactic movement of PMN *in vivo* and on plastic substrates (Todd and Freyer, 1988; Schmalstieg *et al.*, 1986) is also dependent on the expression of functional CD18 antigens, presumably because these molecules are required for attachment to the substrate.

Movement of leukocytes through the endothelium is essential for both the initiation and maintenance of inflammation. In several animal models, the blockade of CD18 with monoclonal antibodies and the resultant decrease in the cellular infiltrate has produced beneficial anti-inflammatory effects. For example, the blockade of CD11b/CD18 decreased the infarct size by 40% in a canine cardiac ischemia/reperfusion model (Simpson *et al.*, 1988), and the blockade of CD18 prevented breakdown of the blood–brain barrier, cerebral edema, and death in a rabbit model of meningitis (Tuomanen *et al.*, 1989). These studies suggest that the leukocyte integrins play a pivotal role in inflammation, that understanding these molecules is critical to understanding inflammation, and that reagents that block these molecules may be potent anti-inflammatory drugs.

4. The Binding Activity of Leukocyte Integrins is Regulated

The adhesive interactions of PMN with endothelium must be regulated on a very short time scale. The adhesion receptors must be inactive to allow circulation of leukocytes, they must be made active within seconds to enable adhesion as the PMN are carried through an inflammatory site, and they must become inactive in the succeeding seconds to enable the cells to detach from the endothelium in order to move through. Successive cycles of adhesion and release are then necessary for locomotion through tissues. We were the first to recognize and characterize this regulated behavior in studies on the binding of C3bi to CR3 (Wright and Meyer, 1986). We discovered that the binding activity of CR3 is negligible on resting PMN, is dramatically enhanced on exposure of cells to agonists, and declines to background levels thereafter. Similar behavior is observed in measurements of the binding of CR3 either to C3bi (Wright and Meyer, 1986) or to ligands on EC (Lo *et al.*, 1989a). Subsequent studies by other investigators have extended these observations to LFA-1 on lymphocytes (Dustin and Springer, 1989) and β1 integrins on fibroblasts (Danilov and Juliano, 1989). A similar regulation of the β3 integrin, gpIIb/IIIa, has long been appreciated (reviewed in Phillips *et al.*, 1991), and with the recognition that gpIIb/IIIa and CR3 are structural homologues (Hynes, 1987), it has become clear that this type of regulation may be a general property of integrins.

Several stimuli are capable of increasing the avidity of CR3 in a transient manner. These include the chemotactic stimuli C5a (Lo *et al.*, 1989a), formyl peptides, LTB_4 (Wright, unpublished), Ser-NAP-1/IL-8, Ala-NAP-1/IL-8, NAP-2, and melanoma growth-stimulating activity (gro/MGSA) (Detmers *et al.*, 1990, 1991), as well as the proinflammatory cytokines TNF (Lo *et al.*, 1989a), G-CSF, and GM-CSF (P. A. Detmers, personal communication). Since the chemotactic movement of PMN requires attachment to the substrate, and since leukocyte integrins are necessary for this attachment (Todd and Freyer, 1988), the activation of the leukocyte integrins appears to be an integral part of chemotaxis. The transient nature of the adhesion also appears necessary for chemotaxis.

A moving cell must not only adhere to the substrate at the leading edge, but must also be capable of selectively detaching its uropod if locomotion is to occur. The loss of avidity of the leukocyte integrins may enable the uropod to detach and the receptors to be reutilized for fresh rounds of adhesion.

5. Adhesion Molecules May Function Sequentially in a Cascade

The stimuli that activate leukocyte integrins are not limited to soluble chemoattractant factors, but may also include other adhesion molecules. We have shown that the adhesion of PMN to stimulated EC causes a dramatic increase in the binding activity of CR3 on the adherent PMN (Lo et al., 1991). Several lines of evidence indicate that this enhancement is caused by the EC surface glycoprotein, ELAM-1 (endothelial leukocyte adhesion molecule-1). Only EC treated with agonists that promote the expression of ELAM-1 on the cell surface are capable of enhancing CR3 function, and the time course of the acquisition of CR3-activating ability corresponds well to the time course of expression of ELAM-1. Purified recombinant ELAM-1 on plastic substrates causes the activation of CR3, and blockade of ELAM-1 on a plastic surface or on EC with anti-ELAM-1 antibody prevents the activation of CR3 on PMN. These observations suggest that the adhesion molecule ELAM-1 not only serves to attach PMN to the endothelium, but also recruits the function of additional adhesion molecules, the leukocyte integrins. The ability of ELAM-1 to induce a response in PMN suggests that the ligand(s) for ELAM-1 may be more appropriately termed "counter-receptors," since they are apparently capable of initiating response in the cells that bear them. We thus predict that at least one of the ligands recognized by ELAM-1 will be a receptor that is able to generate intracellular signals.

Our data suggest that ELAM-1 and the leukocyte integrins function in a sequence or cascade. Cytokines elaborated at an inflammatory locus induce the expression of ELAM-1 on the EC. A PMN entering an inflamed vascular bed is likely to bind first to the ELAM-1 on the EC. The function of leukocyte integrins induced by the binding of PMN to ELAM-1 may then serve in the next event, the movement of PMN between EC into the surrounding tissue. The sequential nature of the function of these molecules may explain why antibodies against the β chain of the leukocyte integrins completely block movement of PMN into inflammatory sites, despite the presence of ELAM-1 on the EC lining vessels at such sites (e.g., see Tuomanen et al., 1989). Recent studies have described additional adhesion cascades employing a variety of receptors (reviewed in Schweighoffer and Shaw, 1992): for example, the ligation of PECAM (CD31) on lymphocytes may result in the enhanced activity of β1 integrins (Tanaka et al., 1992), and the ligation of CD3/TCR on T cells may result in the enhanced activity of β2 integrins (Dustin and Springer, 1989). While the particular receptors vary, to date all adhesion cascades appear to exploit the ability of integrins to modulate their adhesion-promoting capacity.

6. The Mechanism of Regulation of the Binding Activity of CD18 Molecules

Several results indicate that the activation of leukocyte integrins for binding occurs through alterations in the properties of existing cell surface receptors, not through changes in their number. For example, changes in the avidity of LFA-1 on lymphocytes (Dustin and Springer, 1989) and PMN (Lo *et al.*, 1989a) are not accompanied by changes in the number of cell surface receptors. Agonists that enhance the binding activity of CR3 on PMN do cause a two- to threefold rise in the expression of CR3 through the exocytosis of specific granules, but this increase in receptor expression appears unnecessary for the much larger change in CR3 function in response to these agonists. Cytoplasts of PMN created by centrifuging cytochalasin-treated cells on a Ficoll step gradient are devoid of nuclei, mitochondria, and all intracellular granules. Cytoplasts cannot alter the number of cell surface receptors because they lack an intracellular store, but they nonetheless exhibit a transient increase in CR3-dependent adhesion to EC (Lo *et al.*, 1989a; Detmers *et al.*, 1990; Vedder and Harlan, 1988) or to C3bi-coated erythrocytes (Detmers *et al.*, 1990).

Additional studies have shown that not only β2, but also β1 (Danilov and Juliano, 1989; Shimizu *et al.*, 1990) and β3 integrins (Phillips *et al.*, 1991; Shattil *et al.*, 1985; Coller, 1986; Plow and Ginsberg, 1989) exhibit striking changes in their ability to promote adhesion without corresponding changes in the number of receptors at the cell surface. Several results suggest that conformational changes in integrins underlie their regulated binding behavior. The stimulation of platelets causes a coordinate increase in the avidity of gpIIb/IIIa (αIIbβ3) and the induction of a neoepitope on the gpIIb/IIIa molecule, suggesting that a conformational change accompanies activation (Shattil *et al.*, 1985; Coller, 1986; Plow and Ginsberg, 1989). The avidity of β1 (Neugebauer and Reichardt, 1991), β2 (Keizer *et al.*, 1988; van Kooyk *et al.*, 1991; Robinson *et al.*, 1992), and β3 (Kouns *et al.*, 1990; O'Toole *et al.*, 1990) integrins can be strongly enhanced by incubating cells with certain monoclonal antibodies against the corresponding integrins, suggesting that the binding of these monoclonals can induce conformational changes associated with activation. Most importantly, the avidity of purified gpIIb/IIIa can be enhanced by pretreating the purified receptor with peptide ligands (Du *et al.*, 1991; Parise *et al.*, 1987), the avidity of purified αvβ3 can be altered by reconstituting the purified receptor in liposomes of different lipid compositions (Conforti *et al.*, 1990), and our studies (see below) have shown that the avidity of purified CR3 can be altered by adding a specific lipid (Hermanowski-Vosatka *et al.*, 1991). In these experiments, avidity changes were observed with purified receptors, thus eliminating the possible contribution of cellular processes.

7. Integrin Modulating Factor (IMF-1)

We have recently described a mechanism by which the conformational changes of CR3 may be driven in cells (Hermanowski-Vosatka *et al.*, 1991). Stimulated PMN contain a novel lipid, IMF-1, which can induce a large increase in the avidity of the $\beta 2$ integrins CR3 and LFA-1. The change in avidity can be observed whether the purified lipid is added to living cells or to purified CR3 adsorbed to a plastic surface. The ability of the lipid to affect purified CR3 indicates that cellular components or metabolism are not required to observe avidity changes in CR3, and that the lipid must interact directly with CR3. The ability of the lipid to affect the avidity of CR3 in living cells suggests that it is not simply correcting an artifactual blockade of CR3 activity caused by the *in vitro* conditions. Further studies show that adding IMF-1 changes the avidity of CR3, not just for C3bi, but for other distinct ligands (fibrinogen, LPS, and structures on EC).

The addition of IMF-1 is not absolutely necessary for the binding of ligand by purified CR3. Binding of C3bi-coated erythrocytes to CR3-coated surfaces can be observed in the absence of IMF-1 if very high densities of both receptor and ligand are employed (Van Strijp *et al.*, 1993). This result suggests that IMF-1 raises the avidity of CR3, and that in the absence of IMF-1, the avidity of CR3 is low but measurable. We cannot, however, rule out the possibility that a sub-population of CR3 molecules exhibits the "active" conformation without the intentional addition of IMF-1.

Several observations suggest that IMF-1 may represent a physiological regulator of CR3 activity in living cells. Resting PMN express CR3 in a low-avidity state (Wright and Meyer, 1986), and IMF-1 cannot be isolated from these cells (Hermanowski-Vosatka *et al.*, 1991). Within minutes of stimulation with any of several agonists, CR3 is rapidly "activated," and we have observed a coincident rise in the amount of IMF-1 that can be extracted from the cells. Further incubation results in a loss of CR3 activity, and the amount of IMF-1 in the cells declines in a synchronous fashion. Additional studies have shown that cells without integrins, such as erythrocytes, neither contain nor can be stimulated to manufacture IMF-1, while cells such as macrophages, which express constitutively active integrins, show constitutively high levels of IMF-1 (A. Hermanowski-Vosatka and SDW, unpublished observations).

It appears that IMF-1 is rapidly metabolized or inactivated after its manufacture. The IMF-1 content of cells declines to baseline by 30 min after stimulation (Hermanowski-Vosatka *et al.*, 1991). Moreover, the addition of exogenous IMF-1 causes a transient rise in CR3 activity that also declines to baseline by 20 min. The decline in CR3 activity is unlikely to be caused by "desensitization" of the cells, since successive additions of fresh IMF-1 each cause a transient rise in CR3 activity. In addition, the desensitization of PMN with agonists such as PMA does not affect their ability to respond to exogenously added IMF-1.

Our studies support the following model for the control of CR3 function. In resting cells, CR3 exists in a low-avidity state. Upon stimulation, PMN elaborate the lipid IMF-1 that binds CR3 and causes a conformational change to a high-avidity state. Destruction of the IMF-1 then enables CR3 to relax to a low-avidity state. The rapid production of IMF-1 is consonant with the rapid rise in adhesivity seen when the cells are stimulated, and the labile nature of IMF-1 may allow CR3 to be inactivated quickly. The rapid destruction of IMF-1 may also permit the formation of gradients of IMF-1 within a cell, which enables adhesion at the leading edge and release at the uropod. The above model posits that IMF-1 represents a principal physiologic modulator of the activity of $\beta 2$ integrins in neutrophils. We hasten to point out, however, that our studies do not rule out the possibility that alternative or additional mechanisms may also serve to drive conformational changes in CR3 or other integrins.

7.1. Specificity

Recent studies with platelets (Smyth *et al.*, 1992) have provided evidence that lipids may also play a role in regulating $\beta 3$ integrins. A lipid that altered the avidity of purified gpIIb/IIIa was found in extracts of stimulated, but not unstimulated, platelets. These studies confirm the generality of the scheme described above, and raise the question of whether IMF-1 modulates all integrins or a subset of integrins. We have recently found that IMF-1 is not active in regulating the avidity of either $\beta 1$ or $\beta 3$ integrins (A. Hermanowski-Vosatka, W. J. Swiggard, R. L. Silverstein, and S. D. Wright, unpublished), and IMF-1 does not affect the avidity of purified gpIIb/IIIa (S. S. Smyth, L. V. Parise, and S. D. Wright, unpublished observations). These data suggest that IMF-1 may be restricted to the regulation of $\beta 2$ integrins. We would thus suggest that structurally distinct IMFs may exist to regulate other integrins. If this is the case, cells may activate their $\beta 2$ integrins by producing IMF-1, yet leave their $\beta 1$ integrins in an inactive state. Since platelets produce a distinct lipid with the capacity to modulate a second class of integrins, it is possible that several IMFs may exist to allow independent regulation of the several classes of integrins that may exist on a single cell. Recent studies have shown, in fact, that different stimuli may differentially activate $\beta 1$ or $\beta 2$ integrins on a single cell (Tanaka *et al.*, 1992).

7.2. Structure

A preliminary chemical analysis of IMF-1 has revealed that it is an anionic, hydrophobic molecule of $M_r \sim 340$ Da (Hermanowski-Vosatka et al., 1991). Additional studies on the susceptibility of IMF-1 to degradative procedures suggest that IMF-1 does not contain ester, amide, glycosidic, or peptide bonds, and that it does not contain phosphate, sialic acid, or a chromophore. Ozonoloysis causes quantitative destruction of IMF-1, suggesting the presence of double bonds. These studies narrow the list of potential candidates, but a very large number of consistent structures can still be drawn. We currently favor the hypothesis that IMF-1 is an unsaturated, hydroxylated fatty acid. A large body of work has demonstrated the rapid, stimulus-dependent release of arachidonic acid from membrane phospholipids and its rapid metabolism to prostaglandins and leukotrienes. The enzymes necessary for the release of fatty acids and for their subsequent metabolism are present in a wide variety of cell types, and PMN are known to release large amounts of cellular fatty acids in response to the agonists we have employed. Work is currently under way to determine the structure of IMF-1 by unambiguous chemical means.

ACKNOWLEDGMENTS. Supported by USPHS grants AI22003 and AI30556. I thank Dr. Patricia Detmers for critical reading of the manuscript.

References

Altieri, D. C., and Edgington, T. S., 1988, The saturable high affinity association of factor X to ADP-stimulated monocytes defines a novel function of the Mac-1 receptor, J. Biol. Chem. 263:7007–7015.

Altieri, D. C., Bader, R., Mannucci, P. M., and Edgington, T. S., 1988, Oligospecificity of the cellular adhesion receptor MAC-1 encompasses an inducible recognition specificity for fibrinogen, J. Cell Biol. 107:1893–1900.

Arfors, K. E., Lundberg, C., Lindblom, L., Lundberg, K., Beatty, P. G., and Harlan, J. M., 1987, A monoclonal antibody to the membrane glycoprotein complex CD18 inhibits polymorphonuclear leukocyte accumulation and plasma leakage in vivo, Blood 69:338–340.

Coller, B. S., 1986, Activation affects access to the platelet receptor for adhesive glycoproteins, J. Cell Biol. 103:451–456.

Conforti, G., Zanetti, A., Pasquali-Ronchetti, I., Quaglino, D., Jr., Neyroz, P., and Dejana, E., 1990, Modulation of vitronectin receptor binding by membrane lipid composition, J. Biol. Chem. 265:4011–4019.

Danilov, Y. N., and Juliano, R. L., 1989, Phorbol ester modulation of integrin-mediated cell adhesion: A postreceptor event, J. Cell Biol. 108:1925–1933.

Detmers, P. A., Lo, S. K., Olsen-Egbert, E., Walz, A., Baggiolini, M., and Cohn, Z. A., 1990, NAP-1/IL-8 stimulates the binding activity of the leukocyte adhesion receptor CD11b/CD18 on human neutrophils, J. Exp. Med. 171:1155–1162.

Detmers, P. A., Powell, D. E., Walz, A., Clark-Lewis, I., Baggiolini, M., and Cohn, Z. A., 1991,

Differential effects of neutrophil-activating peptide 1/IL-8 and its homologues on leukocyte adhesion and phagocytosis, *J. Immunol.* **147**:4211–4217.

Diamond, M. S., Staunton, D. E., de Fougerolles, A. R., Stacker, J., Garcia-Aguilar, J., Hibbs, M. L., and Springer, T. A., 1990, TA ICAM-1 (CD54): A counter-receptor for Mac-1 (CD11b/CD18), *J. Cell Biol.* **111**:3129–3139.

Du, X., Plow, E. F., Frelinger, A. L., O'Toole, T. E., Loftus, J. C., and Ginsberg, M. H., 1991, Ligands "activate" integrin αIib-β3 (platelet GPIIa-IIa), *Cell* **65**:409–416.

Dustin, M. L., and Springer, T. A., 1989, T-cell receptor cross-linking transiently stimulates adhesiveness through LFA-1, *Nature* **341**:619–624.

Harlan, J. M., Killen, P. D., Senecal, F. M., Schwartz, B. R., Yee, E. K., Taylor, F. R., Beatty, P. G., Price, T. H., and Ochs, H. D., 1985, The role of neutrophil membrane glycoprotein GP-150 in neutrophil adhesion to endothelium *in vitro*, *Blood* **66**:167–178.

Hermanowski-Vosatka, A., Swiggard, W. J., Silverstein, R. L., and Wright, S. D. (unpublished).

Hermanowski-Vosatka, A., Van Strijp, J. A. G., Swiggard, W. J., and Wright, S. D., 1991, Integrin modulating factor-1: A lipid that alters the function of adhesion receptors, *Cell* **68**:341–352.

Hernandez, L. A., Grisham, M. B., Twohig, B., Arfors, K. E., Harlan, J. M., and Granger, D. N., 1987, Role of neutrophils in ischemia-reperfusion-induced microvascular injury, *Am. J. Physiol.* **253**:H699–H703.

Hynes, R. O., 1987, Integrins: A family of cell surface receptors, *Cell* **48**:549–554.

Keizer, G. D., Visser, W., Vliem, M., and Figdor, C. G., 1988, A monoclonal antibody (NKI-L16) directed against a unique epitope of the α-chain of human leukocyte function-associated antigen 1 induces homotypic cell–cell interactions, *J. Immunol.* **140**:1393–1400.

Kouns, W. C., Wall, C. D., White, M. M., Fox, C. F., and Jennings, K., 1990, A conformation-dependent epitope of human platelet glycoprotein IIIa, *J. Biol. Chem.* **265**:20594–20601.

Lo, S. K., Detmers, P. A., Levin, S. M., and Wright, S. D., 1989a, Transient adhesion of neutrophils to endothelium, *J. Exp. Med.* **169**:1779–1793.

Lo, S. K., Van Seventer, G. A., Levin, S. M., and Wright, S. D., 1989b, Two leukocyte receptors (CD11a/CD18 and CD11b/CD18) mediate transient adhesion to endothelium by binding to different ligands, *J. Immunol.* **143**:3325–3329.

Lo, S. K., Lee, S., Ramos, R. A., Lobb, R., Rosa, M., Chi-Rosso, G., and Wright, S. D., 1991, ELAM-1 stimulates the adhesive activity of leukocyte integrin CR3 (CD11b/CD18, Mac-1, $\alpha_m\beta_2$) on human neutrophils, *J. Exp. Med.* **173**:1493–1500.

Loike, J. D., Sodeik, B., Cao, L., Leucona, S., Weitz, J. I., Detmers, P. A., Wright, S. D., and Silverstein, S. C., 1991, CD11c/CD18 on neutrophils recognizes a domain at the N-terminus of the Aα chain of fibrinogen, *Proc. Natl. Acad. Sci. USA* **88**:1044–1048.

Lundberg, C., and Wright, S. D., 1990, Relation of the CD11/CD18 family of leukocyte antigens to the transient neutropenia caused by chemattractants, *Blood* **76**:1240–1245.

Neugebauer, K. M., and Reichardt, L. F., 1991, Cell-surface regulation of β1-integrin activity on developing retinal neurons, *Nature* **350**:68–71.

O'Toole, T. E., Loftus, J. C., Du, X., Glass, A. A., Ruggieri, Z. M., Shattil, S. J., Plow, E. F., and Ginsberg, M. H., 1990, Affinity modulation of the $\alpha_{11b}\beta_3$ integrin (platelet GP11b-111a) is an intrinsic property of the receptor, *Cell Regul.* **1**:883–893.

Parise, L. V., Helgerson, S. L., Steiner, B., Nannizzi, L., and Phillips, D. R., 1987, Synthetic peptides derived from fibrinogen and fibronectin change the conformation of purified platelet glycoprotein 11b-111a, *J. Biol. Chem.* **262**:12597–12602.

Phillips, D. R., Charo, I. F., and Scarborough, R. M., 1991, GPIIb-IIIa: The responsive integrin, *Cell* **65**:359–362.

Plow, E. F., and Ginsberg, M. H., 1989, Cellular adhesion: gp11b/111a as a prototypic adhesion receptor, *Prog. Hemostasis Thromb.* **9**:117–156.

Postigo, A. A., Corbi, A. L., Sanchez-Madrid, F., and de Landazuri, M. O., 1991, Regulated expression and function of CD11c/CD18 integrin on human B lymphocytes. Relation between attachment to fibrinogen and triggering of proliferation through CD11c/CD18, *J. Exp. Med.* **174:**1313–1322.

Robinson, A. K., Andrew, D., Rosen, H., Brown, D., Ortlepp, S., Stephens, P., and Butcher, E. C., 1992, Antibody against the Leu-CAM β-chain (CD18) promotes both LFA-1 and CR3-dependent adhesion events, *J. Immunol.* **148:**1080–1085.

Rosen, H., and Gordon, S., 1987, Monoclonal antibody to the murine type 3 complement receptor inhibits adhesion of myelomonocytic cells *in vitro* and inflammatory cell recruitment *in vivo, J. Exp. Med.* **166:**1685–1701.

Schmalstieg, F. C., Rudloff, H. E., Hillman, G. R., and Anderson, D. C., 1986, Two-dimensional and three-dimensional movement of human polymorphonuclear leukocytes: Two fundamentally different mechanisms of locomotion, *J. Leukocyte Biol.* **40:**677–691.

Schweighoffer, T., and Shaw, S., 1994, Concepts in adhesion regulation, in: *Handbook of Immunopharmacology: Adhesion Molecules* (C. D. Wegner, ed.), Academic Press, New York, in press.

Shattil, S. J., Hoxie, J. A., Cunningham, M., and Brass, L., 1985, Changes in the platelet membrane glycoprotein IIb/IIIa complex during platelet activation, *J. Biol. Chem.* **260:**11107–11114.

Shimizu, Y., van Seventer, G., Horgan, K. J., and Shaw, S., 1990, Regulated expression and function of three VLA (β1) integrin receptors on T cells, *Nature (London)* **345:**250.

Simpson, P. J., Todd, R. F., III, Fantone, J. C., Mickelson, J. K., Griffin, J. D., and Lucchesi, B. R., 1988, Reduction of experimental canine myocardial reperfusion injury by a monoclonal antibody (anti-Mo1, anti-CD11b) that inhibits leukocyte adhesion, *J. Clin. Invest.* **81:**624–629.

Smith, C. W., Rothlein, R., Hughes, B. J., Mariscalco, M. M., Rudloff, H. E., Schmalstieg, F. C., and Anderson, D. C., 1988, Recognition of an endothelial determinant for CD18-dependent human neutrophil adherence and transendothelial migration, *J. Clin. Invest.* **82:**1746–1756.

Smyth, S. S., Hillery, C. A., and Parise, L. V., 1992, Fibrinogen binding to purified platelet glycoprotein IIb-IIIa (integrin $\alpha_{IIb}\beta_3$) is modulated by lipids, *J. Biol. Chem.* **267:**15568–15577.

Tanaka, Y., Albelda, S. M., Horgan, K. J., van Seventer, G. A., Shimizu, Y., Newman, W., Hallam, J., and Newman, P. J., 1992, CD31 expressed on distinctive T cell subsets is a preferential amplifier of β1 integrin-mediated adhesion, *J. Exp. Med.* **176:**245–253.

Todd, R. F., III, and Freyer, D. R., 1988, The CD11/CD18 leukocyte glycoprotein deficiency, in: *Hematology/Oncology Clinics of North America* (J. T. Curnutte, ed.), Saunders, Philadelphia, pp. 13–31.

Tuomanen, E. I., Saukkonen, K., Sande, S., Cioffe, C., and Wright, S. D., 1989, Reduction of inflammation, tissue damage, and mortality in bacterial meningitis in rabbits treated with monoclonal antibodies against adhesion-promoting receptors of leukocytes, *J. Exp. Med.* **170:**959–968.

van Kooyk, Y., Weder, P., Hogervorst, F., Verhoeven, A. J., van Seventer, G., te Velde, A. A., Borst, J., Keizer, G. D., and Figdor, C. G., 1991, Activation of LFA-1 through a calcium-dependent epitope stimulates lymphocyte adhesion, *J. Cell Biol.* **112:**345–354.

Van Strijp, J. A. G., Russell, D. G., Tuomanen, E., Brown, E. J., and Wright, S. D., 1993, Ligand specificity of purified complement receptor type three (CD11b/CD18, Mac-1); indirect effects of an Arg-Gly-Asp (RGD) sequence, *J. Immunol.* **151:**3324–3336.

Vedder, N. B., and Harlan, J. M., 1988, Increased surface expression of CD11b/CD18 (Mac-1) is not required for stimulated neutrophil adherence to cultured endothelium, *J. Clin. Invest.* **81:**676.

Wright, S. D., 1992, Receptors for complement and the biology of phagocytosis, in: *Inflammation* (J. I. Gallin, I. M. Goldstein, and R. Snyderman, eds.), Raven Press, New York, Chapter 25.

Wright, S. D., and Detmers, P. A., 1991, Receptor-mediated phagocytosis in human leukocytes, in: *The Lung: Scientific Foundations* (R. G. Crystal and J. B. West, eds.), Raven Press, New York, pp. 539–551.

Wright, S. D., and Meyer, B. C., 1986, Phorbol esters cause sequential activation and deactivation of complement receptors on polymorphonuclear leukocytes, *J. Immunol.* **136:**1759.

Wright, S. D., Rao, P. E., Van Voorhis, W. E., Craigmyle, L. S., Iida, K., Talle, M. A., Westberg, E. F., Goldstein, G., and Silverstein, S. C., 1983, Identification of the C3bi receptor on human monocytes and macrophages by using monoclonal antibodies, *Proc. Natl. Acad. Sci. USA* **80:**5699.

Wright, S. D., Weitz, J. I., Huang, A. J., Levin, S. M., Silverstein, S. C., and Loike, J. D., 1988, Complement receptor type three (CR3, CD11b/CD18) of human polymorphonuclear leukocytes recognizes fibrinogen, *Proc. Natl. Acad. Sci. USA* **85:**7734.

3

Structure–Function Aspects of Selectin–Carbohydrate Interactions

LAURENCE A. LASKY, LEONARD G. PRESTA, and DAVID V. ERBE

1. Introduction

A major parameter in the migration of leukocytes to sites of acute or chronic inflammation involves the adhesive interactions between these cells and the endothelium. This specific adhesion is the initial event in the cascade initiated by inflammatory insults, and is, therefore, of paramount importance to the regulated defense of the organism. While the lack of such adhesive responses to inflammatory stimuli results in disastrous consequences for the individual (Anderson and Springer, 1987), there are many instances in which such inflammatory reactions result in tissue damage rather than protection. Such chronic instances of inflammatory cell-mediated damage are seen in a variety of diseases including arthritis, psoriasis, inflammatory bowel disease, and other diverse examples. In addition to the chronic diseases, there are also a number of more life-threatening acute instances of inappropriate inflammation. These include the inflammation of various organs by neutrophils during acute responses to septic shock and trauma. It is obvious, therefore, that drugs that act as inhibitors of leukocyte adhesion during chronic or acute inflammation would be widely used by the medical community.

In order to develop such drugs, it is critically important to understand the molecules involved with cell adhesion during inflammation. There are currently four types of cell adhesion molecules involved in the interaction between white

LAURENCE A. LASKY, LEONARD G. PRESTA, and DAVID V. ERBE • Department of Immunology, Genentech, Inc., South San Francisco, California 94080.

Cellular Adhesion: Molecular Definition to Therapeutic Potential, edited by Brian W. Metcalf, Barbara J. Dalton, and George Poste. Plenum Press, New York, 1994.

cells and the endothelium during an inflammatory response: the integrins, their ligands the immunoglobulin superfamily (IgG) members, the selectins, and their carbohydrate and glycoprotein ligands (Springer, 1990; Lasky, 1992). While other chapters in this volume will deal with the integrins, the IgG superfamily members, and their roles in inflammation, this chapter will discuss the molecular details of some of the interactions between the selectins and their carbohydrate ligands.

As has been recently summarized (Lasky, 1992), the leukocyte or L-selectin, the endothelium or E-selectin, and the platelet/endothelium or P-selectin glycoproteins mediate regional cell adhesive interactions by recognizing cell-specific plasma membrane carbohydrates by the calcium-dependent lectin, or carbohydrate binding, motif found in this family of adhesion molecules. These three adhesive glycoproteins are expressed on different cell types, and their expression is regulated by a diversity of inflammatory mediators (Table I). They appear to mediate the initial low-affinity, or "rolling," interaction between leukocytes and the endothelium adjacent to an inflammatory site. This low affinity interaction appears to be an absolute pre-requisite to the higher affinity binding that is mediated by the Integrin-IgG superfamily type of adhesion (Lasky, 1992). Thus, the selectins are among the first participants in a cascade of events that ultimately leads to the migration of inflammatory cells from the bloodstream to the tissues.

Because of their involvement in the initial events that precede inflammation, inhibitors of the selectins may be excellent anti-inflammatory reagents. A large number of drug companies are currently using diverse strategies to develop antagonists of selectin-mediated adhesion. These strategies include developing small-molecule blockers based on the known structures of naturally occurring carbohydrate ligands for the selectins. An understanding of the molecular details involved in the interactions between the selectins and their carbohydrate ligands may allow for a more directed approach to the design of such drug candidates. This chapter will discuss our current understanding of these interactions, and relate these data to the ultimate development of selectin antagonists.

2. Selectin Carbohydrate Ligands

Information regarding the role of carbohydrate–protein interactions in immune cell adhesion was first derived from work on the adhesive interaction between lymphocytes and the high endothelial venules of peripheral lymph nodes (Stoolman, 1989). These early data demonstrated that this type of binding could be inhibited by certain monomeric and polymeric carbohydrates that were negatively charged. These data were supported by subsequent experiments, which demonstrated that removing sialic acid, a negatively charged carbohydrate con-

Table I. Selectin Structure and Function

Structure[a]	Name	Location	Expression	Adherent cell types	Proposed function
	L-Selectin	Leukocytes (constitutive)	Decreases upon cell activation	PLN[b] endothelium Endothelium adjacent to inflammatory sites (rolling)	Lymphocyte recirculation through PLN Neutrophil (+ other leukocyte?) inflammation
	E-Selectin	Endothelium (transcriptionally activated)	Increases upon inflammatory activation (IL-1, TNF, LPS) (~ hours)	Monocytes Neutrophils (rolling) T cell subsets (cutaneous?)	Leukocyte inflammation
	P-Selectin	Platelets (α-granules) Endothelium (Weibel Palade bodies)	Increases upon thrombin activation, histamine, substance P, peroxide (~ minutes)	Monocytes Neutrophils (rolling) T cell subsets (cutaneous?)	Leukocyte inflammation

[a]L, lectin; E, epidermal growth factor-like; C, complement binding proteinlike.
[b]PLN, peripheral lymph node.

stituent of the cell surface, abolished the ability of lymphocytes to adhere to peripheral lymph node endothelium *in vitro* and *in vivo* (Rosen *et al.*, 1985). The data were ultimately unified when the carbohydrate binding protein on the lymphocyte surface that recognized these diverse negatively charged carbohydrate ligands turned out to be the L-selectin adhesion molecule (Lasky *et al.*, 1989). These data thus provided conclusive evidence that a negatively charged carbohydrate ligand could mediate cell adhesion by binding to a cell surface, lectin-containing adhesion molecule, L-selectin.

The cDNA cloning of E- and P-selectin demonstrated that these molecules were highly homologous to L-selectin, and it seemed likely that they, too, were involved with cell adhesion by carbohydrate recognition (Stoolman, 1989). Initial data concerning the calcium dependence of adhesion mediated by E- and P-selectin were consistent with a role for the calcium-dependent lectin domains of these glycoproteins in cell binding. In order to more fully characterize this interaction, a number of investigators used a diversity of techniques to demonstrate the nature of the carbohydrate ligands for E- and P-selectin. Lowe *et al.* (1990) demonstrated that transfection of a single fucosyl transferase into mammalian cells resulted in E-selectin binding. These data were consistent with the possibility that a carbohydrate ligand was involved with E-selectin binding, and that this carbohydrate ligand contained fucose in a specific linkage. Previous characterization of this glycosylating enzyme suggested that it added fucose to N-acetylglucosamine in an $\alpha(1-3)$ linkage, suggesting that the carbohydrate ligand recognized by E-selectin contained fucose specifically linked to this sugar. These data were supported by two other laboratories, which demonstrated that the binding of E-selectin to myeloid cells could be inhibited by an antibody directed against a myeloid-specific carbohydrate, sialyl Lewis[x] (Phillips *et al.*, 1990; Walz *et al.*, 1990). Data supporting a role for sialyl Lewis[x] in E-selectin-mediated adhesion were obtained by demonstrating that a glycolipid form of this carbohydrate could inhibit the E-selectin-mediated adhesion of neutrophils. These data proved to be entirely consistent with the transfection work of Lowe *et al.*, since the structure of sialyl Lewis[x], sialic acid $\alpha(2-3)$ galactose-$\beta(1-3)$ [fucose-$\alpha(1-3)$]-N-acetylglucosamine, contained a fucose residue in the predicted linkage [$\alpha(1-3)$] with N-acetylglucosamine. Further confirmation of these data were obtained by mass spectral analysis of glycolipids from neutrophils that conferred E-selectin binding (Tiemeyer *et al.*, 1991). These results demonstrated that a sialyl Lewis[x]-related carbohydrate, recognized by the VIM 2 antibody, could also mediate E-selectin adhesion, although with much lower affinity than sialyl Lewis[x]. Finally, the inclusion of a sialic acid in a specific 2–3 linkage to galactose in sialyl Lewis[x] was reminiscent of the data concerning the requirement for this negatively charged carbohydrate in L-selectin-mediated adhesion. Subsequent work has clearly revealed that E-selectin-mediated adhesion also requires sialic acid (Tyrrell *et al.*, 1991). Figure 1 illustrates the energy-minimized structure of sialyl Lewis[x].

An analysis of P-selectin-mediated adhesion has proved somewhat more confusing. A sialyl Lewis[x]-like carbohydrate, LNF III, inhibited P-selectin-mediated adhesion, although high concentrations of the carbohydrate were required (Larsen *et al.*, 1990). An antibody against Lewis[x], an unsialated form of the E-selectin carbohydrate ligand, reacted with LNF III and also inhibited P-selectin-mediated adhesion. Additional data have suggested that sulfatide (galactose-4-sulfate ceramide) can also serve as a ligand for P-selectin (Aruffo *et al.*, 1991), although the fact that this interaction is calcium independent argues against a physiological role for this glycolipid. Other data have demonstrated that P-selectin-mediated adhesion is sensitive to treatment of cells with sialidase, suggesting that sialic acid is required for adhesion mediated by all three selectins (Corrall *et al.*, 1990). This proposal was supported by data demonstrating that antibody to sialyl Lewis[x] inhibited P-selectin-mediated adhesion, as did a glycolipid form of this carbohydrate in solution (Polley *et al.*, 1991). This latter result is consistent with the idea of sialyl Lewis[x] as a ligand for P-selectin. However, this carbohydrate alone seems insufficient for high-affinity adhesion. For example, Chinese hamster ovary cells expressing sialyl Lewis[x] showed low-affinity, nonsaturable binding to P-selectin, while neutrophils showed high-affinity, saturable binding (Moore et al., 1991). This result may be explained by the results of protease digestion experiments, which revealed that P-selectin-mediated adhesion of neutrophils is protease sensitive, while E-selectin-mediated adhesion is not (Larsen *et al.*, 1990). These results are consistent with the idea of a glycoprotein contribution to the ligand activity of P-, but not E-, selectin-mediated adhesion, and a candidate neutrophil surface glycoprotein that binds P-selectin has recently been described (Moore *et al.*, 1992). In summary, these data suggest that sialyl Lewis[x] is a component of the P selectin ligand, but the carbohydrate alone appears insufficient to mediate high-affinity adhesion.

Another interesting confirmation of the importance of the conformation of the sialyl Lewis[x] ligand to selectin-mediated adhesion has come from the analysis of another ligand for E-selectin, sialyl Lewis[a] (Berg *et al.*, 1991; Tyrrell *et al.*, 1991). This nonmyeloid carbohydrate has the same composition as sialyl Lewis[x], but the fucose and *N*-acetylglucosamine are found in different linkages: sialic acid-α(2–3)galactose-β(1–3)[fucose-α(1–4)]*N*-acetylglucosamine. This carbohydrate also shows ligand activity for E-selectin-mediated binding, suggesting that although it may not be a naturally occurring ligand, it can still function as a high-affinity ligand for E-selectin under *in vivo* conditions. As Fig. 1 shows, energy-minimized structural analysis of sialyl Lewis[a] reveals that its conformation is virtually identical to sialyl Lewis[x]. This conformation reveals that the sialic acid and fucose residues both point to the same "face" of the carbohydrate ligand. Since these two sugars are critical for selectin-mediated binding, it seems obvious that the similar conformations of these two selectin ligands have important structural implications for recognition by the lectin domain.

3. Structure–Function Analysis of E-Selectin by Mutagenesis

We began our analysis of the details of the interactions between E-selectin and its naturally occurring ligand, sialyl Lewis^x, with a mutagenesis approach (Erbe et al., 1992). We combined the mapping of epitopes recognized by a panel of blocking and nonblocking monoclonal antibodies (MAbs) to E-selectin with an investigation of sialyl Lewis^x binding. In order to fully characterize these interactions, we developed a new type of assay system that allowed us to rapidly analyze the effects of mutations on the structure and function of the E-selectin lectin domain. To do this, we converted a portion of the extracellular domain of E-selectin into an antibodylike molecule by joining it to the hinge, CH2 and CH3 domains of human IgG1 (Watson et al., 1991). This antibody chimera provided a number of advantages, including the ability to quantitate the amount of mutant chimera produced from each transfection and the ability to readily assay for binding of a given mutant to a MAB- or carbohydrate-coated solid phase (Foxall et al., 1992). Microtiter wells were coated either with a MAb to E-selectin or with sialyl Lewis^x glycolipid, after which a given mutant E-selectin–IgG chimera was incubated in the well. Binding of the IgG chimera was assayed by an antihuman IgG1 antibody conjugated to peroxidase. The absence of binding either to a given MAb or to sialyl Lewis^x would indicate that a mutation had an effect on the epitope recognized by the antibody or on a region of the lectin domain that was involved with carbohydrate recognition, respectively.

Mutagenesis of the E-selectin lectin domain was driven by two considerations. The first was provided by earlier data on the mapping of the L-selectin epitope recognized by the adhesion blocking MAb, MEL 14. These data suggested that this antibody bound to the N-terminus of the L-selectin lectin domain

---→

Figure 1. Energy-minimized solution structures of sialyl Lewis^x and sialyl Lewis^a. Photo courtesy of Dr. B. N. Narasinga Rao, Glycomed, Inc.

Figure 2. Models of the E- and P-selectin lectin domains. Illustrated are ribbon diagrams of the E-selectin (left) and P-selectin (right) lectin domains produced by modeling based on the X-ray coordinates of the mannose binding protein (Weis et al., 1991). Residues are colored as follows: for P-selectin, K8 light green, K84 light brown, S97 white, and top to bottom, Y48–K113–K111 yellow; for E-selectin, E8 dark green, R84 brown, R97 red, and top to bottom, Y48–K113–K111 yellow. The purple (residues 43–48) are dark blue (residues 94–100) loops in both models show two insertions in the selectin lectin domains relative to the mannose binding protein lectin. In each model, the single bound calcium is depicted as a light blue ball.

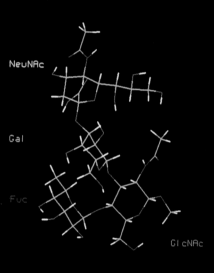

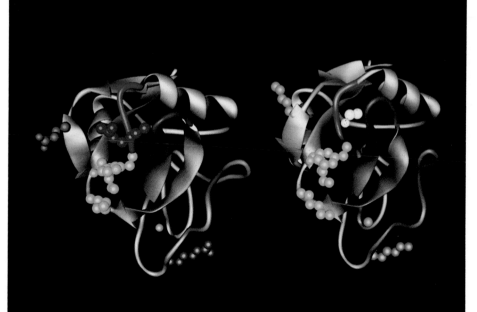

(Bowen *et al.*, 1990), and further unpublished analysis revealed that this antibody recognized a site within the first eight residues of the lectin motif (D. Erbe, unpublished data). These results suggested that, by analogy, the N-terminus of E-selectin might be involved with the binding of adhesion-blocking MAbs, sialyl Lewis[x], or both. Thus, the N-terminus of the E-selectin lectin was targeted for mutagenesis. The second consideration was derived from data concerning the charge nature of ligands for the selectins. Most previously published information concerning the carbohydrate ligands for these molecules suggested that they all contained a component that was anionic. Thus, it seemed likely that cationic residues, such as arginine and lysine, in the selectin lectin domain may have been involved with ligand recognition, perhaps through an electrostatic interaction. This supposition led us to mutagenize various cationic amino acids in the lectin domain to alanine.

Mutagenized E-selectin–IgG chimeras were initially analyzed for binding to a panel of blocking and nonblocking MAbs to E-selectin, and three results were obtained (Table II). Some of the mutants showed a lack of binding to the entire panel of E-selectin MAbs. These mutants were assumed to be structurally unstable proteins that were either unfolded or proteolytically degraded. A second subset of the antibodies no longer bound to certain E-selectin mutants. These mutants thus defined a region that appeared to be involved with the epitope recognized by a given MAb. In the case of the blocking MAbs, this information served to highlight regions of the E-selectin lectin domain that were potentially involved with carbohydrate recognition. In the case of loss of recognition by nonblocking MAbs, this type of mutant revealed regions of E-selectin that were probably not directly involved with ligand recognition. The third type of result demonstrated mutants that still bound all of the antibodies.

As Table II shows, specific regions of E-selectin appeared to be recognized by blocking and nonblocking antibodies. For example, the binding of the adhesion-blocking antibodies ENA1, BBA2, 8E4, and 7H5 were all disrupted by mutations in the N-terminus of E-selectin. This result helped to confirm the previous mapping of the blocking MAb MEL 14 to the N-terminus of L-selectin, and suggested a role for the N-terminus of E-selectin in carbohydrate ligand recognition and cell adhesion. Another blocking MAB, 3B7, was found to map to residues at positions 98 and 101, a region distant from the N-terminus. The potentially folded state of the lectin domain was highlighted by the fact that the blocking MAb 7H5, which mapped to an epitope in the N-terminus, showed a loss of binding by mutation of the lysine at position 113. This residue is near the C-terminus of the protein, and this result is consistent with the possibility that the N- and C-termini of the lectin domain are closely apposed to one another. Finally, the binding of a number of the nonblocking MAbs was disrupted by a mutation at position 74, which is consistent with the location of this residue at a site that is spatially distinct from the carbohydrate binding site. In summary, the

Table II. E-Selectin–IgG Mutant Binding Summary

E-Selectin–IgG Mutant	MAB binding[a]															sLEX binding
	BBA1	BBA2	ENA1	3B7	8E4	7H5	9H9	1B3	11G5	1E5	14G2	4D9	1D6	9A1	7E10	
S2T N4H T5Y	+	+	+	+	+	+	+	+	+	+	+	+	+	+	+	+
T7A A9N	+	+	+	+	-	-	+	+	+	+	+	+	+	+	+	ND
E8A	+	-	-	+	+	+	+	+	+	+	+	+	+	+	+	+++
M10Y T11S Y12W	+	+	+	+	+	+	+	+	+	+	+	+	+	+	+	+
K32A	+	+	+	+	+	+	+	+	+	+	+	+	+	+	+	+
K67A	+	+	+	+	+	+	+	+	+	+	+	+	+	+	+	+
K74A	+	+	+	+	+	+	+	+	+	+	+	+	-	+	+	+
R84A K86A	+	+	+	+	+	+	+	-	+	+	±	+	+	+	+	+
R97A	+	+	+	-	+	+	+	+	+	+	+	+	+	+	+	-
E98P V101T	+	+	+	+	+	±	+	+	+	+	+	+	+	+	+	ND
K99A	+	+	+	±	+	+	+	+	+	+	+	+	+	+	+	±
D100A	+	+	+	+	+	+	+	+	+	+	+	+	+	+	+	+
V101A	+	+	+	+	+	+	+	+	+	+	+	+	+	+	+	+
K113A	+	+	+	+	+	-	+	+	+	+	+	+	+	+	+	-
M10A	-	-	-	-	-	-	-	-	-	-	-	-	-	-	-	-
Y12A	-	-	-	-	-	-	-	-	-	-	-	-	-	±	-	-
E14A								Not expressed								
I93A	ND	ND	ND	-	-	-	-	-	-	-	-	-	-	±	±	ND
Y94A	ND	ND	ND	-	-	-	-	-	-	-	-	-	-	-	-	ND
I95A	ND	ND	ND	-	-	-	-	-	-	-	-	-	-	±	-	ND
K96A	-	-	-	-	-	-	-	-	-	-	-	-	-	±	±	-
E98A	ND	ND	ND	±	±	±	±	±	±	±	±	±	±	±	±	ND
M103A	±	±	±	-	-	-	-	-	-	-	-	-	-	±	±	ND
E14I, A15S, S16R, A17K	ND	-	-	-	-	-	-	-	-	-	-	-	-	-	-	-
Y18A, Q20A, R22A	-	-	-	-	-	-	-	-	-	-	-	-	-	-	-	-
N39A, Y44A	-	-	-	-	-	-	-	-	-	-	-	-	-	±	±	-
R54A, K55A, N57A	-	-	-	-	-	-	-	-	-	-	-	-	-	-	-	-
K96A, R97A, K99A	-	-	-	-	-	-	-	-	-	-	-	-	-	-	-	-
K111S, K112A, K113A	-	-	-	-	-	-	-	-	-	-	-	-	-	-	-	-

[a] Key: +, binding; ±, partial binding; -, no binding; ND, not determined.

mutagenesis analysis of MAb binding to E-selectin revealed that the N- and C-termini and a region encompassing residues 98–101 appeared to be involved with blocking antibody binding, while a region that contained residue 74 was involved with nonblocking MAb binding.

Mutants whose overall structure was not destroyed were subsequently analyzed for sialyl Lewis^x binding in the solid-state assay described above. Of course, the most interesting mutants to investigate first were those that disrupted the binding of blocking MAbs: the three results obtained from this analysis are summarized in Table II. The most interesting mutation was at position 113, a site that was shown to be involved with the binding of the blocking MAb 7H5. This mutant was completely unable to bind to the immobilized carbohydrate ligand. The comapping of a blocking antibody binding site with a site involved with carbohydrate recognition is consistent with this region's direct involvement with sialyl Lewis^x binding and cell adhesion. Mutation of the R at position 97, a region near residues 98–101 that is another site involved with the binding of a blocking antibody (3B7), also showed a complete loss of sialyl Lewis^x recognition. This result was consistent with the involvement of this site in cell adhesion as well. A second result demonstrated that many mutations had no effect on carbohydrate recognition. Perhaps most interesting was the K at position 74, whose mutation to A abolished the binding of a number of nonblocking MAbs, but which had no effect on the binding of immobilized carbohydrate. This result was consistent with this region being distant from the carbohydrate binding site. Finally, a third type of result was obtained with the N-terminal mutation at position 8. This change affected the binding of a number of blocking MAbs, but enhanced, rather than decreased, the binding of E-selectin to immobilized sialyl Lewis^x. In conclusion, the carbohydrate binding analysis of E-selectin mutants was consistent with the data derived from the MAb binding studies, and suggested that the N-terminus, a region surrounding residues 98–101, and the C-terminus were involved in carbohydrate recognition.

The data from a mutagenic analysis of E-selectin suggested that certain regions of the lectin domain were involved with carbohydrate recognition and cell adhesion. These results would be more interesting, however, if they could be put into a more three-dimensional context. While the crystallography of E-selectin is currently being attempted, a structure of this molecule is not yet available. However, a crystal structure for a related type C lectin, the mannose binding protein, has been reported (Weis *et al.*, 1991). As initially proposed by Weis *et al.* (1991), we have used this structure to construct a model of the E-selectin domain. Imposing the mutations described above on this model has revealed a number of interesting findings (Fig. 2). As predicted from the epitope mapping of the 7H5 MAb, the N- and C-termini of the E-selectin lectin domain are found to be very closely apposed to each other. This apposition is, in part, accomplished by a disulfide bond that is formed between these two domains of the protein. On the same face of the E-selectin model is found the 98–101 region that was shown to bind another blocking MAb, and whose mutation resulted in a loss of sialyl Lewis[x] binding. The partial effect of a mutation at this site on the binding of the 7H5 MAb was also consistent with the close proximity of this region to the N- and C-termini in the model. This region thus seems to define a face of E-selectin that appears to be critically involved with the recognition of sialyl Lewis[x] and cell adhesion. The model also reveals that the mutation at amino acid 74, which resulted in the loss of the binding of nonblocking MAbs, is found on the side opposite the one conjectured here to interact with the carbohydrate ligand. This result is consistent with the proposition that the face we have defined here is principally involved with cell adhesion by recognition of sialyl Lewis[x].

One of the most powerful aspects of a structural model such as the one that we have proposed is that it can predict other mutations that can be tested. The successful prediction of such mutations provides supportive evidence for the validity of the model. Examination of the model shows that a small loop (S43 to Y48; in purple, Fig. 2) is found within the region defined by the blocking MAb and sialyl Lewis[x] binding studies. In addition, a K at position 111, adjacent to the critical K at position 113, was also hypothesized to point into the apparent carbohydrate binding site. Mutation of the K at position 111 completely abolished carbohydrate binding, but had no effect on MAb binding. Moreover, the mutation of residues within the S43-to-Y48 loop also showed decreased sialyl Lewis[x] binding (Table II). Perhaps most intriguing was the mutation of a Y to an F at position 48. This mutation, which was identical to the wild-type protein with the exception of a loss of the tyrosine hydroxyl group, was completely unable to bind immobilized carbohydrate, suggesting a critical role for this hydroxyl group in carbohydrate recognition. These results provided further support for the model and for the conclusion that the region defined here is intimately involved with sialyl Lewis[x] recognition.

4. Structure–Function Analysis of P-Selectin: Conservation of a Carbohydrate Recognition Site

Although a number of the residues conserved between the selectins are found in all calcium-dependent mammalian lectins (Drickamer, 1988), there is also conservation of many of the residues that were found to be critical for E-selectin-mediated binding (Erbe et al., 1992). This conservation, together with the data concerning a role for sialyl Lewis[x] in P-selectin-mediated binding, suggested that a similar region of P-selectin may be involved in carbohydrate recognition. In order to analyze this possibility, a mutagenesis approach similar to that undertaken for E-selectin was performed on P-selectin. An additional assay involving the ability of mutants to bind to neutrophils or HL 60 cells was added to the P-selectin analysis. This assay consisted of incubating these cells with a P-selectin–IgG chimera and then analyzing the cells for P-selectin binding by fluorescence-activated cell sorting (FACS) (S. Watson, unpublished). The binding of P-selectin in this assay was calcium dependent, sensitive to sialidase (Corrall *et al.*, 1990) and protease (Larsen *et al.*, 1990), and blocked by the carbohydrates mannose-1-phosphate and dextran sulfate. Thus, this assay appeared to mimic the binding of neutrophils to P-selectin on the endothelial or platelet cell surface.

P-selectin appears to have a broader range of carbohydrate recognition than E- or L-selectin, at least with respect to naturally occurring cell surface sugars. For example, previous data have indicated that sialyl Lewis[x] with sialic acid in either a 2–3 or 2–6 linkage to galactose (2–3 and 2–6 sialyl Lewis[x], respectively) and sulfatide (galactose-4-sulfate ceramide) may be potential biological ligands for P-selectin. We analyzed all three of these ligands in the solid-state binding assay described above, and found that a P-selectin–IgG chimera is capable of binding all three (Fig. 3). However, the binding is not identical. Thus, the binding to the 2–3 form of sialyl Lewis[x] appears to have much higher affinity that the binding to the 2–6 form of the carbohydrate. In addition, P-selectin binds with considerably lower affinity to 2–3 sialyl Lewis[x] than does E-selectin. Figure 3 also illustrates that the carbohydrate specificity of P-selectin resides in the lectin domain, since an IgG chimera containing the P-selectin lectin domain and the E-selectin EGF-like and complement binding protein-like domains has the same carbohydrate specificity as the P-selectin–IgG chimera. In addition, while the binding of P-selectin to sulfatide appears to be of relatively high affinity, this binding is almost completely independent of calcium. Since all lectins in the calcium-dependent lectin family have an absolute requirement for calcium bound to the lectin domain, it may be concluded that the binding of P-selectin to sulfatide is nonphysiological.

In contrast to the large panel of MAbs specific for E-selectin, there was a much smaller array of MAbs to P-selectin. However, one antibody, termed AK-6, is capable of blocking cell adhesion mediated by P-selectin. In addition, two nonblocking P-selectin MAbs were also obtained (AC 1.2 and CRC 81). An analysis of antibody binding to the construct containing only the P-selectin lectin domain (see above) demonstrated that the AK-6 blocking MAb recognized an epitope in the lectin domain, while the two nonblocking antibodies recognized uncharacterized epitopes outside of the lectin domain. Finer level mapping of the epitope recognized by the AK-6 MAb was then executed by incorporating mutations known to affect the binding of blocking MAbs in E-selectin into homologous sites of P-selectin contained in a P-selectin–IgG chimera. While these mutations were made to investigate whether there was conservation in the car-

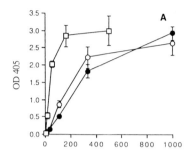

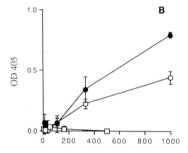

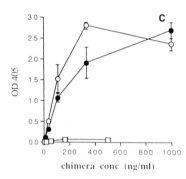

chimera conc (ng/ml)

Figure 3. Binding of selectin chimeras to immobilized glycolipids. P-selectin–IgG (open circles), E-selectin–IgG (open squares), or the PE–IgG chimera (closed circles) were tested at the indicated concentrations for binding to (A) 2–3 sialyl Lewis^x, (B) 2–6 sialyl Lewis^x, and (C) sulfatides by an ELISA assay. Results are ±SD of triplicate determinations.

bohydrate binding sites of E- and P-selectin, they were also produced because these identical mutations had little effect on the overall structure of E-selectin; it therefore seemed unlikely that they would have a disruptive effect on P-selectin structure. Figure 4 shows that when two of the mutations that affected the binding of blocking MAbs to E-selectin were incorporated into homologous sites, they affected the binding of the P-selectin-blocking MAb. Thus, when mutations at residues 8 and 113, which were shown to disrupt binding of some of the E-selectin-blocking MAbs, were incorporated into homologous sites of E-se-

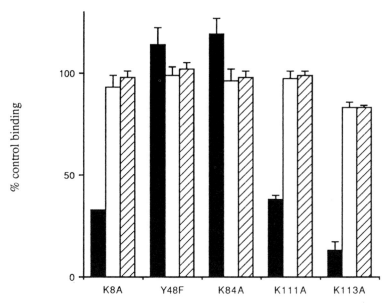

Figure 4. Reactivity of anti-P-selectin MAbs with mutant chimeras. P-selectin–IgG chimeras with the indicated substitutions were tested for capture by the blocking MAb (solid bars) and the nonblocking MAbs AC 1.2 (open bars) or CRC 81 (hatched bars). Results are ±SD of duplicate determinations, and are shown as percentages of wild-type P-selectin–IgG binding.

lectin, they also disrupted the recognition of P-selectin by the AK-6-blocking MAb. In addition, another mutation, a conserved K residue at position 111, also affected the binding of this MAb to P-selectin. As expected, none of these mutations had an effect on the binding of the nonblocking P-selectin antibodies, which is consistent with the hypothesis that these mutations did not have a gross effect on the structure or stability of the P-selectin–IgG chimera. Finally, a mutation at residue 48, which was found to have a profound effect on the recognition of sialyl Lewis^x by E-selectin, had no effect on the binding of either the blocking or nonblocking MAbs, which is again consistent with a lack of structural disruption by this mutation. In summary, these mapping results were consistent with the conservation of E- and P-selectin regions that were recognized by blocking MAbs.

These mutations were then analyzed for carbohydrate binding ability. As previously discussed, P-selectin shows binding to 2–3 and 2–6 sialyl Lewis^x as well as to sulfatide (Fig. 3). Thus, each mutant was tested for binding to these carbohydrates immobilized onto plastic. As Fig. 5 shows, various mutants affected the binding to different carbohydrates in different manners. In a manner analogous to E-selectin, mutations at Y48, K111, and K113 all affected the

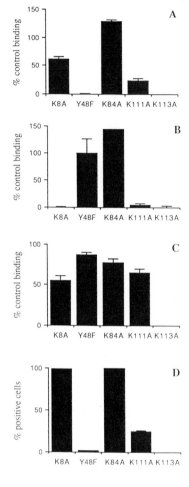

Figure 5. Binding of P-selectin–IgG mutants to immobilized glycolipids and cells. P-selectin–IgG chimeras with the indicated substitutions were tested for binding to immobilized 2–3 sialyl Lewis^x glycolipid (A), 2–6 sialyl Lewis^x glycolipid (B), sulfatides (C) using the ELISA shown in Fig. 3 for staining of HL 60 cells (D) by FACS analysis.

binding of P-selectin to 2–3 sialyl Lewis^x. A mutation at residue 8 had a mild effect on 2–3 sialyl Lewis^x recognition, and a mutation at residue 84 had no effect on carbohydrate recognition. In contrast, mutations at residue 8, in addition to 111 and 113, had strong effects on 2–6 sialyl Lewis^x recognition, while a mutation at residue 48, which had a powerful inhibitory effect on 2–3 sialyl Lewis^x binding, had no effect on binding of the 2–6 form of sialyl Lewis^x. These results suggest that some features of the carbohydrate recognition face of P-selectin recognize both the 2–3 and 2–6 forms of sialyl Lewis^x, but the loop containing residue 48 appears to be involved only with the recognition of the 2–3-linked form of the ligand. Finally, only the mutation at residue 113 appears to inhibit the recognition of sulfatide, while mutations at 111 and 48 have no effect on the recognition of this ligand. This result suggests that sulfatide binds to the same face of P-selectin as do 2–3 and 2–6 sialyl Lewis^x, which is consistent with the ability of sulfatide to block P-selectin-mediated adhesion of cells.

The differential recognition of these three carbohydrates by the various P-selectin mutants allowed an analysis of the nature of the carbohydrate ligand on the neutrophil cell surface, using the previously described P-selectin–IgG cell binding FACS assay. Various investigators have proposed that both 2–6 sialyl Lewis^x (Larsen *et al.*, 1990) as well as sulfatide (Aruffo *et al.*, 1991) contribute to the binding of neutrophils mediated by P-selectin. Figure 5 clearly shows that cell binding by the P-selectin mutations appears to segregate only with binding to 2–3 sialyl Lewis^x. Thus, the binding of mutants at residues 48, 111, and 113, which were either completely or almost completely incapable of 2–3 sialyl Lewis^x recognition, is completely abolished, suggesting that, as with E-selectin, these residues are critical for the cell adhesion mediated by P-selectin. The binding of P-selectin mutant K8A, which results in a complete lack of 2–6 sialyl Lewis^x recognition, shows wild-type binding to cells, while the binding of Y48F, which has no effect on sulfatide binding, shows a complete lack of cell adhesion in this assay. These results are consistent with the hypothesis that 2–3 sialyl Lewis^x is a major component of cell recognition by P-selectin, but they are inconsistent with a critical role for either 2–6 sialyl Lewis^x or sulfatide in this process.

Figure 2 shows a P-selectin model similar to that previously described for E-selectin. It is clear that the proposed structures of E- and P-selectin are remarkably similar. In addition, the residues that have been suggested to be of functional significance are seen in homologous domains in both proteins. Thus, K111, K113, and Y48 are seen to map to a conserved face of both E- and P-selectin. These data are consistent with the proposal that both of these lectins recognize carbohydrates and cell surfaces by similar means.

5. Conclusions

The data reported here suggest that homologous faces of E- and P-selectin mediate cell adhesion by recognition of cell surface carbohydrate, particularly sialyl Lewis^x. While these data are interesting in themselves, they raise some additional questions for the future. While biological and biochemical experiments suggest that E- and P-selectin both recognize similar cell types, these two selectins clearly bind to these cells via different mechanisms. Data supporting this contention include the protease sensitivity of P-, but not E-, selectin binding, and the apparent higher affinity of P-selectin binding to neutrophil surfaces in our FACS assays. In addition, a recently described protein component of the P-selectin ligand is consistent with a glycoprotein scaffold scheme homologous to that previously suggested for an endothelial glycoprotein ligand for L-selectin. The question of whether P-selectin binding requires protein contact in addition to carbohydrate recognition therefore remains open, although it seems clear from the mutagenesis analysis reported here that sialyl Lewis[x] is a critical component of P-selectin-mediating binding, and homologous regions of the E- and P-selectin lectin domains appear to be required for cell adhesion. An additional question raised by these data is whether they will provide sufficient information to aid in the design of E- and P-selectin antagonists. It is hoped that the models described here, together with the delineation of potential ligand binding sites, will allow for the design of sialyl Lewis[x] analogues that can be tested for inhibitory activity in cell adhesion and *in vivo* inflammation assays. If this proves to be the case, then a new class of drugs based on the structures of naturally occurring carbohydrate ligands may ultimately be added to the arsenal of current anti-inflammatory compounds.

ACKNOWLEDGMENT. The authors thank Dr. B. N. Narasinga Rao for providing the figure showing the solution structures of sialyl Lewis[x] and sialyl Lewis[a].

References

Anderson, D. C., and Springer, T. A., 1987, Leukocyte adhesion deficiency: An inherited defect in the Mac-1, LFA-1, and p150, 95 glycoproteins, *Annu. Rev. Med.* **38:**175–194.

Aruffo, A., Kolanus, W., Walz, G., Fredman, P., and Seed, B., 1991, CD62/P-selectin recognition of myeloid and tumor cell sulfatides, *Cell* **67:**35–44.

Berg, E. L., Robinson, M. K., Mansson, O., Butcher, E. C., and Magnani, J. L., 1991, A carbohydrate domain common to both sialyl Le[a] and sialyl Le[x] is recognized by the endothelial cell leukocyte adhesion molecule ELAM-1, *J. Biol. Chem.* **265:**14869–14872.

Bowen, B., Fennie, C., and Lasky, L. A., 1990, The Mel-14 antibody binds to the lectin domain of the murine peripheral lymph node homing receptor, *J. Cell Biol.* **107:**1853–1862.

Corrall, L., Singer, M., Macher, B., and Rosen, S., 1990, Requirement for sialic acid on neutrophils in a gmp140(PADGEM) mediated adhesive interaction with activated platelets, *Biochem. Biophys. Res. Commun.* **172:**1349–1352.

Drickamer, K., 1988, Two distinct classes of carbohydrate-recognition domains in animal lectins, *J. Biol. Chem.* **263:**9557–9560.

Erbe, D. V., Wolitzky, B. A., Presta, L. G., Norton, C. R., Ramos, R. J., Burns, D. K., Rumberger, J. M., Rao, B. N. N., Foxall, C., Brandley, B. K., and Lasky, L. A., 1992, Identification of an E-selectin region critical for carbohydrate recognition and cell adhesion, *J. Cell Biol.* **119:**215–227.

Foxall, C., Watson, S., Dowbenko, D., Fennie, C., Lasky, L., Kiso, M., Hasegawa, A., Asa, D., and Brandley, B., 1992, The three members of the selectin receptor family recognize a common carbohydrate epitope, the sialyl Lewis x oligosaccharide, *J. Cell Biol.* **117:**895–902.

Larsen, E., Palabrica, T., Sajer, S., Gilbert, G. E., Wagner, D. D., Furie, B. C., and Furie, B., 1990, PADGEM-dependent adhesion of platelets to monocytes and neutrophils is mediated by a lineage-specific carbohydrate, LNF III (CD15), *Cell* **63:**467–474.

Lasky, L. A., 1992, The selectins: Interpreters of cell-specific carbohydrate information during inflammation, *Science* **258:**964–969.

Lasky, L. A., Singer, M. S., Yednock, T. A., Dowbenko, D., Fennie, C., Rodriquez, J., Hguyen, T., Stachel S., and Rosen, S. D., 1989, Cloning of a lymphocyte homing receptor reveals a lectin domain, *Cell* **56**:1045–1055.

Lowe, J. B., Stoolman, L. M., Nair, R. P., Larsen, R. D., Berhend, T. L., and Marks, R. M., 1990, ELAM-1 dependent cell adhesion to vascular endothelial determined by a transfected human fucosyltransferase cDNA, *Cell* **63**:475–484.

Moore, K. L., Varki, A., and McEver, R. P., 1991, GMP-140 binds to a glycoprotein receptor on human neutrophils: Evidence for a lectin-like interaction, *J. Cell Biol.* **112**:491–499.

Moore, K. L., Stults, N. C., Diaz, S., Smith, D. F., Cummings, R. D., Varki, A., and McEver, R. P., 1992, Identification of a specific glycoprotein ligand for P-selectin (CD62) on myeloid cells, *J. Cell Biol.* **118**:445–456.

Phillips, M. L., Nudelman, E., Gaeta, F. C. A., Perez, M., Singhai, A. K., Hakomori, S., and Paulson, J. C., 1990, ELAM-1 mediates cell adhesion by recognition of a carbohydrate ligand, sialyl-Le^x, *Science* **250**:1130–1132.

Polley, M. J., Phillips, M. L., Wayner, E., Nucelman, E., Singhai, A. K., Hakomori, S., and Paulson, J. C., 1991, CD62 and endothelial cell–leukocyte adhesion molecule 1 (ELAM-1) recognize the same carbohydrate ligand, sialyl-Lewis x, *Proc. Natl. Acad. Sci. USA* **88**:6224–6228.

Rosen, S. D., Singer, M. S., Yednock, T. A., and Stoolman, L. M., 1985, Involvement of sialic acid on endothelial cells in organ-specific lymphocyte recirculation, *Science* **228**:1005–1007.

Springer, T. A., 1990, Adhesion receptors of the immune system, *Nature* **346**:425–434.

Stoolman, L. M., 1989, Adhesion molecules controlling lymphocyte migration, *Cell* **56**:907–910.

Tiemeyer, M., Swiedler, S. J., Ishihara, M., Moreland, M., Schweingruber, H., Hirtzer, P., and Brandley, B. K., 1991, Carbohydrate ligands for endothelial–leukocyte adhesion molecule 1, *Proc. Natl. Acad. Sci. USA* **88**:1138–1142.

Tyrrell, D., James, P., Rao, N., Foxall, C., Abbas, S., Dasgupta, F., Nashed, M., Hasegawa, A., Kiso, M., Asa, D., Kidd, J., and Brandley, B. K., 1991, Structural requirements for the carbohydrate ligand of E-selectin, *Proc. Natl. Acad. Sci. USA* **88**:10372–10376.

Walz, G., Aruffo, A., Kolanus, W. E., Bevilacqua, M., and Seed, B., 1990, Recognition by ELAM-1 of the sialyl-Le^x determinant on myeloid and tumor cells, *Science* **250**:1132–1135.

Watson, S. R., Imai, Y., Fennie, C., Geoffroy, J., Singer, M., Rosen, S. D., and Lasky, L. A., 1991, The complement binding-like domains of the murine homing receptor facilitate lectin activity, *J. Cell Biol.* **115**:235–243.

Weiss, W., Kahn, R., Fourme, R., Drickamer, K., and Hendrickson, W., 1991, Structure of the calcium-dependent lectin domain from a rat mannose-binding protein determined by MAD phasing, *Science* **254**:1608–1615.

4

Recognition of Cell Surface Carbohydrates by C-Type Animal Lectins

WILLIAM I. WEIS

1. Introduction

Protein-carbohydrate interactions often mediate recognition events at the cell surface. Examples include cell–cell adhesion, such as the targeting of certain T lymphocytes to their sites of action (Springer and Lasky, 1991), and the binding of sperm to egg (Wassarman, 1987); the clearance of glycoproteins from the circulation (Ashwell and Harford, 1982); host recognition of pathogens (Ezekowitz, 1991); and the binding of certain viruses (Paulson, 1985) and toxins (van Heyningen, 1983) to their target cells. Each of these cases involves the recognition of specific carbohydrate structures. Recent progress in physical and structural characterization has greatly increased our understanding of these interactions at the molecular level.

1.1. Carbohydrates as Informational Molecules

Information is encoded in carbohydrate structure by the tissue-specific expression of glycosyl transferases and glycosidases, which gives rise to characteristic glycolipids and glycoproteins on the cell surface and on secreted mole-

WILLIAM I. WEIS • Department of Biochemistry and Molecular Biophysics and Howard Hughes Medical Institute, Columbia University, New York, New York 10032. *Present address:* Department of Cell Biology, Stanford University School of Medicine, Stanford, California 94305.
Cellular Adhesion: Molecular Definition to Therapeutic Potential, edited by Brian W. Metcalf, Barbara J. Dalton, and George Poste. Plenum Press, New York, 1994.

cules. In contrast to proteins, a given oligosaccharide contains only a few distinct building blocks. However, the ability of carbohydrates to form linkages at different ring positions and in different anomeric configurations, and to form branched structures by the linkage of a single monosaccharide to several sugars, permits an enormous number of chemically distinct structures to be made from a few monosaccharides. Amino acids, on the other hand, usually form only linear polymers with one kind of linkage. This distinction can be appreciated by considering that only 24 distinct tetrapeptides can be formed from four amino acids, while over 35,000 tetrasaccharides can be made from four hexose sugars (Sharon and Lis, 1989). Modifications such as sulfation add further complexity to carbohydrate structure.

The concept of carbohydrate structure as a carrier of information is well illustrated by influenza virus, which uses cell-surface sialic acid as its receptor (Paulson, 1985). The hemagglutinin (HA) glycoprotein of the viral membrane is the molecule that binds to the receptor. HAs from different strains of influenza discriminate between sialic acids in $\alpha(2,6)$ linkage and $\alpha(2,3)$ linkage to the penultimate sugar on a carbohydrate chain; a particular strain will bind only those cells bearing sialic acid in the preferred linkage (Rogers et al., 1983). Thus, a difference of a single glycosidic linkage produces profoundly different interactions between the virus and the host cell.

1.2. Lectins as Recognition Molecules

The proteins that decode the information contained in carbohydrate structures are lectins, defined as nonenzymatic carbohydrate-binding proteins. A characteristic feature of lectins is their extremely weak intrinsic affinity for carbohydrates, with K_d typically on the order of 10^{-5} to 10^{-3} M for monovalent ligands. Continuing with the example of influenza virus, HAs from virus strains that preferentially recognize $\alpha(2,6)$-linked sialic acids bind to sialated trisaccharides bearing this linkage with a K_d of 2.0 mM, and to $\alpha(2,3)$-linked sialic acids with a K_d of 3.2 mM (Sauter et al., 1989). Conversely, HAs from strains that prefer $\alpha(2,3)$ linkages bind to $\alpha(2,3)$-linked sialic acids with a K_d of 2.9 mM, and to $\alpha(2,6)$-linked sialic acids with a K_d of 5.9 mM (Sauter et al., 1989). Note that in each instance the affinity is very low, and the difference between binding to ligands containing the preferred and nonpreferred linkages is energetically very small. Despite this intrinsically low affinity, lectin-mediated binding at the cell surface is extremely strong, and lectins discriminate among various carbohydrates with great specificity. This apparent discrepancy seems to be a consequence of multivalent binding, whereby multiple weak interactions acting together produce strong binding. Multivalent binding amplifies the energetically small difference between binding to preferred and nonpreferred ligands to produce the observed discrimination.

These properties of lectins lead to two basic structural and mechanistic questions. First, how is weak, specific binding to monovalent ligands achieved? Second, what are the spatial requirements for multivalent binding? The following sections review recent progress in addressing these questions in one particular system.

2. C-Type Lectins

2.1. Definition, Functions, and Primary Structures

The largest and most diverse group of animal lectins is the calcium-dependent or C-type lectins (Drickamer, 1988). This family includes endocytic receptors such as the asialoglycoprotein receptors responsible for clearance of serum glycoproteins (Ashwell and Harford, 1982); selectin cell adhesion molecules, which target leukocytes to lymphoid tissues and sites of inflammation (Stoolman, 1989); macrophage receptors involved in the phagocytosis of pathogens (Taylor *et al.*, 1990); proteoglycan core proteins (Halberg *et al.*, 1988); and soluble proteins, including pulmonary surfactant apoproteins (White *et al.*, 1985) (Fig. 1). Each family member contains one or more C-type carbohydrate-recognition domains (CRDs) that are combined with effector domains responsible for the physiological function of the molecule. C-type CRDs are approximately 120 amino acids long, and display a characteristic sequence motif of 14 invariant amino acids and another 18 amino acids conserved in character (Drickamer, 1988). Despite their shared dependence on Ca^{2+} for activity, the C-type lectins display a wide variety of carbohydrate specificities.

Sequence alignments of C-type CRDs reveal that the most closely related CRDs are those attached to similar effector domains, implying that the C-type lectins diverged after juxtaposition of the CRD with various effector domains (Bezouska *et al.*, 1991). The organization of C-type lectin genes is consistent with this hypothesis (Bezouska *et al.*, 1991; Kim *et al.*, 1992). The C-type lectins can be grouped into several families according to the overall organization of their domains and the evolutionary distance of their CRDs (Figs. 1, 2). It is important to note that there is not a strong correlation between the carbohydrate specificities of particular C-type lectins and the sequence relatedness of their CRDs. For example, two type II membrane receptors, rat hepatic lectin and chicken hepatic lectin, are closely related in overall sequence and yet have nonoverlapping carbohydrate specificities: the mammalian receptor binds to galactose- or *N*-acetylgalactosamine-terminated oligosaccharides while the avian molecule binds to oligosaccharides that terminate in *N*-acetylglucosamine. This suggests that while the conserved residues define a Ca^{2+}-dependent carbohydrate binding domain, comparatively few amino acids determine carbohydrate specificity.

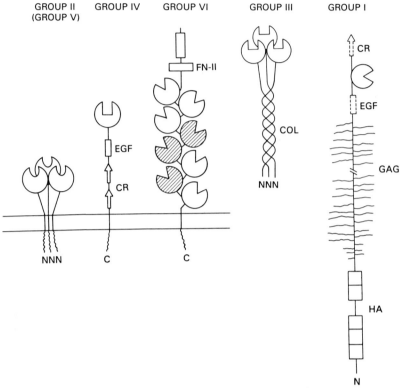

GROUP II   GROUP IV   GROUP VI   GROUP III   GROUP I
(GROUP V)

Figure 1. Organization of C-type lectins. The CRD is represented as a circle, with cutouts of different shapes representing the various carbohydrate specificities attached to different effector domains. Membrane-bound C-type lectins include group II, the type II membrane receptors such as the asialoglycoprotein receptor (hepatic lectins); group IV, the selectin cell adhesion molecules; and group VI, the macrophage mannose receptor. Group III, the collectins, are soluble C-type lectins that contain a collagenous domain such as the MBPs. Group I, the proteoglycan core proteins, forms part of the extracellular matrix. Abbreviations: EGF, epidermal growth factor-like domain; CR, complement regulatory domain; FN-II, fibronectin type II repeat; COL, collagenlike domain; GAG, glycosaminoglycan attachment sites; HA, hyaluronic acid-binding domain. N and C denote the amino- and carboxyl-termini. (From Weis *et al.*, 1992b.)

The shared Ca^{2+} dependence of the C-type lectins manifests itself in certain biochemical properties. In the absence of Ca^{2+}, the CRD becomes extremely sensitive to proteases, suggesting a conformational change in the protein that renders it unable to bind to carbohydrates. In two cases, it has been shown that two Ca^{2+} bind to a CRD, based on the second-order Ca^{2+} dependence of ligand binding and resistance to proteolysis (Loeb and Drickamer, 1988; Weis *et al.*, 1991a). Ca^{2+} and carbohydrate binding also change as a function of

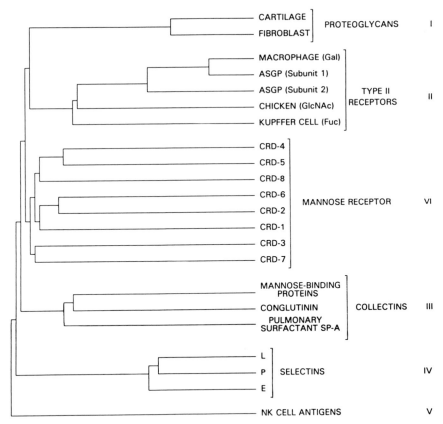

Figure 2. Dendrogram of sequence similarity among C-type CRDs. Similarities are based on amino acid sequence comparisons (Bezouska *et al.*, 1991; Kim *et al.*, 1992). The groups are summarized in Fig. 1. The horizontal scale is arbitrary, and is not meant to imply specific times of evolutionary divergence. The CRDs of the mannose receptor have been segregated from the other proteins for clarity. (From Weis *et al.*, 1992b.)

pH, a characteristic important in the functioning of endocytic receptors (see below).

2.2. Mannose-Binding Proteins

Mammalian mannose-binding proteins (MBPs) are serum C-type lectins that mediate immunoglobulin-independent host defense against pathogens. MBPs function by binding to high-mannose structures found on the surfaces of bacterial and fungal pathogens, and kill these organisms by complement-mediated cell lysis (Ikeda *et al.*, 1987) or opsonization (Kuhlman *et al.*, 1989). Stress response elements are found in the promoter region of the MBP gene (Taylor *et al.*, 1989), and levels of MBP mRNA have been observed to be elevated during the acute phase response (Ezekowitz *et al.*, 1988). Low serum levels of MBP, which are the result of a point mutation in the MBP gene (Sumiya *et al.*, 1991), are associated with an opsonic defect that leads to severe and recurrent bacterial infections (Super *et al.*, 1989). MBPs may confer a primitive form of immunological self versus nonself discrimination by recognizing carbohydrate structures absent from mammalian cell surfaces (Ezekowitz, 1991).

The primary structure of MBPs consists of a cysteine-rich N-terminal region, a collagenous domain of 18–20 Gly-X-Y repeats, a "neck" region of about 40 amino acids, and a C-terminal CRD (Drickamer *et al.*, 1986) (Fig. 3). MBPs are found as higher oligomers of a trimeric building block formed by a collagenlike triple helix (Drickamer *et al.*, 1986); the formation of higher oligomers may result from disulfide bond formation in the N-terminal region. The similarity in domain organization between MBP and complement protein C1q, which also contains a collagenous domain, suggests that the ability of MBP to activate the complement cascade resides in the collagenous domain, and that the MBP CRD functions analogously to the Fc-binding domain of C1q (Drickamer *et al.*, 1986; Ikeda *et al.*, 1987; Lu *et al.*, 1990).

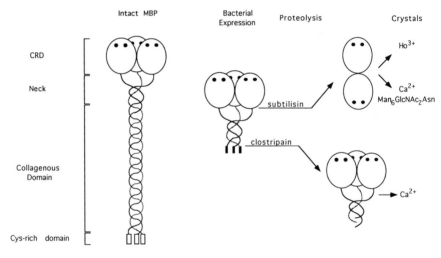

Figure 3. Summary of mannose-binding protein structure, expression, and crystallization. The domain structure of a single trimer of the intact MBP is shown on the left. The C-terminal CRD is represented as an oval, with the two black dots representing the two Ca^{2+} sites. The bacterially expressed trimer contains the C-terminal 161 amino acids of the intact polypeptide, plus seven extra amino acids at the N-terminus that can be removed by treatment with clostripain. Treatment with subtilisin produces the dimeric fragment. Crystals of the different fragments have been obtained in the presence of the cations and oligosaccharide, as indicated.

2.3. Structure of a Mannose-Binding Protein Carbohydrate-Recognition Domain

Structural analysis of C-type lectins is most advanced in mannose-binding protein A from rat. In order to obtain sufficient amounts of protein for structural studies, the C-terminal portion of MBP-A containing the neck and CRD was overproduced in *E. coli* (Weis *et al.*, 1991a). The fragment expressed contains seven extra amino acids introduced by the expression construct, which can be removed by limited proteolysis with clostripain. Gel filtration and chemical cross-linking analyses indicate that this molecule (with or without the seven extra amino acids) is a trimer (Fig. 3). Treatment of the bacterially expressed protein with subtilisin produces a fragment containing the C-terminal 115 residues of MBP-A, which retains specific, Ca^{2+}-dependent carbohydrate binding. Interestingly, this fragment is a dimer (Weis *et al.*, 1991a) (Fig. 3). Therefore, interactions within the neck region contribute to the stability of the MBP trimer.

The dimeric MBP-A fragment yields extremely well-diffracting crystals in the presence of Ca^{2+} and a $Man_6GlcNAc_2Asn$ asparaginyl-oligosaccharide isolated from ovalbumin (Weis *et al.*, 1991) (Fig. 3). In order to solve the crystallographic phase problem, the Ca^{2+} dependence of the CRD was exploited by replacing Ca^{2+} with trivalent lanthanide ions. It was demonstrated that lanthanides substitute functionally for Ca^{2+}, as assessed by their ability to support protease resistance of the CRD as well as carbohydrate binding (Weis *et al.*, 1991a). The structure of the dimer was first determined with Ho^{3+} substituted for Ca^{2+} (Weis *et al.*, 1991b); this model was then used to determine the structure of the Ca^{2+}-containing complex with $Man_6GlcNAc_2Asn$ (Weis *et al.*, 1992a).

The CRD protomer adopts an unusual fold that contains a large proportion of nonregular secondary structure (Fig. 4A). All of the regular secondary structure elements, consisting of two α helices and five β strands, lie in the lower two-thirds of the domain. Although hydrogen bonding occurs between pairs of strands, there are no extended β-sheet regions. The upper one-third of the molecule contains several loops and extended regions that are stabilized by the two required cations.

The N- and C-termini of the MBP-A CRD are paired as antiparallel β strands, such that the polypeptide begins and ends in the same vicinity. Since the CRD lies in different positions in the primary structure of the various C-type lectins (Fig. 1), this observation indicates that the CRD is a module that can be placed in different orientations in the primary structure, yet can still maintain the same spatial orientation with respect to the rest of the protein. For example, although the extracellular CRD lies at the C-terminus of the asialoglycoprotein receptors and at the N-terminus of the selectins (Fig. 1), the structure suggests that in both cases the CRD would be disposed similarly with respect to the cell membrane.

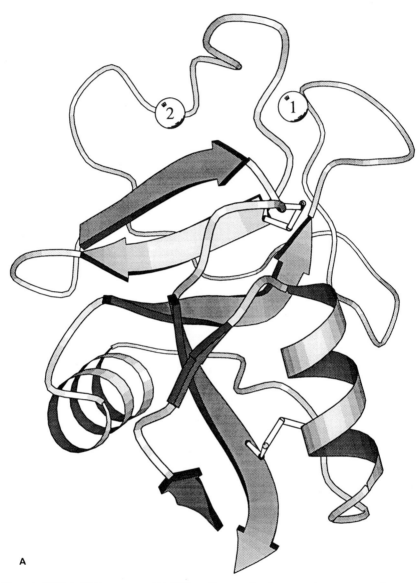

A

Figure 4. MBP-A CRD structure. (A) Ribbon diagram showing secondary structure elements. Spheres marked 1 and 2 represent the two calcium ions. (B, C) Side chains of conserved residues are shown in black on the α-carbon backbone of the CRD. Conserved disulfide bonds and calcium ligands are shown in B, while the other conserved, mostly hydrophobic core residues are shown in C. (From Weis *et al.*, 1992b.)

The MBP CRD structure largely explains the pattern of conserved residues in the C-type lectins. These residues appear to stabilize the basic fold, and fall mostly into four groups: (1) four cysteines that form two disulfide bonds; (2) amino acids that serve as Ca^{2+} ligands; (3) residues that form two hydrophobic cores; and (4) several turns (Fig. 4B,C) (Weis *et al.*, 1991b). Conversely, the insertions and deletions needed to align the CRD sequences map to five regions, all in surface turns, of the CRD (Weis *et al.*, 1991b). Therefore, it is clear that the basic fold of the MBP CRD is conserved among the C-type CRDs.

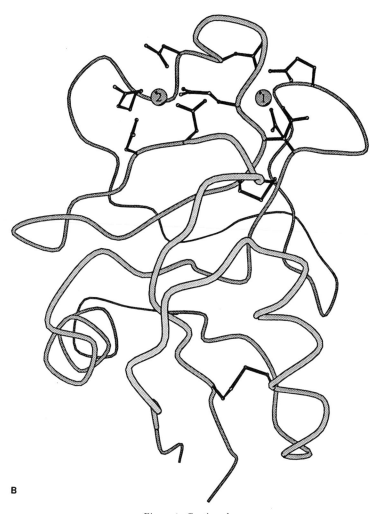

B

Figure 4. Continued.

The MBP CRD dimer interface is composed largely of hydrophobic residues from the paired β1 and β5 strands, and the ends of the two helices. In the Ho^{3+}-complex crystal structure, the conformation of the β1 strands is different in the two protomers; their conformation is the same in the Ca^{2+}-containing structure. Moreover, the protomers bear a different rotational relationship with respect to one another in the two different crystal forms. These observations support the notion that the dimer interface is an artifact produced by the subtilisin treatment, perhaps by exposing hydrophobic residues that are buried within a single protomer in the natural molecule.

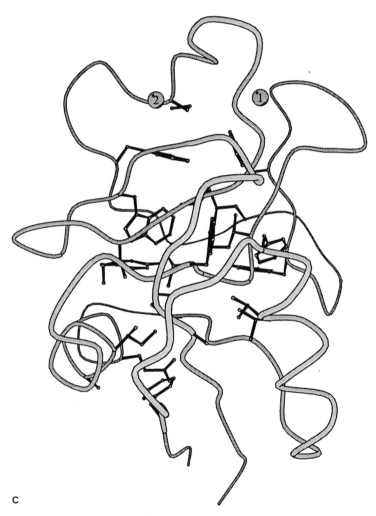

C

Figure 4. Continued.

3. Carbohydrate Binding by C-Type Lectins

3.1. Structure of a C-Type Carbohydrate-Recognition Domain–Oligosaccharide Complex

The structure of the Ca^{2+}-containing CRD complexed with $Man_6GlcNAc_2Asn$ (Fig. 5) has been determined to 1.7 Å resolution (Weis *et al.*, 1992a). The terminal mannose residues of two of the three branches of a single oligosaccharide molecule bind to different protomers of neighboring CRD dimers in the crystal lattice, thereby cross-linking the dimers within the crystal (Fig. 5). A model of the interaction between one terminal mannose and a CRD is shown in Fig. 6.

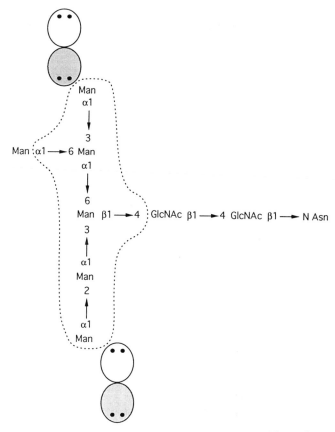

Figure 5. Oligosaccharide structure and crystal cross-linking. The dashed line encloses residues and bonds that are visible in the electron density map. The CRD is represented as in Fig. 3. The carbohydrate links one protomer (white) of one dimer to the other protomer (shaded) of another dimer in the crystal.

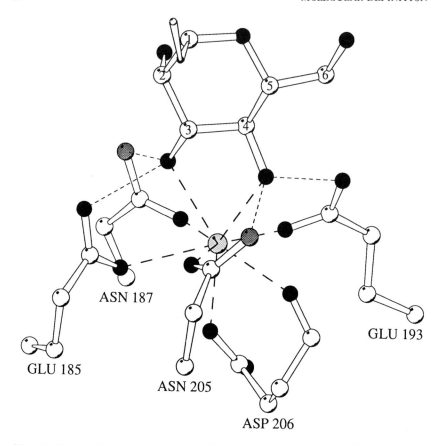

ASN 187

GLU 193

GLU 185

ASN 205

ASP 206

Figure 6. Mannose binding at calcium site 2. The calcium ion is represented as a light gray sphere. White, dark gray, and black spheres represent carbon, nitrogen, and oxygen atoms, respectively. Calcium coordination bonds are shown as thick dashed lines, while thin dashed lines represent hydrogen bonds. Numbers on the mannose carbon atoms represent ring positions. The α-glycosidic bond to the next sugar of the oligosaccharide (at carbon 1) has been cut off for clarity. (From Weis *et al.*, 1992a.)

The key feature of the mannose–CRD interaction is the direct ligation of calcium ion 2 (sites are numbered in Fig. 4) by the 3- and 4-hydroxyl (OH) groups of mannose to form an eight-coordinate calcium ion complex. These two OH groups displace a water molecule that serves as the seventh Ca^{2+} 2 ligand in the sugar-free structure, but no changes in protein structure are observed upon sugar binding. Both sugar hydroxyl groups utilize their full complement of hydrogen bonding potential. Specifically, one lone electron pair forms a calcium coordination bond, the other lone pair accepts a proton from an asparagine side chain amide group, and the hydroxyl proton donates to a carboxylate oxygen atom from a glutamic acid. The remaining oxygen atom of each asparagine and glutamic acid side chain that participates in these interactions is a ligand for the Ca^{2+}, thereby producing an intimate ternary complex of Ca^{2+}, protein, and sugar.

The mode of binding observed between the MBP CRD and mannose contrasts sharply with other protein–carbohydrate complexes that have been studied crystallographically. There are few van der Waals interactions between the oligosaccharide and the protein. The only significant contacts are between the exocyclic C6 of mannose and the side chain of Ile 207, the mannose C4 and Cβ of His 189, and the O2 of mannose and Cδ2 of His 189. Also, the direct ligation of Ca^{2+} by the sugar differs from the role played by Ca^{2+} and Mn^{2+} in the legume lectins; in the latter, the cations position amino acid side chains for binding to the ligand, but do not interact directly with the sugar.

Biochemical and mutagenesis data support the model of the mannose–CRD interaction shown in Fig. 6. Lee and co-workers studied the binding of a variety of derivatized sugars to MBP, and concluded that, as seen in the crystal structure, the equatorial 3- and 4-OH groups of the D-pyranose ring are the primary determinants of carbohydrate specificity (Lee *et al.*, 1991). Analysis of single-site mutants in the MBP CRD generated by random cassette mutagenesis indicates that changes in amino acids over much of the MBP CRD surface have no effect on carbohydrate binding (Quesenberry and Drickamer, 1992), which is consistent with the limited interaction surface observed in the crystal. An especially interesting mutant contains the single change Asn 187 $\rightarrow$ Asp. According to the model, this change would remove the hydrogen bond-donating amide group and replace it with a carboxylate oxygen unable to satisfy the hydrogen bonding requirements of the mannose 3-OH group (Fig. 6). However, the other side chain oxygen would still be present to ligate Ca^{2+} 2. Therefore, this mutant would be expected to bind Ca^{2+} but not carbohydrate. This is in fact the phenotype: the mutant binds Ca^{2+} with affinity indistinguishable from the wild-type protein, but does not bind to carbohydrates (Quesenberry and Drickamer, 1992; Weis *et al.*, 1992a).

3.2. Modulation of Activity by pH

Most C-type CRDs are unable to bind carbohydrates at mildly acid pH (5.0–5.5). This property is central to those C-type lectins that function as endocytic receptors, which bind to their glycoprotein ligands at the cell surface, are taken up in coated vesicles, release their ligand in the endosome, and return to the cell surface for another round of binding. The CRD becomes sensitive to proteases at low pH, and Ca^{2+} and carbohydrate binding to the CRD have similar pH dependencies (Loeb and Drickamer, 1988). These observations have been interpreted as a reversible loss of Ca^{2+} at low pH that renders the CRD unable to bind carbohydrate. The intimate linkage of Ca^{2+} and sugar binding observed in the crystal structure suggests that the effect of pH may be to directly titrate the acidic groups that ligate Ca^{2+} 2. Alternatively, the carbohydrate binding site could be disrupted by titration of the acid ligands of Ca^{2+} site 1, since this site is linked directly to site 2 through Glu 193 and a water bridge to Asp 206 (Weis *et al.*, 1991b).

3.3. Specificity of Mannose-Binding Proteins and Other C-Type Lectins

MBPs bind to D-mannose, N-acetyl-D-glucosamine, D-glucose, and L-fucose, but not to D-galactose and its derivatives. Model building indicates that this specificity can be simply explained by the binding scheme shown in Fig. 6. Glucose and N-acetylglucosamine differ from mannose only at the C2 position, where they have equatorial instead of axial substituents. When models of these sugars are superimposed on the mannose model by matching ring positions, the equatorial substituents at C2 do not clash sterically with the protein (Fig. 7A). The binding scheme of Fig. 6 can also be preserved with L-fucose, since the 3- and 2-OH groups of the L-pyranose ring can also be superimposed respectively on the 3- and 4-OH groups of the D-pyranose ring. The naturally occurring α-glycosidic linkages between fucose and other sugars can be accommodated on such a model with no steric clashes with the protein (Fig. 7B). In contrast, there is no way to preserve the binding geometry with D-galactose, since it has an axial 4-OH, and the equatorial 2- and 3-OH groups of the D-pyranose ring are twisted with the opposite handedness with respect to the 3- and 4-OH groups.

The explanation of MBP specificity provided by the crystal structure can be applied to the specificity of other C-type lectins. Chicken hepatic lectin binds to oligosaccharides that terminate in N-acetyl-D-glucosamine. This molecule has exactly the same set of Ca^{2+} 2 ligands as MBP-A. Therefore, it is likely that this receptor binds its carbohydrate ligands in the manner shown in Fig. 7A. Similarly, the selectins bind to carbohydrates containing both terminal fucose and sialic acid. These molecules also have the same set of Ca^{2+} 2 ligands as MBP, so it is likely that fucose is bound at Ca^{2+} 2, as shown in Fig. 7B.

In those C-type lectins that bind mannose or glucose and their derivatives, a Glu/Asn pair is found at the positions corresponding to Glu 185 and Asn 187 of MBP-A, the Ca^{2+} 2 ligands that interact with the 3-OH of mannose (Fig. 6). In C-type lectins that recognize galactose, the corresponding residues are Gln and Asp, effectively reversing the positions of the side chain amide and carboxylate groups while preserving the oxygen atoms that ligate Ca^{2+} 2. In contrast, the other Ca^{2+} 2 ligands, residues Glu 193 and Asn 205, which interact with the mannose 4-OH and Asp 206, are invariant in all of the C-type lectins. Drickamer has constructed the double mutant Glu 185 → Glu/Asn 187 → Asp of the MBP-A CRD, and has shown that this protein preferentially binds galactose and

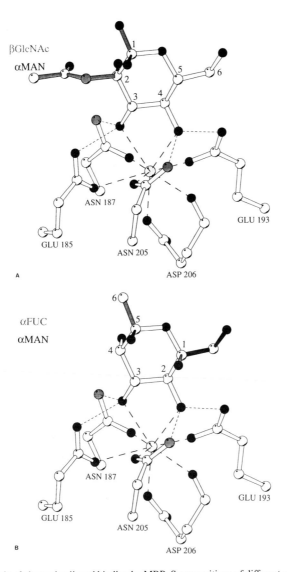

Figure 7. Models of alternative ligand binding by MBP. Superpositions of different monosaccharides on mannose are represented as in Fig. 6. Covalent bonds common to both sugars after the superposition are shown in white; black bonds occupy positions unique to mannose, and gray bonds occupy positions unique to the sugar being modeled. (A) Superposition of β-*N*-acetyl-D-glucosamine (GlcNAc) on α-D-mannose. Numbers on ring carbon atoms are those of both GlcNAc and mannose. (B) Superposition of α-L-fucose on α-D-mannose. Numbers on ring carbon atoms are those of fucose.

its derivatives and discriminates against mannose and glucose (Drickamer, 1992). The dramatic reversal in specificity resulting from this simple change suggests that the amino acids in these positions are the primary determinants of C-type lectin specificity. However, many of the mannose/glucose-binding C-type lectins bind to distinct subsets of sugars, so other residues in the CRD must contribute to the fine specificity of these proteins. Interestingly, those family members with more unusual specificities have other residues at the positions equivalent to 185 and 187 of MBP-A (Drickamer, 1992).

4. Multivalent Binding

In order to understand the structural basis of the multivalent binding of lectins to carbohydrates at the cell surface, it is necessary to consider both the ability of carbohydrates to form branched structures and present multiple copies of the ligand to one or more lectin domains, and the clustering of lectin domains to permit simultaneous binding to multiple carbohydrate ligands. For example, the intact hexamer and the isolated monomeric CRD of chicken hepatic lectin bind to monosaccharides with similar affinities, but the hexamer binds to a multivalent ligand with 25-fold greater affinity than does the monomer (Loeb and Drickamer, 1988). In this case, the isolated CRD also binds to synthetic multivalent ligands more strongly than it binds to monosaccharides, suggesting that multiple binding sites, which contribute to the overall valency of binding, may exist within the CRD (Lee *et al.*, 1989).

4.1. Oligosaccharide Multivalency

Several recent crystal structures have provided detailed pictures of branched oligosaccharides: the N-linked carbohydrate of *Erythrina corallodendron* lectin (EcoRL) (Shaanan *et al.*, 1991); a complex between wheat germ agglutinin (WGA) and a dibranched glycopeptide containing terminal sialic acids, one of the sugars recognized by this lectin (Wright, 1992); an octasaccharide–*Lathyrus ochrus* isolectin I (LOLI) complex (Bourne *et al.*, 1992); and the MBP CRD–Man$_6$GlcNAc$_2$Asn complex (Weis *et al.*, 1992a). In the EcoRL, WGA, and MBP structures, the oligosaccharide forms an integral part of the crystal lattice, with a single conformation from the ensemble that exists in solution incorporated into the crystal. Since oligosaccharides exhibit some conformational heterogeneity (Brisson and Carver, 1983; Homans *et al.*, 1986), there is concern that the conformation "trapped" in the lattice may not represent a conformation present in the ensemble that exists in solution. However, in the cases of EcoRL and the MBP CRD–oligosaccharide complex, NMR data from the same or similar oligosaccharide structures confirm that the structure seen in the crystal corresponds to an energetically allowed solution conformer, demonstrating that the lattice does not impose a nonnative structure on the oligosaccharide. In the LOLI–octasaccharide complex, the carbohydrate does not participate in any lattice interactions but is nonetheless extremely well-ordered (Bourne *et al.*, 1992).

In the MBP-A CRD–oligosaccharide complex, the terminal mannose of each visible branch interacts similarly with the CRD. In the WGA–glycopeptide complex, which contains four similar, but not identical, carbohydrate-binding sites within a single polypeptide, one subsite binds to the α(2,3) branch, while a different subsite of a neighboring molecule in the crystal lattice binds to the α(2,6) branch (Wright, 1992). The nature of the interactions is similar at each site. In both of these structures, the penultimate sugar of each oligosaccharide branch interacts differently with the protein, mostly through ordered water molecules (Weis *et al.*, 1992a; Wright, 1992). Ordered water molecules are also observed to mediate interactions between LOLI and oligosaccharides (Bourne *et al.*, 1990, 1992), although in the octasaccharide complex both branches interact with a single protomer. These observations suggest that the large number of hydroxyl groups on sugars can be used in combination with ordered water molecules to enlarge the interaction surface between lectin and ligand in a nonspecific manner. This nonspecific binding appears to complement the direct binding of particular sugars on nonequivalent branches, and thus permits the specific recognition of ligands presented in different contexts on the cell surface.

4.2. Oligomeric Structure of Lectins

Lectins generally exist as multimeric proteins. The crystal structures of a number of plant and viral lectins in their natural oligomeric states are known, so the relationship of the binding sites in these proteins is clear. No structural information yet exists concerning the three-dimensional relationships of the carbohydrate binding sites in C-type lectins. As noted above, the MBP CRD dimer forms upon subtilisin treatment of the bacterially expressed trimeric fragment, and is probably an artifact of its preparation. Since the natural trimer must have a different spatial relationship among the carbohydrate-binding sites than the dimer, knowledge of the crystal structure of the trimer will be necessary to understand the natural relationships of the CRDs in MBP.

It is clear that knowledge of the oligomeric structure of the C-type lectins will be of fundamental importance in understanding how these molecules function at the molecular level. The carbohydrate-binding sites in the C-type lectins appear to be clustered in three basic modes. Molecules such as MBP form soluble oligomers (Drickamer *et al.*, 1986), while the type II membrane proteins oligomerize in the membrane (Loeb and Drickamer, 1987; Verry and Drickamer, 1993) (Fig. 1). In both cases the oligomerization appears to be driven by regions of the protein outside of the CRD, as the isolated CRD is monomeric, or, as in MBP, an artifactual dimer. The macrophage mannose receptor represents another mode of clustering, as it contains eight tandemly repeated CRDs in a single polypeptide (Taylor et al., 1990) (Fig. 1). Studies using in vitro expression of various portions of this receptor have shown that while CRD 4 alone mimics the binding of monosaccharides to the intact receptor, CRDs 4 through 7 are needed to achieve high-affinity binding to a multivalent ligand (Taylor et al., 1992).

Putting the oligomeric structures of lectins together with their multivalent carbohydrate ligands has proven difficult. In the MBP CRD and WGA–oligosaccharide complexes, multivalent binding leads to cross-linking of lectins by the carbohydrate in the crystal lattice. While these pictures clearly illustrate general principles that are likely to govern multivalent binding, it is unlikely that they represent the situation on a cell surface, since the relative disposition of carbohydrate-binding sites in the crystal lattice is largely the result of the requirements of protein–protein packing interactions. In the case of the MBP CRD, the situation is compounded by the use of a nonnatural oligomer. Hopefully, a more realistic picture will be achieved by having a multivalent complex form a single asymmetric unit of a crystal.

5. Future Directions

Our current knowledge of the three-dimensional structures of lectins interacting with their carbohydrate ligands is limited to a few systems, but recent advances in carbohydrate purification and analysis should improve this situation. The successful overproduction and purification of several C-type CRDs should permit detailed molecular characterization of the mechanism of carbohydrate specificity in this system. Although the unusual mode of carbohydrate binding displayed by the MBP CRD demonstrates that generalizations about protein–carbohydrate interactions based on the structures of a limited number of proteins should be viewed with some caution, the lessons about multivalent binding learned from this system should be applicable to the general problem of carbohydrate recognition at cell surfaces.

The elucidation of the molecular mechanisms of carbohydrate recognition will form the basis for the design of drugs that would interfere with processes dependent on protein–carbohydrate interactions, such as inflammation and viral infectivity. The development of these molecules will depend on understanding the mechanism of multivalent binding, since it is unlikely that a monovalent carbohydrate analogue will be able to compete successfully with the natural multivalent interactions at the cell surface, even if it binds with an intrinsically higher affinity. The presentation of such compounds on a microsurface or highly branched polymer (Matrosovich, 1989; Weis et al., 1988) will almost certainly be needed in order to obtain a compound that could successfully compete with the natural interactions.

ACKNOWLEDGMENTS. The structural studies of C-type lectins were carried out in the laboratory of Wayne A. Hendrickson, in collaboration with Kurt Drickamer, and I am grateful for their continuing collaboration and support. This work was supported by grants from the National Institutes of Health, the National Science Foundation, and by the Howard Hughes Medical Institute. The author was a Life Sciences Research Foundation Fellow of the Howard Hughes Medical Institute.

References

Ashwell, G., and Harford, J., 1982, Carbohydrate-specific receptors of the liver, *Annu. Rev. Biochem.* **51**:531–554.

Bezouska, K., Crichlow, G. V., Rose, J. M., Taylor, M. E., and Drickamer, K., 1991, Evolutionary conservation of intron position in a subfamily of genes encoding carbohydrate-recognition domains. *J. Biol. Chem.* **266**:11604–11609.

Bourne, Y., Rougé, P., and Cambillau, C., 1990, X-ray structure of a (α-Man(1-3)β-Man(1-4)GlcNac)–lectin complex at 2.1-Å resolution, *J. Biol. Chem.* **265**:18161–18165.

Bourne, Y., Rougé, P., and Cambillau, C., 1992, X-ray structure of a biantennary octasaccharide–lectin complex refined at 2.3-Å resolution, *J. Biol. Chem.* **267**:197–203.

Brisson, J.-R., and Carver, J. P., 1983, Solution conformation of asparagine-linked oligosaccharides: α(1-6)-linked moiety, *Biochemistry* **22**:3680–3686.

Drickamer, K., 1988, Two distinct classes of carbohydrate-recognition domains in animal lectins, *J. Biol. Chem.* **263**:9557–9560.

Drickamer, K., 1992, Engineering galactose-binding activity into a C-type mannose-binding protein, *Nature* **360**:183–186.

Drickamer, K., Dordal, M. S., and Reynolds, L., 1986, Mannose-binding proteins isolated from rat liver contain carbohydrate-recognition domains linked to collagenous tails, *J. Biol. Chem.* **261**:6878–6886.

Ezekowitz, R. A. B., 1991, Ante-antibody immunity, *Curr. Biol.* **1**:60–62.

Ezekowitz, R. A. B., Day, L. E., and Herman, G. A., 1988, A human mannose-binding protein is an acute-phase reactant that shares sequence homology with other vertebrate lectins, *J. Exp. Med.* **167**:1034–1046.

Halberg, D. F., Proulx, G., Doege, K., Yamada, Y., and Drickamer, K., 1988, A segment of the cartilage proteoglycan core protein has lectin-like activity, *J. Biol. Chem.* **263**:9486–9490.

Homans, S. W., Dwek, R. A., Boyd, J., Mahmoudian, M., Richards, W. G., and Rademacher, T. W., 1986, Conformational transitions in N-linked oligosaccharides, *Biochemistry* **25**:6342–6350.

Ikeda, K., Sannoh, T., Kawasaki, N., Kawasaki, T., and Yamashina, I., 1987, Serum lectin with known structure activates complement through the classical pathway. *J. Biol. Chem.* **262**:7451–7454.

Kim, S. J., Ruiz, N., Bezouska, K., and Drickamer, K., 1992, Organization of the gene encoding the human macrophage mannose receptor, *Genomics* **14**:721–727.

Kuhlman, M., Joiner, K., and Ezekowitz, R. A. B., 1989, The human mannose-binding protein functions as an opsonin, *J. Exp. Med.* **169**:1733–1745.

Lee, R. T., Rice, K. G., Rao, N. B. N., Ichikawa, Y., Barthel, T., Piskarev, V., and Lee, Y.C.,

1989, Binding characteristics of N-acetylglucosamine-specific lectin of the isolated chicken hepatocytes: Similarities to mammalian hepatic galactose/N-acetylgalactosamine-specific lectin, *Biochemistry* **28**:8351–8358.

Lee, R. T., Ichikawa, Y., Fay, M., Drickamer, K., Shao, M.-C., and Lee, Y. C., 1991, Ligand-binding characteristics of rat serum-type mannose-binding protein (MBP-A), *J. Biol. Chem.* **266**:4810–4815.

Loeb, J. A., and Drickamer, K., 1987, The chicken receptor for endocytosis of glycoproteins contains a cluster of N-acetylglucosamine-binding sites, *J. Biol. Chem.* **262**:3022–3029.

Loeb, J. A., and Drickamer, K., 1988, Conformational changes in the chicken receptor for endocytosis of glycoproteins, *J. Biol. Chem.* **263**:9752–9760.

Lu, J., Thiel, S., Wiedemann, H., Timpl, R., and Reid, K. B. M., 1990, Binding of the pentamer/hexamer forms of mannan-binding protein to zymosan activates the proenzyme C1r2C1s2 complex, of the classical pathway of complement, without involvement of C1q, *J. Immunol.* **144**:2287–2294.

Matrosovich, M. N., 1989, Toward the development of antimicrobial drugs acting by inhibition of pathogen attachment to host cells: A need for polyvalency, *FEBS Lett* **252**:1–4.

Paulson, J. C., 1985, Interactions of animal viruses with cell surface receptors, in: *The Receptors,* Volume 2 (P. M. Conn, ed.), Academic Press, New York.

Quesenberry, M. S., and Drickamer, K., 1992, Role of conserved and nonconserved residues in the Ca^{2+}-dependent carbohydrate-recognition domain of a rat mannose-binding protein, *J. Biol. Chem.* **267**:10831–10841.

Rogers, G. N., Paulson, J. C., Daniels, R. S., Skehel, J. J., Wilson, I. A., and Wiley, D. C., 1983, Single amino acid substitutions in influenza haemagglutinin change receptor binding specificity, *Nature* **304**:76–78.

Sauter, N. K., Bednarski, M. D., Wurzburg, B. A., Hanson, J. E., Whitesides, G. M., Skehel, J. J., and Wiley, D. C., 1989. Hemagglutinins from two influenza virus variants bind to sialic acid derivatives with millimolar dissociation constants: A 500-MHz proton nuclear magnetic resonance study, *Biochemistry* **28**:8388–8396.

Shaanan, B., Lis, H., and Sharon, N., 1991, Structure of a legume lectin with an ordered N-linked carbohydrate in complex with lactose, *Science* **254**:862–866.

Sharon, N., and Lis, H., 1989, Lectins as cell recognition molecules, *Science* **246**:227–234.

Springer, T. A., and Lasky, L. A., 1991, Sticky sugars for selectins, *Nature* **349**:196–197.

Stoolman, L. M., 1989, Adhesion molecules controlling lymphocyte migration, *Cell* **56**:907–910.

Sumiya, M., Super, M., Tabona, P., Levinsky, R. J., Arai, T., Turner, M. W., and Summerfield, J. A., 1991, A point mutation in the human mannose binding protein gene resulting in immunodeficiency, *Lancet* **337**:1569–1570.

Super, M., Thiel, S., Lu, J., Levinsky, R. J., and Turner, M. W., 1989, Association of low levels of mannan-binding protein with a common defect of opsonisation, *Lancet* **2**:1236–1239.

Taylor, M. E., Brickell, P. M., Craig, R. K., and Summerfield, J. A., 1989, Structure and evolutionary origin of the gene encoding a human serum mannose-binding protein, *Biochem. J.* **262**:763–771.

Taylor, M. E., Conary, J. T., Lennartz, M. R., Stahl, P., and Drickamer, K., 1990, Primary structure of the mannose receptor contains multiple motifs resembling carbohydrate-recognition domains, *J. Biol. Chem.* **265**:12156–12162.

Taylor, M. E., Bezouska, K., and Drickamer, K., 1992, Contribution to ligand binding by multiple carbohydrate-recognition domains in the macrophage mannose receptor, *J. Biol. Chem.* **267**:1719–1726.

van Heyningen, S., 1983, The interaction of cholera toxin with gangliosides and the cell membrane, in: *Current Topics in Membranes and Transport,* Volume 18 (A. Kleinzeller and B. R. Martin, eds.), Academic Press, New York.

Verry, F., and Drickamer, K., 1993, Determinants of oligomeric structure in the chicken liver glycoprotein receptor, *Biochem. J.* **292**:149–155.

Wassarman, P. M., 1987, The biology and chemistry of fertilization, *Science* **235**:553–560.

Weis, W., Brown, J. H., Cusack, S., Paulson, J. C., Skehel, J. J., and Wiley, D. C., 1988, Structure of the influenza virus haemagglutinin complexed with its receptor, sialic acid, *Nature* **333**:426–431.

Weis, W. I., Crichlow, G. V., Murthy, H. M. K., Hendrickson, W. A., and Drickamer, K., 1991a, Physical characterization and crystallization of the carbohydrate-recognition domain of a mannose-binding protein from rat, *J. Biol. Chem.* **266**:20678–20686.

Weis, W. I., Kahn, R., Fourme, R., Drickamer, K., and Hendrickson, W. A., 1991b, Structure of the calcium-dependent lectin domain from a rat mannose-binding protein determined by MAD phasing, *Science* **254**:1068–1615.

Weis, W. I., Drickamer, K., and Hendrickson, W. A., 1992a, Structure of a C-type mannose-binding protein complexed with an oligosaccharide, *Nature* **360**:127–134.

Weis, W. I., Quesenberry, M. S., Taylor, M. E., Bezouska, K., Hendrickson, W. A., and Drickamer, K., 1992b, Molecular mechanisms of complex carbohydrate recognition at the cell surface, *Cold Spring Harbor Symp. Quant. Biol.* **52**:281–289.

White, R. T., Damm, D., Miller, J., Spratt, K., Schilling, J., Hawgood, S., Benson, B., and Cordell, B., 1985, Isolation and characterization of the human pulmonary surfactant apoprotein gene, *Nature* **317**:361–363.

Wright, C. S., 1992, Crystal structure of a wheat germ agglutinin/glycophorin–sialoglycopeptide receptor complex, *J. Biol. Chem.* **267**:14345–14352.

5

The Uvomorulin–Catenin Complex
Insights into the Structure and Function of Cadherins

MARTIN RINGWALD

1. Introduction

Cadherins are a protein family of Ca^{2+}-dependent transmembrane-type cell adhesion molecules. They mediate Ca^{2+}-dependent cell–cell adhesion by homophilic binding. All solid tissues in vertebrates express some cadherins. Very importantly for their function, the different cadherin subtypes show distinct developmental and tissue-specific patterns of expression and display corresponding binding specificities. In this way, cadherins are thought to influence cell sorting, morphogenesis, and the maintenance of adult tissues (Takeichi, 1988, 1991; Kemler *et al.*, 1989).

Molecular cloning and sequence comparison has led to the characterization of a gene superfamily, composed of a highly homologous group of "classical cadherins" and more distantly related members, including the desmocollins and desmogleins, proteins of the desmosomal dense plaque, and the *Drosophila* tumor suppressor gene *fat* (Kemler, 1992a; Buxton and Magee, 1992).

The controlled interaction of cadherins with the actin filament network and with additional submembranous elements seems to be of fundamental importance to the function of these cell adhesion molecules. The study of uvomorulin has led to the identification of three proteins, catenin-α, -β, and -γ, which complex with

MARTIN RINGWALD • Max-Planck Institute for Immunobiology, Division of Molecular Embryology, W-7800 Freiburg, Germany. *Present address:* The Jackson Laboratory, Bar Harbor, Maine 04609.

Cellular Adhesion: Molecular Definition to Therapeutic Potential, edited by Brian W. Metcalf, Barbara J. Dalton, and George Poste. Plenum Press, New York, 1994.

the intracellular protein domain of cadherins and which are probably involved in this control. The molecular structure of catenin-α and -β has now been determined.

This chapter will concentrate on the structure and function of classical cadherins, and specifically on the epithelial cadherin uvomorulin (= E-cadherin) (Ringwald et al., 1987; Nagafuchi et al., 1987), which will be used to elucidate the structure–function relationships of the cadherin proteins. Possible implications for catenin and cadherin function will be discussed.

2. Uvomorulin-Mediated Cell Adhesion—A Multistep Process

Uvomorulin is expressed in epithelial cells during mouse development and in adult tissues (Vestweber and Kemler, 1984a; Takeichi, 1988). Uvomorulin provides epithelial cells with adhesion specificity. In addition, it is important for the induction of epithelial polarization (McNeill et al., 1990), and for the formation of epithelial junctional complexes (Gumbiner et al., 1988; Mege et al., 1988). In polarized epithelial cells, uvomorulin is concentrated in the adherens junctions (Boller et al., 1985). The redistribution of uvomorulin to the basolateral cell surface and its concentration in the zonula adherens seems to modulate both the cell-adhesive function of uvomorulin and the transition from a nonpolarized to a polarized epithelial cell. This can be seen best in vivo in the compaction process of the mouse morula, which is mediated by uvomorulin (Kemler et al., 1977; Hyafil et al., 1980) and which is the first calcium-dependent adhesive process during embryogenesis (Fig. 1). Before compaction, uvomorulin is found uniformly distributed on the surface of all blastomeres. During compaction, however, it is progressively removed from the outer surface of the embryo and becomes concentrated in membrane domains involved in the cell–cell contact of adjacent outer blastomeres. These cells then maximize their cell contacts, polarize, and establish junctional complexes to become the epitheliumlike trophectodermal cells of the blastocysts, on which uvomorulin is predominantly located in the adherens junctions. Redistribution of uvomorulin is developmentally regulated in preimplantation embryos. It occurs only in cells committed to epithelial cell differentiation. In the inner blastomeres of compact morulae or on inner-cell-mass cells of the blastocysts, uniform cell surface distribution of uvomorulin is maintained (Vestweber et al., 1987).

This short scenario already shows that uvomorulin-mediated cell adhesion is a complex process. Recent studies of the structure and function of uvomorulin have provided insights into the molecular mechanism of this process. These analyses have been performed in three stages: (1) the comparison of uvomorulin and other cadherins at the amino acid sequence level; (2) the identification of motifs/domains that are conserved in the cadherin protein family; (3) a functional

2-cell embryo   4-cell embryo   8-cell embryo   morula   blastocyst

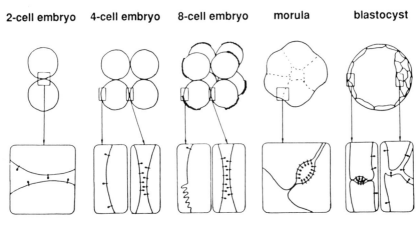

 uvomorulin

Figure 1. Schematic representation of uvomorulin cell-surface localization during mouse preimplantation development. Uvomorulin is present on the cell surface throughout all stages of preimplantation development. Uvomorulin is uniformly distributed on the cell surface of the egg and the two-cell-stage embryo. On four-cell-stage embryos, the concentration of uvomorulin first occurs in membrane areas that are in close contact with adjacent blastomeres. During compaction, uvomorulin vanishes from the outer surface of the embryo and is found predominantly in membrane domains that are involved in cell–cell contact. In compact morulae, uvomorulin is no longer present on the outer surface of the embryo; it is located exclusively on the basolateral membranes of the outer cells, and is concentrated in their adherens junctions. Uvomorulin remains uniformly distributed on inner cells. Once this differential distribution has been established during compaction, it is maintained and is also found in the blastocyst. (After Kemler *et al.*, 1990, by permission of John Wiley and Sons.)

analysis of these units in uvomorulin. Therefore, the results summarized schematically in Fig. 2 and discussed in the following section go beyond a specific analysis of uvomorulin and reflect those features common to all cadherin molecules.

3. Structure–Function Analysis

3.1. The Extracellular Protein Domain

Figure 2 shows a schematic view of the primary structure of the uvomorulin protein and, at the same time, the basic structure of all classical cadherins. With the exception of T-cadherin, all of these molecules are transmembrane type 1 glycoproteins, with the N-terminal domain located extracellularly and the C-terminal domain located in the cytoplasm (Kemler, 1992a). [The truncated T-cad-

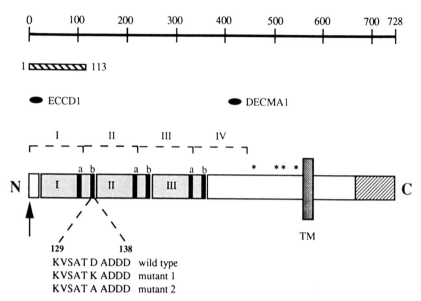

Figure 2. Schematic representation of the primary structure of the mature uvomorulin protein, as it appears on the cell surface. The precursor region of the protein is left out. (Like all classical cadherins, uvomorulin is synthesized as a larger preproprotein which is intracellularly processed by endoproteolytic processing.) The arrow indicates the conserved cleavage site for the removal of the propeptide. Three or four internally repeated domains can be identified in the extracellular protein part; both of these alternatives are shown. a and b, putative Ca^{2+}-binding units; the sequences of the wild-type and mutant peptides, used to demonstrate the Ca^{2+}-binding capability of unit b in repeat I, are indicated (for details see text). Asterisks, conserved cysteine residues; TM, transmembrane region; hatched box at the C-terminus, catenin-binding domain. Upper part: The N-terminal domain of 113 amino acid residues mediates the binding selectivity of uvomorulin with respect to P-cadherin. Closed ovals, epitopes recognized by the adhesion-blocking monoclonal antibodies ECCD1 and DECMA1.

herin (Ranscht *et al.*, 1991) is exceptional in this respect; it is the only cadherin so far described which is missing a cytoplasmic domain and is anchored to the cell membrane by phosphatidylinositol.]

The extracellular portion of cadherins is composed largely of repeating domains, each approximately 110 amino acid residues in length. Depending on the definition of where they start, three or four repeats can be identified (Ringwald *et al.*, 1987; Hatta *et al.*, 1988). Both alternatives are indicated in Fig. 2. Further research at the protein level is needed to clarify where the real boundaries of the respective protein domains are located. The homology of the corresponding repeats in different cadherins is higher than the internal homology of the respective domains in each molecule (Kemler *et al.*, 1989; Ringwald and Kemler, unpublished results). This suggests that each repeated domain exhibits not only common, but also unique functional properties, and that both of these have been conserved in all vertebrate cadherins since their divergence from a common ancestral protein.

Two major characteristics of cadherins are that they express their adhesive function only in the presence of Ca^{2+}, and that Ca^{2+} protects the extracellular part of these proteins from proteolytic degradation (Takeichi, 1977, 1988; Hyafil *et al.*, 1981). As with uvomorulin, each repeated domain contains putative Ca^{2+} binding sites (two each in the three-repeat model; Ringwald *et al.*, 1987). This proposed function has recently found two confirmations. First, a synthetic peptide corresponding to the sequence of a Ca^{2+}-binding motif can complex Ca^{2+}. A single amino acid substitution in this peptide abolishes complex formation. The same mutation introduced into uvomorulin leads to protease sensitivity near the site of amino acid replacement and completely destroys the cell-adhesive properties of uvomorulin (Ozawa *et al.*, 1990a). Second, for all members of the cadherin gene superfamily (defined by the presence of "cadherin-like repeats," described above), the putative Ca^{2+}-binding motifs originally defined for uvomorulin are well conserved in all repeats and are located at analogous positions in all proteins from *Drosophila* to humans (Mahoney *et al.*, 1991; Amagai *et al.*, 1991; Mechanic *et al.*, 1991; Wheeler *et al.*, 1991; Parker *et al.*, 1991; Collins *et al.*, 1991; Koch *et al.*, 1991). The latter finding suggests that Ca^{2+} plays the same molecular role for all of these molecules, reflecting the essential importance of Ca^{2+} for cell–cell adhesion in all animal species.

Cadherins mediate cell adhesion mostly by homophilic binding; i.e., identical molecules from neighboring cells selectively bind to each other. This has been shown by transfection experiments: L-cells normally have little cadherin activity, but L-cells expressing E- or P-cadherin show Ca^{2+}-dependent adhesion activity that can be blocked by monoclonal antibodies specific for the respective cadherin subclass. In a mixture, E-cadherin transfectants segregate from P-cadherin transfectants (Nose *et al.*, 1988), excluding the possibility that cadherins bind in a heterophilic manner to some ligands present on L-cells. In addition, the N-cadherin of chicken neural retina cells has been shown to bind to N-cadherin immobilized on tissue culture dishes (Balsamo *et al.*, 1991), further confirming a homophilic binding of cadherins.

The molecular mechanism of homophilic binding is still not completely understood. For example, it is not clear whether the molecules interact along their length in a head-to-tail manner, or whether only specific parts of the respective extracellular domains act as active centers in homophilic binding. The epitopes of five adhesion-blocking monoclonal antibodies, directed against different cadherins, are localized either at the N-terminal region or close to the membrane-spanning domain (Nose *et al.*, 1990; Geiger *et al.*, 1990; Ozawa *et al.*, 1991). The cleavage site for the removal of the propeptide is well conserved in all cadherins. Analysis of the proteolytic processing of the uvomorulin precursor polypeptide reveals that correct cleavage is necessary for the activation of uvomorulin function. In the case of incorrect processing, even a small residual stretch of precursor peptide exhibits inhibitory effects, further confirming the importance of the N-terminal domain for adhesive binding (Ozawa and Kemler, 1990). The membrane proximal part of classical cadherins harbors four conserved cysteine residues. As shown for uvomorulin, reductive cleavage of these disulfide bonds by DTT inhibits the formation of close cell–cell contacts (Ozawa et al., 1991). Interpreting these results is difficult, because it is not yet possible to differentiate between direct and conformational blocking effects. Further studies are needed to reveal the details of the interplay between cadherin conformation and cadherin activity.

When one considers that cadherins descended from a common ancestral protein, it seems clear that the ability for homophilic binding must have evolved first. Thereafter, the binding specificity of cadherins could be generated by the divergence to different cadherin subtypes. An analysis of chimeric proteins between E- and P-cadherin reveals that the N-terminal domain of 113 amino acid residues mediates the selectivity of binding (Nose et al., 1990). It would be interesting to extend such analysis to other cadherin combinations to see whether the N-terminal domain determines the subtype specificity of cadherins in general.

3.2. The Cytoplasmic Protein Domain

The cytoplasmic domain is the most highly conserved protein portion of classical cadherins (Kemler et al., 1989; Takeichi, 1990). This finding points to an important biological function for this protein domain common to all cadherins. The enrichment of uvomorulin in the zonula adherens of epithelial cells and the colocalization of cadherins and actin bundles (Boller et al., 1985; Hirano et al., 1987; Volk and Geiger, 1986) suggests that one functional role of the cytoplasmic domain could be to connect cadherins to cytoskeletal structures. Recent research has confirmed this hypothesis, unraveling the molecular components of cytoplasmic anchorage and highlighting the importance of these molecules for the regulation of cadherin function. Again, most information comes from studies on uvomorulin/E-cadherin.

Three independent proteins, termed catenin-α, -β, and γ, size 102, 88, and 80 kDa, respectively, have been found to associate with the cytoplasmic domain of uvomorulin. These proteins had originally been detected in cell lysates from uvomorulin-expressing cells by coprecipitation with antiuvomorulin antibodies (Vestweber and Kemler, 1984b; Peyrieras et al., 1985). A breakthrough in their molecular analysis was the finding that catenins are also expressed in uvomorulin-negative fibroblasts. Upon transfection with uvomorulin cDNA, the catenins of the host cell associate with the introduced uvomorulin; they thus become detectable in immunoprecipitation with antiuvomorulin antibodies (Ozawa *et al.*, 1989). Therefore, fibroblasts offered the model of choice to study the molecular interaction of uvomorulin and catenins. As already mentioned, the ectopic expression of uvomorulin in fibroblasts confers Ca^{2+}-dependent cell-adhesion activity to these cells. Moreover, uvomorulin expression induces the redistribution of Na^+/K^+-ATPase and fodrin to sites of uvomorulin-mediated cell–cell contacts (McNeill *et al.*, 1990), and leads to the formation of adherens junctions and gap junctions (Mege *et al.*, 1988). Thus, fibroblasts can also be used to assess the molecular role of uvomorulin and the catenins in cell adhesion and cell polarization.

An analysis of deletion mutants of uvomorulin and of chimeric proteins between uvomorulin and H-2K^d, upon transfection in L-cells, reveals that the C-terminal 72-amino-acid domain of uvomorulin, which corresponds almost exactly to the coding part of exon 16 of the uvomorulin gene (Ringwald *et al.*, 1991), is necessary and sufficient for complex formation with catenins (Fig. 2; Ozawa *et al.*, 1990b). In cell-aggregation assays, only the normal and mutant forms of uvomorulin that are able to associate with catenins express full adhesion activity. Deletion of the catenin-binding domain correlates with the loss of adhesiveness (Ozawa *et al.*, 1990b; Nagafuchi and Takeichi, 1989). Moreover, uvomorulin mutants without the catenin-binding domain are not able to induce the redistribution of uvomorulin, Na^+/K^+-ATPase, and fodrin to sites of cell–cell contact (McNeill *et al.*, 1990). Therefore, catenins seem to be important both for cell polarization and for uvomorulin-mediated cell adhesion.

4. The Uvomorulin–Catenin Complex—Structure and Function of Catenins

Further biochemical studies have elucidated the molecular role of catenins in cell adhesion and cell polarization (Ozawa *et al.*, 1990b). In phase-separation experiments, 30–40% of the uvomorulin protein is normally found in the insoluble cytoskeletal and nuclear fraction. This is also true of uvomorulin expressed in transfected L-cells. However, when the catenin-binding domain is deleted, it leads to an exclusive separation of the mutant uvomorulin into the detergent-soluble fraction. This indicates that catenins connect uvomorulin to the cytoskeleton.

As already mentioned, uvomorulin and actin bundles are found colocalized in the zonula adherens of epithelial cells. Biochemical evidence that catenins connect uvomorulin to the actin filament network has been obtained by experiments with DNase I–Sepharose. DNase I has specific affinity for G-actin. Therefore, molecular complexes containing actin bind to DNase I–Sepharose. Uvomorulin is found in the DNase I-bound fraction of cell lysates, but only if it is complexed with catenins. The selective removal of catenin-α from the uvomorulin–catenin complex destroys its ability to bind to DNase I and to separate into the detergent-insoluble cytoskeletal fraction. Thus, catenin-α seems to play a key role in the cytoplasmic anchorage of uvomorulin to the actin microfilament network.

Figure 3A summarizes our present knowledge about the molecular organization of the uvomorulin–catenin complex (Ozawa and Kemler, 1992). Based on phase-separation and cross-linking experiments, catenin-β binds directly to uvomorulin even in the absence of catenin-α. Catenin-α and -β probably interact directly with each other. It is not yet known whether catenin-α can also interact directly with uvomorulin. Catenin-γ seems to be rather loosely associated, since it is detected with some, but not all, analytical methods used. Catenin-α mediates linkage to actin. It is not clear, however, whether this linkage is direct, or whether additional molecules are involved.

The uvomorulin–catenin complex is assembled very rapidly in the cell (Fig. 3B). Pulse-chase experiments show that catenin-β associates immediately after uvomorulin synthesis, and that the association of catenin-α takes place at the

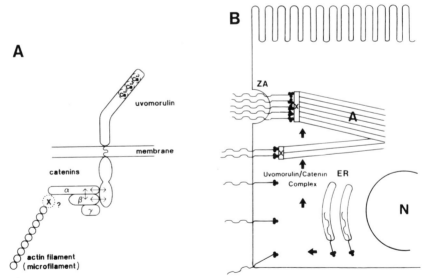

Figure 3. (A) Diagram of the molecular organization of the uvomorulin–catenin complex. For details see text. (From Kemler, 1992b, by permission of Elsevier Science Publishers B.V.) (B) Simplified model of assembly and intracellular transport of the uvomorulin–catenin complex and of the formation of functional adhesive compartments at the zonula adherens. A, actin bundles; ER, endoplasmic reticulum; N, nucleus; ZA, zonula adherens. (From Kemler *et al.*, 1989, by permission of Current Biology, Ltd.)

time of endoproteolytic processing of the uvomorulin precursor (Ozawa and Kemler, 1992). In MDCK cells, precursor processing is observed shortly after the addition of complex carbohydrates in the late Golgi complex (Shore and Nelson, 1991). In these polarized epithelial cells, the uvomorulin–catenin complex is transported both to the apical and to the basal-lateral cell surface immediately after the induction of cell–cell contact. At the apical surface, however, the complex is rapidly internalized and is probably degraded thereafter. In contrast, the uvomorulin–catenin complex has a much longer residence time at the basal-lateral membrane (Wollner *et al.*, 1992). This is probably the result of the integration of the complex into a larger submembranous network (Nelson *et al.*, 1990), which occurs at the basal-lateral cell surface. The Ca^{2+}-dependent homophilic binding of the extracellular part of uvomorulin is a prerequisite for this process. This binding finally leads to a lateral clustering of the uvomorulin–catenin complex, its enrichment in the zonula adherens, and its concomitant linkage to the actin filament network. All of these steps are necessary for the formation of a strong and stable adhesive complex.

Recent insights into the primary structure of catenin-α and -β suggest that catenins are not only important for building up the intermolecular chain that connects uvomorulin to the cytoskeleton (Latin: *catena* = chain), but may also be directly involved in the process of lateral assembly.

Sequence comparison reveals that catenin-α shows homology to vinculin (Fig. 4A; Herrenknecht *et al.*, 1991; Nagafuchi *et al.*, 1991). Vinculin is a cytoskeletal protein found both in adherens junctions and in focal contacts (Geiger, 1979; Burridge *et al.*, 1988). By interacting with talin (Otto, 1983; Burridge and Mangeat, 1984) and α-actinin (Belkin and Koteliansky, 1987; Wachsstock *et al.*, 1987), vinculin is thought to participate in the linkage of integrins to the actin microfilament system (Burridge *et al.*, 1988). The homology between catenin-α and vinculin is restricted to three major regions, one each in the N-terminal, central, and C-terminal parts of both proteins. The strongest degree of homology is found in the C-terminal domain. For vinculin, this region has been described as exhibiting self-association properties (Bendori *et al.*, 1989). By analogy, catenin-α could associate with another catenin-α molecule via this domain. It is also possible that catenin-α and vinculin interact with each other and are part of a multiprotein complex. In the N-terminal vinculin-homologous region, catenin-α contains a short pentapeptide that is evolutionarily conserved in vinculin; this is located in a region that is believed to regulate the binding of vinculin to talin (Jones *et al.*, 1989). Talin is not found in adherens junctions, but catenin-α may exhibit multiple biological properties in different cell types, since it is also found, for example, in cadherin-negative L-cells.

Catenin-β is homologous to plakoglobin, a protein found in the adherens junctions and in the desmosomal dense plaque (Cowin *et al.*, 1986). Both proteins are highly homologous in a large central region, while the N- and C-termi-

VINCULIN

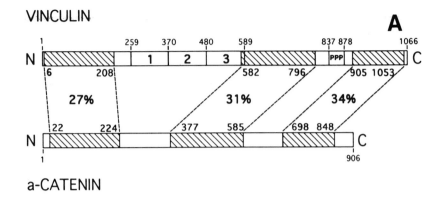

a-CATENIN

ß-CATENIN

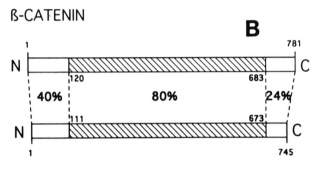

PLAKOGLOBIN

Figure 4. (A) Schematic comparison of mouse catenin-α and chicken vinculin. Homologous regions are marked with hatched boxes; the percentage of amino acid identity is indicated. The repeated units (1,2,3) and the proline-rich sequence (PPP), both characteristic for vinculin, are absent in catenin-α. (After Herrenknecht *et al.*, 1991.) (B) Schematic comparison of mouse catenin-β and human plakoglobin. The percentage of amino acid identity is indicated. Hatched box, highly homologous central region. Numbering of catenin-β according to Butz *et al.* (1992); numbering of human plakoglobin according to Franke *et al.* (1989).

nal sequences diverge more (Fig. 4B; Butz *et al.*, 1992; McCrea *et al.*, 1991). It has been shown that plakoglobin molecules can form dimers (Kapprell *et al.*, 1987); the strong homology implies that catenin-β may have the same capacity. Thus, both catenin-α and -β could participate directly in the lateral clustering of the uvomorulin–catenin complex.

While catenin-β binds to uvomorulin strongly, plakoglobin associates very weakly with the complex, if at all (Butz *et al.*, 1992; Peifer *et al.*, 1992; Knudsen and Wheelock, 1992). In contrast, plakoglobin interacts with desmoglein in the desmosomes (Korman *et al.*, 1989), where no catenin-β can be detected. The specific binding capabilities of catenin-β and plakoglobin could be mediated by their less conserved and specific terminal protein domains.

Interestingly, the *Drosophila* segment polarity gene *armadillo* also belongs to the new catenin-β/plakoglobin protein family (Peifer and Wieschaus; 1990; Peifer *et al.*, 1992). The armadillo protein is even more homologous to catenin-β than to plakoglobin, and anti-armadillo antibodies recognize catenin-β. This surprising finding raises a lot of exciting questions concerning cell communication and signaling. However, the question of whether armadillo is the true species homologue of catenin-β, or whether even more closely related molecules exist in vertebrates, cannot be answered at the present time.

5. Multiple Cadherin–Catenin Complexes

Antibodies against catenin-α and -β recognize analogous proteins in human, mouse, chicken, and *Xenopus* (Herrenknecht *et al.*, 1991; Butz *et al.*, 1992). This confirms earlier results obtained by peptide pattern analysis (Ozawa *et al.*, 1989), showing that catenins are evolutionarily highly conserved. In immunofluorescence and Western blot analysis, catenin-α and -β are expressed in a wide variety of cell types, most of which are known to be negative for uvomorulin. Moreover, catenin-α and -β are localized in membrane areas of cell–cell contact in almost all tissues in which they are expressed. This suggests that catenins may also interact with other cadherins. Indeed, catenin-α has been associated with all cadherins tested; i.e., with mouse N- and P-cadherin, chicken A-CAM, and *Xenopus* U-cadherin (Herrenknecht *et al.*, 1991). More indirect data indicate that the same is true for catenin-β. Therefore, one selective pressure for the high conservation of the cadherin cytoplasmic domain seems to be complex formation with catenins; the data described above for uvomorulin will probably be representative of all cadherins.

Several tissues express more than one cadherin. In addition, a second subtype of catenin-α has recently been described. It is expressed mainly in the nervous system, where it associates with N-cadherin, and is thus termed αN-catenin. (According to this new nomenclature, the catenin-α described above should be termed αE-catenin.) The important role of αN-catenin for cell aggregation and polarization has been convincingly demonstrated. Upon transfection, αN-catenin can substitute for αE-catenin, and can interact with uvomorulin as well (Hirano *et al.*, 1992). Thus, different combinations of cadherins and catenins seem to be expressed in different cell types. Each combination should have distinct chemical affinities and biological properties that could be used for regulating and fine-tuning specific cadherin activities. In keeping with this hypothesis, the ectopic expression of different combinations and concentrations of cadherins in murine sarcoma cells results in distinct sorting behavior (Friedlander *et al.*, 1989). In other experiments, the overexpression of the N-cadherin cytoplasmic domain in *Xenopus* embryos inhibits catenin binding to uvomorulin and abolishes cell adhesion (Kintner, 1992). In the latter case, the cadherin/catenin system of the affected cells is probably completely out of balance.

The perturbation of cadherin function has often been observed to correlate with the increase of metastatic activity of tumor cells, both *in vitro* and *in vivo*. In several cases this is the result of a reduction of cadherin expression (Behrens *et al.*, 1992; Takeichi, 1991). However, there are also invasive carcinomas in which cadherin function seems to be impaired by another mechanism (Shimoyama and Hirohashi, 1991; Eidelman *et al.*, 1989). Recently, a correlation of suppression of cadherin function and tyrosine-specific phosphorylation of catenin-β was reported in v-src-transformed cells (Matsuyoshi *et al.*, 1992). Specific phosphorylation and dephosphorylation of components of the cadherin–catenin complex might thus be a further important regulatory mechanism of Ca^{2+}-dependent cell adhesion.

6. Conclusion

Two observations—that compaction of the mouse morula correlates with the redistribution of uvomorulin at the cell membrane, and that the cytoplasmic domain of cadherins is conserved—led to the discovery of an elaborate intracellular system necessary for uvomorulin-mediated cell–cell adhesion. The formation of a fully functional adhesive complex involves multiple steps, including the assembly of cell-type-specific cadherin–catenin complexes, their lateral clustering, and their linkage to the actin filament network, as well as their integration into a larger submembranous protein network. The sophisticated process of Ca^{2+}-dependent cell adhesion allows a variety of interacting regulatory mechanisms, all of which should be considered important both for normal development and for disorders like cancer, especially tumor metastasis.

The molecular cloning of catenin-α and -β has revealed the existence of new protein families, and has given us interesting hints about the molecular evolution of junctional complexes. The cadherin superfamily consists of classical cadherins located in the adherens junctions, and of desmogleins and desmocollins, which are both components of the desmosomes. Catenin-α, localized in adherens junctions, is homologous to vinculin, a component of adherens junctions and focal contacts. Catenin-β, plakoglobin, and the armadillo protein are members of a third protein family. While catenin-β is localized in adherens junctions, plakoglobin is found in both adherens junctions and the desmosomal dense plaque. Thus, different members of the same protein families are used specifically to assemble the different junctional complexes. The most reasonable explanation is that these specialized cell junctions have evolved from a common ancestral protein complex. An interesting picture is thus emerging about the coevolution and conservation of protein complexes for the construction of adhesive cellular compartments.

ACKNOWLEDGMENTS. I thank Rolf Kemler, Kurt Herrenknecht, Stefan Butz, Heinz Hoschützky, and Stephen Wood for helpful discussions, and Randy Cassada for critically reading the manuscript.

References

Amagai, M., Klaus-Kovtun, V., and Stanley, J. R., 1991, Autoantibodies against a novel epithelial cadherin in Pemphigus vulgaris, a disease of cell adhesion, *Cell* **67**:869–877.

Balsamo, J., Thiboldeaux, R., Swaminathan, N., and Lilien, J., 1991, Antibodies to the retina N-acetylgalactosaminylphosphotransferase modulate N-cadherin-mediated adhesion from the actin-containing cytoskeleton, *J. Cell Biol.* **113**:429–436.

Behrens, J., Frixen, U., Schipper, J., Weidner, M., and Birchmeier, W., 1992, Cell adhesion in invasion and metastasis, *Semin. Cell Biol.* **3**:169–178.

Belkin, A. M., and Koteliansky, V. E., 1987, Interaction of iodinated vinculin, metavinculin and α-actinin with cytoskeletal proteins, *FEBS Lett.* **220**:291–294.

Bendori, R., Salomon, D., and Geiger, B., 1989, Identification of two distinct functional domains on vinculin involved in its association with focal contacts, *J. Cell Biol.* **108**:2383–2393.

Boller, K., Vestweber, D., and Kemler, R., 1985, Cell adhesion molecule uvomorulin is localized in the intermediate junctions of adult intestinal epithelial cells, *J. Cell Biol.* **100**:327–332.

Burridge, K., and Mangeat, P., 1984, An interaction between vinculin and talin, *Nature* **308**:744–746.

Burridge, K., Fath, K., Kelly, T., Nuckolls, G., and Turner, C., 1988, Focal adhesion: Transmembrane junctions between the extracellular matrix and the cytoskeleton, *Annu. Rev. Cell Biol.* **4**:487–525.

Butz, S., Stappert, J., Weissig, H., and Kemler, R., 1992, Plakoglobin and β-catenin: Distinct but related, *Science* **257**:1142–1144.

Buxton, R. S., and Magee, A. I., 1992, Structure and interaction of desmosomal and other cadherins, *Semin. Cell Biol.* **3**:157–167.

Collins, J. E., Legan, P. K., Kenny, T. P., MacGarvie, J., Holton, J. L., and Garrod, D. R., 1991, Cloning and sequence analysis of desmosomal glycoprotein 2 and 3 (desmocollins): Cadherinlike desmosomal adhesion molecules with heterogeneous cytoplasmic domains, *J. Cell Biol.* **113**:381–391.

Cowin, P., Kapprell, H. P., Franke, W. W., Tamkun, J., and Hynes, R. O., 1986, Plakoglobin: A protein common to different kinds of intercellular adhering junctions, *Cell* **46**:1063–1073.

Eidelman, S., Damsky, C. H., Wheelock, M. J., and Damjanov, I., 1989, Expression of cell–cell adhesion glycoprotein cell CAM 120/80 in normal human tissues and tumors, *Am. J. Pathol.* **135**:101–110.

Franke, W. W., Goldschmidt, M. D., Zimbelmann, R., Mueller, H. M., Schiller, D. L., and Cowin, P., 1989, Molecular cloning and amino acid sequence of human plakoglobin, the common junctional plaque protein, *Proc. Natl. Acad. Sci. USA* **86**:4027–4031.

Friedlander, D. R., Mege, R. M., Cunningham, B. A., and Edelman, G. M., 1989, Cell sorting-out is modulated by both the specificity and amount of different cell adhesion molecules (CAMs) expressed on cell surfaces, *Proc. Natl. Acad. Sci. USA* **86:**7043–7047.

Geiger, B., 1979, A 130K protein from chicken gizzard: Its localization at the termini of microfilament bundles in cultured chicken cells, *Cell* **18:**193–205.

Geiger, B., Volberg, T., Sabanay, I., and Volk, T., 1990, A-CAM: An adherens junction-specific cell adhesion molecule, in: *Morphoregulatory Molecules* (G. M. Edelman, B. A. Cunningham, and J. P. Thiery, eds.), Wiley, New York, pp. 57–79.

Gumbiner, B., Stevenson, B., and Grimaldi, A., 1988, The role of the cell adhesion molecule uvomorulin in the formation and maintenance of the epithelial junctional complex, *J. Cell Biol.* **107:**1575–1587.

Hatta, K., Nose, A., Nagafuchi, A., and Takeichi, M., 1988, Cloning and expression of cDNA encoding a neural calcium-dependent cell adhesion molecule: Its identity in the cadherin gene family, *J. Cell Biol.* **106:**873–881.

Herrenknecht, K., Ozawa, M., Eckerskorn, C., Lottspeich, F., Lenter, M., and Kemler, R., 1991, The uvomorulin-anchorage protein α catenin is a vinculin-homologue, *Proc. Natl. Acad. Sci. USA* **88:**9156–9160.

Hirano, S., Nose, A., Hatta, K., Kawakami, A., and Takeichi, M., 1987, Calcium-dependent cell–cell adhesion molecules (cadherins): Subclass specificities and possible involvement of actin bundles, *J. Cell Biol.* **105:**2501–2510.

Hirano, S., Kimoto, N., Shimoyama, Y., Hirohashi, S., and Takeichi, M., 1992, Identification of a neural α-catenin as a key regulator of cadherin function and multicellular organization, *Cell* **70:**293–301.

Hyafil, F., Morrelo, D., Babinet, C., and Jacob, F., 1980, A cell surface glycoprotein involved in the compaction of embryonal carcinoma cells and cleavage stage embryos, *Cell* **21:**927–934.

Hyafil, F., Babinet, C., and Jacob, F., 1981, Cell–cell interactions in early embryogenesis: A molecular approach to the role of calcium, *Cell* **26:**447–454.

Jones, P., Jackson, P., Price, G. J., Patel, B., Ohanion, V., Lear, A. L., and Critchley, D. R., 1989, Identification of a talin binding site in the cytoskeletal protein vinculin, *J. Cell Biol.* **109:**2917–2927.

Kapprell, H. P., Cowin, P., and Franke, W. W., 1987, Biochemical characterization of the soluble form of the junctional plaque protein, plakoglobin, from different cell types, *Eur. J. Biochem.* **166:**505–517.

Kemler, R., 1992a, Classical cadherins, *Semin. Cell Biol.* **3:**149–155.

Kemler, R., 1992b, Cadherins and catenins, in: *Leucocyte Adhesion: Basic and Clinical Aspects* (C. G. Gahmberg, T. Mandrup-Poulsen, L. Wogensen Bach, and B. Hökfeld, eds.), Excerpta Medica, Amsterdam, pp. 63–78.

Kemler, R., Babinet, C., Eisen, H., and Jacob, F., 1977, Surface antigen and early differentiation, *Proc. Natl. Acad. Sci. USA* **74:**4449–4452.

Kemler, R., Ozawa, M., and Ringwald, M., 1989, Calcium-dependent cell adhesion molecules, *Curr. Opin. Cell Biol.* **1:**892–897.

Kemler, R., Gossler, A., Mansouri, A., and Vestweber, D., 1990, The cell adhesion molecule uvomorulin, in: *Morphoregulatory Molecules* (G. M. Edelman, B. A. Cunningham, and J. P. Thiery, eds.), Wiley, New York, pp. 41–56.

Kintner, C., 1992, Regulation of embryonic cell adhesion by the cadherin cytoplasmic domain, *Cell* **69:**225–236.

Knudsen, K. A., and Wheelock, M. J., 1992, Plakoglobin, or an 83-kD homologue distinct from β-catenin, interacts with E-cadherin and N-cadherin, *J. Cell Biol.* **118:**671–679.

Koch, P. J., Goldschmidt, M. D., Walsh, M. J., Zimbelmann, R., Schmelz, M., and Franke, W. W., 1991, Amino acid sequence of bovine muzzle epithelial desmocollin derived from cloned cDNA: A novel subtype of desmosomal cadherins, *Differentiation* **47:**29–36.

Korman, N. J., Eyre, R. W., Klaus-Kovtin, V., and Stanley, J. R., 1989, Demonstration of an adhering junction molecule (plakoglobin) in the autoantigens of Pemphicus foliaceus and Pemphicus vulgaris, *N. Engl. J. Med.* **321**:631–635.

McCrea, P. D., Turck, C. W., and Gumbiner, B., 1991, A homolog of the armadillo protein in Drosophila (plakoglobin) associated with E-cadherin, *Science* **254**:1359–1361.

McNeill, H., Ozawa, M., Kemler, R., and Nelson, W. J., 1990, Novel function of the cell adhesion molecule uvomorulin as an inducer of cell surface polarity, *Cell* **62**:309–316.

Mahoney, P. A., Weber, U., Onofrechuk, P., Biessmann, H., Bryant, P. J., and Goodman, C. S., 1991, The fat tumor suppressor gene in Drosophila encodes a novel member of the cadherin gene superfamily, *Cell* **67**:853–858.

Matsuyoshi, N, Hamaguchi, M., Taniguchi, S., Nagafuchi, A., Tsukita, S., and Takeichi, M., 1992, Cadherin-mediated cell–cell adhesion is perturbed by v-src tyrosine phosphorylation in metastatic fibroblasts, *J. Cell Biol.* **118**:703–704.

Mechanic, S., Raynor, K., Hill, J. E., and Cowin, P., 1991, Desmocollins form a subset of the cadherin family of cell adhesion molecules, *Proc. Natl. Acad. Sci. USA* **88**:4476–4480.

Mege, R. M., Matsuzaki, F., Gallin, W. J., Goldberg, J. I., Cunningham, B. A., and Edelman, G. M., 1988, Construction of epithelioid sheets by transfection of mouse sarcoma cells with cDNAs for chicken cell adhesion molecules, *Proc. Natl. Acad. Sci. USA* **85**:7274–7278.

Nagafuchi, A., and Takeichi, M., 1989, Transmembrane control of cadherin-mediated cell adhesion: A 94 kDa protein functionally associated with a specific region of the cytoplasmic domain of E-cadherin, *Cell Regul.* **1**:37–44.

Nagafuchi, A., Shirayoshi, Y., Okazaki, K., Yasuda, K., and Takeichi, M., 1987, Transformation of cell adhesion properties by exogenously introduced E-cadherin cDNA, *Nature* **329**:341–343.

Nagafuchi, A., Takeichi, M., and Tsukita, S., 1991, The 102 kD cadherin-associated protein: Similarity to vinculin and post-translational regulation of expression, *Cell* **65**:849–857.

Nelson, W. J., Shore, E. M., Wang, A. Z., and Hammerton, R. W., 1990, Identification of a membrane–cytoskeletal complex containing the cell adhesion molecule uvomorulin (E-cadherin), ankyrin, and fodrin in Madin-Darby canine kidney epithelial cells, *J. Cell Biol.* **110**:349–357.

Nose, A., Nagafuchi, A., and Takeichi, M., 1988, Expressed recombinant cadherins mediate cell sorting in model systems, *Cell* **54**:993–1001.

Nose, A., Tsuji, K., and Takeichi, M., 1990, Localization of specificity determining sites in cadherin cell adhesion molecules, *Cell* **61**:147–155.

Otto, J. J., 1983, Detection of vinculin-binding proteins with an ^{125}I-vinculin gel overlay technique, *J. Cell Biol.* **97**:1283–1287.

Ozawa, M., and Kemler, R., 1990, Correct proteolytic cleavage is required for the cell adhesive function of uvomorulin, *J. Cell Biol.* **111**:1645–1650.

Ozawa, M., and Kemler, R., 1992, Molecular organization of the uvomorulin–catenin complex, *J. Cell Biol.* **116**:989–996.

Ozawa, M., Baribault, H., and Kemler, R., 1989, The cytoplasmic domain of the cell adhesion molecule uvomorulin associates with three independent proteins structurally related in different species, *EMBO J.* **8**:1711–1717.

Ozawa, M., Engel, J., and Kemler, R., 1990a, Single amino acid substitution in one Ca^{2+} binding site of uvomorulin abolish the adhesive function, *Cell* **63**:1033–1038.

Ozawa, M., Ringwald, M., and Kemler, R., 1990b, Uvomorulin–catenin complex formation is regulated by a specific domain in the cytoplasmic region of the cell adhesion molecule, *Proc. Natl. Acad. Sci. USA* **87**:4246–4250.

Ozawa, M., Hoschützky, H., Herrenknecht, K., and Kemler, R., 1991, A possible new adhesive site in the cell-adhesion molecule uvomorulin, *Mech. Dev.* **33**:49–56.

Parker, A. E., Wheeler, G. N., Arnemann, J., Pidsley, S. C., Ataliotis, P., Thomas, C. L., Rees,

D. A., Magee, A. I., and Buxton, R. S., 1991, Desmosomal glycoproteins II and III, *J. Biol. Chem.* **266**:10438–10445.

Peifer, M., and Wieschaus, E., 1990, The segment polarity gene armadillo encodes a functionally modular protein that is the Drosophila homolog of human plakoglobin, *Cell* **63**:1167–1178.

Peifer, M., McCrea, P. D., Green, K. J., Wieschaus, E., and Gumbiner, B. M., 1992, The vertebrate adhesive junction proteins β-catenin and plakoglobin and the Drosophila segment polarity gene armadillo form a multigene family with similar properties, *J. Cell Biol.* **118**:681–691.

Peyrieras, N., Louvard, D., and Jacob, F., 1985, Characterization of antigens recognized by monoclonal and polyclonal antibodies directed against uvomorulin, *Proc. Natl. Acad. Sci. USA* **82**:8067–8071.

Ranscht, B., and Dours-Zimmermann, M. T., 1991, T-cadherin, a novel cadherin cell adhesion molecule in the nervous system lacks the conserved cytoplasmic region, *Neuron* **7**:391–402.

Ringwald, M., Schuh, R., Vestweber, D., Eistetter, H., Lottspeich, F., Engel, J., Dölz, R., Jähnig, F., Epplen, J., Mayer, S., Müller, C., and Kemler, R., 1987, The structure of cell adhesion molecule uvomorulin. Insights into the molecular-mechanism of Ca^{2+}-dependent cell adhesion, *EMBO J.* **6**:3647–3653.

Ringwald, M., Baribault, H., Schmidt, C., and Kemler, R., 1991, The structure of the gene coding for the mouse cell adhesion molecule uvomorulin, *Nucleic Acid Res.* **19**:6533–6539.

Shimoyama, Y., and Hirohashi, S., 1991, Expression of E- and P-cadherin in gastric carcinomas, *Cancer Res.* **51**:2185–2192.

Shore, E. M., and Nelson, W. J., 1991, Biosynthesis of the cell adhesion molecule uvomorulin (E-cadherin) in Madin-Darby canine kidney epithelial cells, *J. Biol. Chem.* **266**:19672–19680.

Takeichi, M., 1977, Functional correlation between cell adhesive properties and some cell surface proteins, *J. Cell Biol.* **75**:464–474.

Takeichi, M., 1988, The cadherins: Cell–cell adhesion molecules controlling animal morphogenesis, *Development* **102**:639–655.

Takeichi, M., 1990, Cadherins: A molecular family important in selective cell–cell adhesion, *Annu. Rev. Biochem.* **59**:237–252.

Takeichi, M., 1991, Cadherin cell adhesion receptors as a morphogenetic regulator, *Science* **251**:1451–1455.

Vestweber, D., and Kemler, R., 1984a, Rabbit antiserum against a purified surface glycoprotein decompacts mouse preimplantation embryos and reacts with specific adult tissues, *Exp. Cell Res.* **152**:169–178.

Vestweber, D., and Kemler, R., 1984b, Some structural and functional aspects of the cell adhesion molecule uvomorulin, *Cell Differ.* **15**:269–273.

Vestweber, D., Gossler, A., Boller, K., and Kemler, R., 1987, Expression and distribution of cell adhesion molecule uvomorulin in mouse preimplantation embryos, *Dev. Biol.* **124**:451–456.

Volk, T., and Geiger, B., 1986, A-CAM: A 135-kD receptor of intercellular adherens junctions. I. Immunoelectron microscopic localization and biochemical studies, *J. Cell Biol.* **103**:1441–1450.

Wachsstock, D. H., Wilkins, J. A., and Lin, S., 1987, Specific interaction of vinculin with α-actinin, *Biochem. Biophys. Res. Commun.* **146**:554–560.

Wheeler, G. N., Parker, A. E., Thomas, C. L., Ataliotis, P., Poynter, D., Arnemann, J., Rutman, A. J., Pidsley, S. C., Watt, F. M., Rees, D. A., Buxton, R. S., and Magee, A. I., 1991, Desmosomal glycoprotein DGI, a component of intercellular desmosome junctions, is related to the cadherin family of cell adhesion molecules, *Proc. Natl. Acad. Sci. USA* **88**:4796–4800.

Wollner, D. A., Krzeminski, K. A., and Nelson, W. J., 1992, Remodeling the cell surface distribution of membrane proteins during the development of epithelial cell polarity, *J. Cell Biol.* **116**:889–899.

6

Cytotactin
A Substrate Adhesion Molecule with Amphitropic Functions in Morphogenesis

KATHRYN L. CROSSIN and ANNE L. PRIETO

1. Introduction

The generation of form during embryologic development is a problem that has occupied biologists since ancient times. A clear understanding of the cellular processes leading to morphogenesis is just beginning to emerge from a combination of genetic, molecular biological, and physiological studies. A clearer understanding of the molecular regulatory properties of cell collectives that lead to macroscopic tissue form should provide deep insights into normal function, as well as into the diagnosis and treatment of disease. A deeper knowledge of these events may help to explain the aberrations of development that lead to birth defects and the alterations in adult organisms that lead to disease and degenerative states.

A large body of data, accumulated over the past several years, implicates cell adhesion as an important mediator of events responsible for the generation of tissue patterns (reviewed in Edelman and Crossin, 1991; Takeichi, 1990). Moreover, recent studies suggest that adhesive forces, as well as counteradhesive events, are important in structuring the embryo (Edelman and Crossin, 1991; Edelman, 1992), particularly its nervous system (reviewed in Keynes and Cook, 1992).

KATHRYN L. CROSSIN and ANNE L. PRIETO • Department of Neurobiology, The Scripps Research Institute, La Jolla, California 92037.

Cellular Adhesion: Molecular Definition to Therapeutic Potential, edited by Brian W. Metcalf, Barbara J. Dalton, and George Poste. Plenum Press, New York, 1994.

This chapter reviews the properties of a novel extracellular matrix glycoprotein, called cytotactin, that contributes in a complex fashion to important functions of embryonic extracellular matrices, including those that regulate cell migration, process extension, and attachment to basement membrane proteins. The expression patterns of cytotactin, observed by immunohistochemistry and by *in situ* hybridization, change dynamically during development. Evidence has now accumulated on the cell biological properties of cytotactin and the control of expression of the cytotactin gene. Cytotactin can mediate cell–substratum adhesion. It also exhibits counteradhesive properties, depending on the presence of other matrix molecules and on the type of cell, and thus is an amphitropic molecule. The combination of these seemingly contradictory functions, together with the tight place-dependent regulation of cytotactin gene expression, can provide powerful constraints on developing morphology in the embryo.

2. Cytotactin/Tenascin: Identification and Biochemistry

Cytotactin (Grumet *et al.*, 1985) was initially discovered as a glycoprotein involved in neuron–glia adhesion. Similar or identical molecules have been discovered using a variety of assays: tenascin (Chiquet-Ehrismann *et al.*, 1986), myotendinous antigen (Chiquet and Fambrough, 1984a), glioma–mesenchymal extracellular matrix protein (Bourdon *et al.*, 1983), hexabrachion (Erickson and Iglesias, 1984), and J1 220/200 (Faissner *et al.*, 1988) (reviewed in Erickson and Bourdon, 1989). As isolated from embryonic chicken brains, the molecule consists of related polypeptides of 220, 200, and 190 kDa when examined on SDS gels in the presence of reducing agents. In the absence of reducing agents, the molecule barely enters a 6% polyacrylamide gel, suggesting that these polypeptides are disulfide-linked in their native state. This is consistent with the six-armed appearance of the molecule in electron micrographs after rotary shadowing, a structure that has been termed a hexabrachion (Erikson and Iglesias, 1984). Cytotactin appears to be composed of six individual polypeptides linked in a hub structure through disulfide bonds.

Analysis of the cDNA sequence of cytotactin reveals a strong homology to three families of proteins (Jones *et al.*, 1988, 1989; Pearson *et al.*, 1988) (Fig. 1A). The 5' end encodes 13.5 EGF-like repeats followed by a series of fibronectin type III repeats [11 in the chicken molecule, 15 or more in the human sequence (Gulcher *et al.*, 1989) and mouse sequences (Weller *et al.*, 1991)], 3 to 6 of which are subject to differential splicing. The 3' end encodes a region similar to fibrinogen and contains a potential calcium binding site (Jones *et al.*, 1989). The different domains of the protein can be mapped onto the hexabrachion structure (Fig. 1B) seen in rotary-shadowed electron microscopic images (Fig. 1C). The central core of the hexabrachion is the highly disulfide-bonded region at

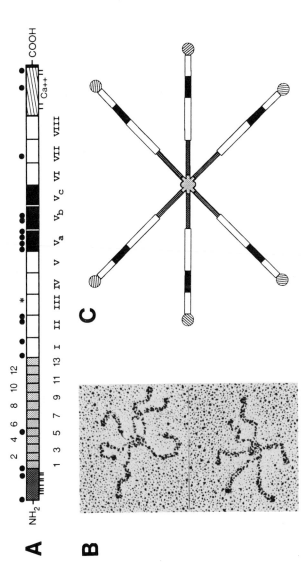

Figure 1. A functional map of cytotactin. (A) Schematic drawing of the cytotactin polypeptides. The EGF-like repeats (lightly stippled boxes) are numbered 1–13, and type III repeats (open boxes) are designated by Roman numerals, with those in the additional insert (black boxes) designated Va, Vb, and Vc. The region similar to the β subunit of fibrinogen (diagonal lines) includes a putative calcium-binding domain (Ca++). Darkly stippled boxes denote regions that have no extensive homology to any known protein; these include the amino-terminal 142 amino acids and the carboxyl-terminal 13 amino acids. Cysteine residues that are not in EGF-like repeats (|), potential asparagine-glycosylation sites (•), and the Arg-Gly-Asp sequence (*) are indicated. (B) Electron micrograph (courtesy of Joseph W. Becker, The Rockefeller University) of chicken brain cytotactin showing two hexabrachions. (C) Proposed model of a cytotactin hexamer. The amino-terminal regions of each polypeptide are disulfide-linked and are represented as a single structure forming the core of the hexabrachion. The EGF-like repeats are stippled; the type III repeats present in both polypeptides are white, whereas those in the insert are black; the fibrinogen-like nodular region is marked with slanted lines. The EGF-like repeats are assumed to make up the thin parts of the arms, and the type III repeats the thicker portions. (From Jones *et al.*, 1989.)

the N-terminus. The arms are composed of the EGF-like repeats and fibronectin type III repeats, with the region of fibrinogen homology forming the terminal knob seen at the end of each arm.

In the nervous system, cytotactin is made by glia, but not neurons (Grumet *et al.*, 1985; Hoffman and Edelman, 1987; Hoffman *et al.*, 1988; Crossin *et al.*, 1989). It is also synthesized by somites, smooth muscle cells, immature chondrocytes, perichondrial cells, myotendinous regions, and various other cells (Crossin *et al.*, 1986; Prieto *et al.*, 1990). A proteoglycan copurifies with cytotactin (Chiquet and Fambrough, 1984b; Hoffman and Edelman, 1987), and it is synthesized by central nervous system neurons as well as by nonneural cells elsewhere in the body (Hoffman *et al.*, 1988). This proteoglycan has been termed the cytotactin-binding (CTB) proteoglycan (Hoffman and Edelman, 1987), since its core protein binds directly to cytotactin in several types of assay. Cytotactin also binds to fibroblasts, to fibronectin (Chiquet and Fambrough, 1984a; Hoffman and Edelman, 1987; Chiquet-Ehrismann *et al.*, 1988; Friedlander *et al.*, 1988), and to other matrix components (Faissner *et al.*, 1990). Cytotactin has been shown to affect cell movement differently than do other extracellular matrix proteins such as fibronectin and laminin. Its most striking functional effect causes cells in culture to round up; it also tends to inhibit cellular migration and neurite invasion in culture. Fibronectin generally supports such migration, but when cytotactin is added, these effects are mitigated. Cytotactin thus appears to be amphitropic; i.e., it carries out both adhesive and repulsive functions at different sites. These notions will be discussed in more detail below.

3. Cytotactin in Developmental Patterning and Cell Migration

The developmental distribution of cytotactin suggests that the cellular effects described above may be important to morphogenesis. In comparison with other matrix molecules such as fibronectin, the expression of the protein is highly transient and site-restricted. Cytotactin differs from other known ECM proteins in that it appears in a series of cephalocaudal waves of expression early in development during gastrulation, neurulation, and somite formation (Crossin *et al.*, 1986; Tan *et al.*, 1987, 1991; Prieto *et al.*, 1990). Moreover, the molecule is present at a number of sites of epithelial–mesenchymal interaction, where it may play a role in signaling and in cell migration critical to morphological development (Crossin *et al.*, 1986; Chiquet-Ehrismann *et al.*, 1986, 1989; Tan *et al.*, 1987; Mackie *et al.*, 1988; Aufderheide and Ekblom, 1988; Stern *et al.*, 1989; Prieto *et al.*, 1990; Kaplony *et al.*, 1991).

Studies of several developmental systems indicate that cytotactin plays an important morphogenetic role in conjunction with cell adhesion molecules, particularly with regard to cell and neurite movement. Three important examples will be described in detail in this section. A striking example of the expression of cytotactin and CTB proteoglycan and their relationship to epithelial–mesenchymal transformations occurs in developing somites (Tan *et al.*, 1987; Crossin *et al.*, 1986). In the epithelial somite, cytotactin protein (Crossin *et al.*, 1986; Tan *et al.*, 1987, 1991; Stern *et al.*, 1989) and mRNA (Tan *et al.*, 1991) are expressed only in the caudal half. Later in development the somite becomes mesenchymal, and neural crest cells and motor axons enter the sclerotome only in the rostral half. This leads to polarity and periodicity of the peripheral nervous system, and positions nerves and ganglia appropriately between the vertebrae. While fibronectin is present uniformly throughout the sclerotome (Duband *et al.*, 1987), cytotactin protein (Tan *et al.*, 1987, 1991; Mackie *et al.*, 1988) and mRNA (Tan *et al.*, 1991) become restricted to the rostral, and CTB proteoglycan to the caudal, half-sclerotome (Tan *et al.*, 1987). We have recently shown, using neural crest extirpation experiments (Tan *et al.*, 1991), that the presence of cytotactin is neither necessary nor sufficient for the appropriate patterning of the neural crest cells. In *in vitro* experiments, however, fibronectin promotes neural crest cell migration, while cytotactin and CTB proteoglycan each inhibit migration (Tan *et al.*, 1987; Mackie *et al.*, 1988) and attachment (Halfter *et al.*, 1989). It is possible that cytotactin synthesized *in vivo* by the rostral somite acts as a substrate that blocks the further migration of neural crest cells, thereby facilitating aggregation and subsequent dorsal root ganglion formation. These findings are consistent with the hypothesis that, in the anterior region of each somite, cytotactin causes cell surface modulation (Edelman, 1976) of migrating neural crest cells, affecting their movement and possibly their subsequent differentiation.

In the developing cerebellum, external granule neurons send neurites out into the subjacent molecular layer and finally translocate on radial glial cells to form the internal granule layer (Rakic, 1971). This process occurs *in vitro* in tissue slices and can be followed by labeling the granule cell with tritiated thymidine; the slice culture system can be perturbed by adding antibodies to various adhesion molecules. The granule cells express both N-CAM and Ng-CAM (Daniloff *et al.*, 1986), and the radial glia synthesize and secrete cytotactin (Grumet *et al.*, 1985; Crossin *et al.*, 1986; Chuong *et al.*, 1987; Prieto *et al.*, 1990). The addition of Fab′ fragments of anti-NgCAM to cerebellar slices inhibits external granule cell migration into the molecular layer (Hoffman *et al.*, 1986; Chuong *et al.*, 1987). In contrast, Fab′ fragments of anticytotactin antibodies cause a pileup in the molecular layer of the granule cells that had already entered that layer (Chuong *et al.*, 1987). Thus, a CAM of neuronal origin and a SAM of glial origin play different roles at different times in the same morphogenetic process. Inasmuch as each of these molecules appears in a sequence prior to these events, there must be a precise coregulation of signals for their expression, as well as coordination of their different binding functions as they affect cell migration.

In the developing mouse cerebral cortex, the somatosensory region contains a highly structured area, called the barrel field, with a characteristic appearance (resembling stacked wine barrels); it is composed of axon-rich areas surrounded by cell-rich areas (Woolsey and Van der Loos, 1970; Van der Loos and Woolsey, 1973). This area represents the projection from the neurons of the vibrissa on the face of the mouse; the cortical pattern is dependent on this vibrissal input (Van der Loos and Woolsey, 1973). Cytotactin (Crossin *et al.*, 1989; Steindler *et al.*, 1989) and CTB proteoglycan (Crossin *et al.*, 1989) are both enriched in the walls of the barrel field from its earliest development. Their expression in the walls peaks at postnatal day 7 and is uniform throughout by day 13. Like the anatomical pattern, the pattern of molecular expression is dependent on peripheral input: when the input is removed by electrocauterization of the whisker follicle, the molecular pattern in the cortex is destroyed (Crossin *et al.*, 1989). These results clearly demonstrate that the molecular pattern does not precede and guide the anatomical events, as has been postulated (Steindler *et al.*, 1989), but rather is coordinate with the developing anatomy; they are both highly dependent on peripheral neural input. This is also supported by more recent results demonstrating that the thalamocortical afferents are segregated prior to the patterned expression of cytotactin and CTB proteoglycan (Jhaveri *et al.*, 1991). Such findings raise the possibility that the expression of cytotactin by glial cells is regulated by neuronal activity, a point that may now be explored, given recent results on the isolation and characterization of the cytotactin promoter.

4. Control of Cytotactin Gene Expression

Given the place-dependent and tissue-specific expression patterns observed for cytotactin, one would expect its synthesis to be differentially responsive to various signals and *trans*-acting elements. Potential signals responsible for the transient and restricted distribution of cytotactin within various tissues have been illuminated by studies of the upstream regulatory region of its gene (Jones *et al.*, 1990). The promoter region of the cytotactin gene shows a very rich series of regulatory motifs, including DNA sequences similar to binding regions for the products of homeotic genes (Jones *et al.*, 1990), which is consistent with tight spatiotemporal expression (reviewed in Erickson and Bourdon, 1989). *In situ* hybridization experiments support these observations and reveal tissue sites where different alternatively-spliced forms appear (Prieto *et al.*, 1990), indicating further levels of control of gene expression.

The postulate that these various motifs may be utilized to control cytotactin expression *in vivo* is supported by the observation that the *in vitro* expression of cytotactin is enhanced by growth factors (Pearson *et al.*, 1988). Moreover, cotransfection experiments with the cytotactin promoter and *hox* genes (Jones *et al.*, 1992) demonstrate that the products of a variety of hox-containing genes may control cytotactin expression, and that this control may, at least in part, be related to control by growth factors.

5. Effects of Cytotactin on Cell Morphology and Physiology

Cytotactin has been shown to support the attachment of a number of primary cells and cell lines (Chiquet and Fambrough, 1984a; Chiquet-Ehrismann *et al.*, 1986, 1988; Friedlander *et al.*, 1988; Lotz *et al.*, 1989). However, interaction with cytotactin can inhibit the spreading of neuronal and fibroblastic cells plated in culture on otherwise permissive substrates (Tan *et al.*, 1987; Lotz *et al.*, 1989; Friedlander *et al.*, 1988; Faissner and Kruse, 1990); it also tends to inhibit cellular migration *in vitro* (Tan *et al.*, 1987; Mackie *et al.*, 1988; Epperlein *et al.*, 1988) and *in vivo* (Riou *et al.*, 1990). As discussed above, some of the effects of cytotactin on cells may be modulated by interaction with other molecules. In the presence of fibronectin, which generally supports cell spreading and migration, the effects of cytotactin are decreased (Tan *et al.*, 1987; Friedlander *et al.*, 1988; Spring *et al.*, 1989); thus, the two interactive molecules may have contrasting effects on cell behavior.

In addition to its effects on cell attachment, morphology, and migration, cytotactin has been shown to affect the proliferation of certain cells. Certain transformed cells from mammary tumors attach to cytotactin-coated substrates and show increased proliferation relative to cells plated on other matrix proteins (Chiquet-Ehrismann *et al.*, 1986). In contrast, normal fibroblastic cells plated in the presence of soluble cytotactin are inhibited in their proliferation for at least one cell cycle time (Crossin, 1991). This inhibition appears to correlate with an inhibition of the increased intracellular pH (Crossin, 1991) that is known to result upon flattening of cells on the substratum or upon growth factor stimulation (reviewed in Moolenaar, 1986; Grinstein *et al.*, 1989).

6. Structure–Function Analysis of Cytotactin Domains

The seemingly contradictory effects that the cytotactin molecule has on cellular processes are now being understood in terms of the multidomain structure of the protein. Early studies of cell attachment to cytotactin-coated surfaces suggested that multiple modes of binding to the molecule existed. For example, fibroblasts bind both to intact cytotactin and to a chymotryptic fragment derived from the C-terminus of the protein (Friedlander *et al.*, 1988). These binding activities can be inhibited by peptides containing the amino acid sequence RGD and by antibodies to specific regions of the cytotactin protein. In contrast to their rounded morphology on intact cytotactin, cells on the chymotryptic fragment exhibit a spread morphology. Using a variety of recombinant fragments of tenascin, a smaller region of the molecule has been identified as a cell binding site, but no spreading has been observed (Spring *et al.*, 1989). In these studies, a fragment in the N-terminal region containing the EGF domains appeared to prevent cell binding to other substrates. Together, these observations suggested that at least two binding activities are present in intact cytotactin: one in the C-terminal half of the protein, mediating cell attachment and flattening, and one in the N-terminal portion, responsible for the so-called antiadhesive effects (Spring *et al.*, 1989) and rounding of cells exposed to the molecule (Friedlander *et al.*, 1988; Chiquet-Ehrismann *et al.*, 1988). Antibody inhibition studies on the effects of cytotactin on neural attachment and neurite outgrowth have suggested the presence of at least one additional interactive site on the molecule (Lochter *et al.*, 1991; Wehrle and Chiquet, 1990; Grierson *et al.*, 1990; Faissner and Kruse, 1990; Crossin *et al.*, 1990; Husmann *et al.*, 1992).

The remarkable pattern of distribution throughout development and its complex multidomain structure of cytotactin are consistent with the idea that it may play multiple roles during morphogenesis. The fact that each type of protein domain in cytotactin appears in several different proteins suggests that individual domains may represent independent functional units. Based on these ideas, we have recently performed a detailed analysis of the binding properties of isolated cytotactin domains using a variety of cell types (Prieto *et al.*, 1992). In contrast to previous mapping studies that identified a single cell binding site and a repulsive site (Friedlander *et al.*, 1988; Spring *et al.*, 1989), at least four nonoverlapping sites on the molecule were found to interact with the cell surface (Prieto *et al.*, 1992). Cell attachment was sensitive to the method of preparation of the single cell suspension; for example, a unique cytotactin binding site was seen in cells prepared by trypsin treatment, but not in cells prepared with EDTA. The results suggest that multiple sites on cytotactin mediate its adhesive and counteradhesive effects on cells (Fig. 2); these effects are likely to be mediated by multiple, specific cell surface receptors, some of which may be differentially expressed on different cell types.

7. Complex Effects of Cytotactin on Neurite Outgrowth

In addition to the effects on neural cell attachment mentioned above, cytotactin has also been shown to affect neurite morphology (Crossin *et al.*, 1990; Wehrle and Chiquet, 1990; Grierson *et al.*, 1990; Faissner and Kruse, 1990; Lochter *et al.*, 1991), extension (Crossin *et al.*, 1990; Faissner and Kruse, 1990), and neuronal migration (Chuong *et al.*, 1987; Halfter *et al.*, 1989; Husmann *et al.*, 1992). However, an estimation of the number of potential cellular receptors

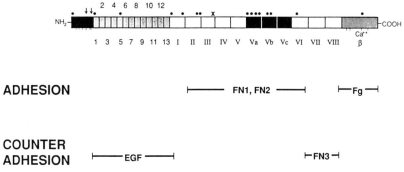

Figure 2. Summary of the cellular binding properties of different regions of cytotactin. Based on the linear sequence of chicken cytotactin (Jones *et al.*, 1989) shown at the top, regions corresponding to the domains of the protein delineated below were prepared in bacterial expression vectors (Prieto *et al.*, 1992). The isolated domains were tested for their ability to support cell attachment when coated onto plastic substrates (adhesive) or to prevent cell attachment and flattening on fibronectin-coated substrates (counteradhesive). (From Prieto *et al.*, 1992.)

for cytotactin becomes even more complicated when one considers the responses of neurons to cytotactin. Chick neurons bind cytotactin only in centrifugation assays, and not in simple assays in which cells are allowed to settle under gravity (Friedlander *et al.*, 1988; Prieto *et al.*, 1992). Inhibition of neuritic outgrowth has been observed for both central (Grierson *et al.*, 1990; Faissner and Kruse, 1990) and peripheral (Crossin *et al.*, 1990) neurites. Rat neurons respond differently to soluble and substratum-bound cytotactin (J1/tenascin) coated onto polyornithine, itself a favorable substrate for neuronal binding (Lochter *et al.*, 1991; Wehrle and Chiquet, 1990). Neurons bind and extend longer neurites on J1/tenascin- and polyornithine-coated substrates, as compared with polyornithine control substrates. This neurite outgrowth promotion is inhibitable by a monoclonal antibody that recognizes an epitope distinct from the cell binding epitope. Moreover, this antibody does not neutralize the antispreading effect of purified J1/tenascin. These results suggest that at least three functional domains exist for cytotactin binding to neurons: one mediates cell attachment, another counters cell spreading, and a third promotes neurite outgrowth. Of course, the possibility that these sites exhibit cooperative activity must be kept in mind. Moreover, several other structurally unrelated molecules have been reported to collapse growth cones and alter cell attachment and migration in the nervous system (Caroni and Schwab, 1988; Cox *et al.*, 1990; Raper and Kapfhammer, 1990; Davies *et al.*, 1990). Thus, a number of different molecular structures may act as "repulsins" for neurite outgrowth and thereby affect neural morphogenesis by maintaining a balance between adhesion and repulsion. With molecules as complex in their molecular interactions as cytotactin, however, this activity is only one of many different activities, and it can be altered by the composition of other molecules in the extracellular matrix. In addition, our recent experiments (Prieto *et al.*, 1992) raise the possibility that different receptors or receptor combinations may be responsible for the diverse effects of cytotactin.

8. Perspectives

The results on the binding of multiple domains of cytotactin prompt the search for a variety of potential receptors (both integrin and nonintegrin) that might mediate these effects on cell attachment, morphology, and physiology. Indeed, recent reports have suggested that cytotactin binds to two members of the integrin family, $\alpha\beta1$ (a novel alpha subunit) (Mendler *et al.*, 1991) and $\alpha v\beta3$ (Mendler *et al.*, 1991; Joshi *et al.*, 1991), and that the binding can be inhibited by peptides containing the RGD sequence. However, the possibility that there are additional non-RGD-related binding sites is supported by our preliminary results and by the observation that the RGD sequence in cytotactin is not conserved among species [e.g., it is absent from the mouse (Weller *et al.*, 1991) and newt (Onda *et al.*, 1991) sequences, but it is present in the human (Gulcher *et al.*, 1989) and chicken (Jones *et al.*, 1989; Spring *et al.*, 1989) sequences]. For example, a recent report indicated that the alternatively-spliced fibronectin type III repeats could disrupt focal contacts (Murphy-Ullrich *et al.*, 1991). This suggests that novel receptors may exist for particular variants of cytotactin generated by alternative splicing. Moreover, cytotactin has recently been shown to bind to sulfated glycolipids (Crossin and Edelman, 1992); several other matrix glycoproteins do the same (Roberts *et al.*, 1986). These may also serve as receptors for cytotactin, or they may modulate the binding of cytotactin to other glycoprotein receptors.

Clearly, the differential expression of domain-specific receptors for cytotactin might alter the response of cells to the presence of the molecule. The activity of the intact cytotactin molecule would represent a summation of its adhesive and counteradhesive activities, providing a rationale for its observed amphitropic properties (Edelman and Crossin, 1991). Further analysis of the cell interactions that mediate these effects will lead to a better understanding of the regulation of cellular processes important to disease and development, including proliferation, migration, neurite extension, and differentiation.

References

Aufderheide, E., and Ekblom, P., 1988, Tenascin during gut development: Appearance in the mesenchyme, shift in molecular forms, and dependence on epithelial–mesenchymal interactions, *J. Cell Biol.* **105**:2341–2349.

Bourdon, M. A., Wikstrand, C. J., Furthmayer, H., Matthews, T. J., and Bigner, D. D., 1983, Human glioma–mesenchymal extracellular matrix antigen defined by monoclonal antibody, *Cancer Res.* **43:**2796–2805.

Caroni, P., and Schwab, M. E., 1988, Two membrane protein fractions from rat central myelin with inhibitory properties for neurite growth and fibroblast spreading, *J. Cell Biol.* **106:**1281–1287.

Chiquet, M., and Fambrough, D. M., 1984a, Chick myotendinous antigen. I. A. monoclonal antibody as a marker for tendon and muscle morphogenesis, *J. Cell Biol.* **98:**1926–1936.

Chiquet, M., and Fambrough, D. M., 1984b, Chick myotendinous antigen. II. A novel extracellular glycoprotein complex consisting of large disulfide-linked subunits, *J. Cell Biol.* **98:**1937–1946.

Chiquet-Ehrismann, R., Mackie, E. J., Pearson, C. A., and Sakakura, T., 1986, Tenascin: An extracellular matrix protein involved in tissue interactions during fetal development and oncogenesis, *Cell* **47:**131–139.

Chiquet-Ehrismann, R., Kalla, P., Pearson, C. A., Beck, K., and Chiquet, M., 1988, Tenascin interferes with fibronectin action, *Cell* **53:**383–390.

Chiquet-Ehrismann, R., Kalla, P., and Pearson, C., 1989, Participation of tenascin and transforming growth factor-β in reciprocal epithelial–mesenchymal interactions of MCF7 cells and fibroblasts, *Cancer Res.* **49:**4322–4325.

Chuong, C.-M., Crossin, K. L., and Edelman, G. M., 1987, Sequential expression and differential function of multiple adhesion molecules during the formation of cerebellar cortical layers, *J. Cell Biol.* **104:**331–342.

Cox, E. C., Muller, B., and Bonhoeffer, F., 1990, Axonal guidance in the chick visual system: Posterior tectal membranes induce collapse of growth cones from the temporal retina, *Neuron* **2:**31–37.

Crossin, K. L., 1991, Cytotactin binding: Inhibition of stimulated proliferation and intracellular alkalinization in fibroblasts, *Proc. Natl. Acad. Sci. USA.* **88:**11403–11407.

Crossin, K. L., and Edelman, G. M., 1992, Specific binding of cytotactin to sulfated glycolipids, *J. Neurosci. Res.* **33:**631–638.

Crossin, K. L., Hoffman, S., Grumet, M., Thiery, J.-P., and Edelman, G. M., 1986, Site-restricted expression of cytotactin during development of the chicken embryo, *J. Cell Biol.* **102:**1917–1930.

Crossin, K. L., Hoffman, S., Tan, S.-S., and Edelman, G. M., 1989, Cytotactin and its proteoglycan ligand mark structural and functional boundaries in somatosensory cortex of the early postnatal mouse, *Dev. Biol.* **136:**381–392.

Crossin, K. L., Prieto, A. L., Hoffman, S., Jones, F. S., and Friedlander, D. R., 1990, Expression of adhesion molecules and the establishment of boundaries during embryonic and neural development, *Exp. Neurol.* **109:**6–18.

Daniloff, J., Chuong, C.-M., Levi, G., and Edelman, G. M., 1986, Differential distribution of cell adhesion molecules during histogenesis of the chick nervous system, *J. Neurosci.* **6:**739–758.

Davies, J. A., Cook, G. M. W., Stern, C. D., and Keynes, R. J., 1990, Isolation from chick somites of a glycoprotein fraction that causes collapse of dorsal root ganglion growth cones, *Neuron* **2:**11–20.

Duband, J.-L., Dufour, S., Hatta, K., Takeichi, M., Edelman, G. M., and Thiery, J.-P., 1987, Adhesion molecules during somitogenesis in the avian embryo, *J. Cell Biol.* **104:**1361–1374.

Edelman, G. M., 1976, Surface modulation in cell recognition and cell growth, *Science* **192:**218–226.

Edelman, G. M., 1992, Morphoregulation, *Dev. Dynamics* **193:**2–10.

Edelman, G. M., and Crossin, K. L., 1991, Cell adhesion molecules: Implications for a molecular histology, *Annu. Rev. Biochem.* **60:**155–190.

Epperlein, H. H., Halfter, W., and Tucker, R. P., 1988, The distribution of fibronectin and tenascin along migratory pathways of the neural crest in the trunk of amphibian embryos, *Development* **103**:743–756.

Erickson, H. P., and Bourdon, M. A., 1989, Tenascin: An extracellular matrix protein prominent in specialized embryonic tissues and tumors, *Annu. Rev. Cell Biol.* **5**:71–92.

Erickson, H. P., and Iglesias, J. L., 1984, A six-armed oligomer isolated from cell surface fibronectin preparations, *Nature* **311**:267–269.

Faissner, A., and Kruse, J., 1990, J1/tenascin is a repulsive substrate for central nervous system neurons, *Neuron* **5**:627–637.

Faissner, A., Kruse, J., Chiquet-Ehrismann, R., and Mackie, E., 1988, The high molecular weight J1 glycoproteins are immunochemically related to tenascin, *Differentiation* **37**:104–114.

Faissner, A., Kruse, J., Kuhn, K., and Schachner, M., 1990, Binding of the J1 adhesion molecules to extracellular matrix constituents, *J. Neurochem.* **54**:1004–1015.

Friedlander, D. R., Hoffman, S., and Edelman, G. M., 1988, Functional mapping of cytotactin: Proteolytic fragments active in cell–substrate adhesion, *J. Cell Biol.* **107**:2329–2340.

Grierson, J. P., Petroski, R. E., Ling, D. S. F., and Geller, H. M., 1990, Astrocyte topography and tenascin/cytotactin expression: Correlation with the ability to support neuritic outgrowth, *Dev. Brain Res.* **55**:11–19.

Grinstein, S., Rotin, D., and Mason, M. J., 1989, Na^+/H^+ exchange and growth factor-induced cytosolic pH changes. Role in cellular proliferation, *Biochim. Biophys. Acta* **988**:73–97.

Grumet, M., Hoffman, S., Crossin, K. L., and Edelman, G. M., 1985, Cytotactin, an extracellular matrix protein of neural and non-neural tissues that mediates glia–neuron interaction, *Proc. Natl. Acad. Sci. USA* **82**:8075–8079.

Gulcher, J. R., Nies, D. E., Marton, L. S., and Stefansson, K., 1989, An alternatively spliced region of the human hexabrachion contains a novel repeat of potential N-glycosylation sites, *Proc. Natl. Acad. Sci. USA* **86**:1588–1592.

Halfter, W., Chiquet-Ehrismann, R., and Tucker, R. P., 1989, The effect of tenascin and embryonic basal lamina on the behavior and morphology of neural crest cells *in vitro*, *Dev. Biol.* **132**:14–25.

Hoffman, S., and Edelman, G. M., 1987, A proteoglycan with HNK-1 antigenic determinants is a neuron-associated ligand for cytotactin, *Proc. Natl. Acad. Sci. USA* **84**:2523–2527.

Hoffman, S., Friedlander, D. R., Chuong, C.-M., Grumet, M., and Edelman, G. M., 1986, Differential contributions of Ng-CAM and N-CAM to cell adhesion in different neural regions, *J. Cell Biol.* **103**:145–158.

Hoffman, S., Crossin, K. L., and Edelman, G. M., 1988, Molecular forms, binding functions, and developmental expression patterns of cytotactin and cytotactin-binding proteoglycan, an interactive pair of extracellular matrix molecules, *J. Cell Biol.* **106**:519–532.

Husmann, K., Faissner, A., and Schachner, M., 1992, Tenascin promotes cerebellar granule cell migration and neurite outgrowth by different domains in the fibronectin type III repeats, *J. Cell Biol.* **116**:1475–1486.

Jhaveri, S., Erzurumlu, R. S., and Crossin, K. L., 1991, Barrel construction in rodent neocortex: Role of thalamic afferents versus extracellular matrix molecules, *Proc. Natl. Acad. Sci. USA* **88**:4489–4493.

Jones, F. S., Burgoon, M. P., Hoffman, S., Crossin, K. L., Cunningham, B. A., and Edelman, G. M., 1988, A cDNA clone for cytotactin contains sequences similar to epidermal growth factor-like repeats and segments of fibronectin and fibrinogen, *Proc. Natl. Acad. Sci. USA* **85**:2186–2190.

Jones, F. S., Hoffman, S., Cunningham, B. A., and Edelman, G. M., 1989, A detailed structural model of cytotactin: Protein homologies, alternative RNA splicing, and binding regions, *Proc. Natl. Acad. Sci. USA* **86**:1905–1909.

Jones, F. S., Crossin, K. L., Cunningham, B. A., and Edelman, G. M., 1990, Identification and characterization of the promoter for the cytotactin gene, *Proc. Natl. Acad. Sci. USA* **87**:6497–6501.

Jones, F. S., Chalepakis, G., Gruss, P., and Edelman, G. M., 1992, Activation of the cytotactin promoter by the homeobox-containing gene, Evx-1, *Proc. Natl. Acad. Sci. USA* **89**:2091–2095.

Joshi, P., Aukhil, I., and Erickson, H. P., 1991, Cell-adhesion to tenascin-endothelial cells adhere to bacterially expressed RGD and fibrinogen domains, but perhaps not to intact hexabrachion, *J. Cell Biol.* **115**:134a.

Kaplony, A., Zimmermann, D. R., Fischer, R. W., Imhof, B. A., Odermatt, B. F., Winterhalter, K. H., and Vaughan, L., 1991, Tenascin Mr 220,000 isoform expression correlates with corneal cell migration, *Development* **112**:605–614.

Keynes, R. J., and Cook, G. M. W., 1992, Repellent cues in axon guidance, *Curr. Opin. Neurobiol.* **2**:55–59.

Lochter, A., Vaughn, L., Kaplony, A., Prochaintz, A., Schachner, M., and Faissner, A., 1991, J1/tenascin in substrate-bound and soluble form displays contrary effects on neurite outgrowth, *J. Cell Biol.* **113**:1159–1171.

Lotz, M. M., Burdsal, C. A., Erickson, H. P., and McClay, D. R., 1989, Cell adhesion to fibronectin and tenascin: Quantitative measurements of initial binding and subsequent strengthening response, *J. Cell Biol.* **109**:1795–1805.

Mackie, E. J., Tucker, R. P., Halfter, W., Chiquet-Ehrismann, R., and Epperlein, H. H., 1988, The distribution of tenascin coincides with pathways of neural crest cell migration, *Development* **102**:237–250.

Mendler, M., Priest, A., and Bourdon, M. A., 1991, Tenascin is recognized by two different integrin receptors, a $\beta 1$-group integrin ($\alpha\beta_1$) and the vitronectin receptor $\alpha_v \beta_3$, *J. Cell Biol.* **115**:137a.

Moolenaar, W. H., 1986, Effects of growth factors on intracellular pH regulation, *Annu. Rev. Physiol.* **48**:363–376.

Murphy-Ullrich, J. E., Lightner, V. A., Aukhil, I., Yan, Y. Z., and Erickson, H. P., 1991, Focal adhesion integrity is downregulated by the alternatively spliced domain of human tenascin, *J. Cell Biol.* **115**:1127–1136.

Onda, H., Poulin, M. L., Tassava, R. A., and Chiu, I., 1991, Characterization of a newt tenascin cDNA and localization of tenascin mRNA during newt limb regeneration by *in situ* hybridization, *Dev. Biol.* **148**:219–232.

Pearson, C. A., Pearson, D., Shibahara, S., Hofsteenge, J., and Chiquet-Ehrismann, R., 1988, Tenascin: cDNA cloning and induction by TGF-β, *EMBO J.* **7**:2977–2982.

Prieto, A. L., Jones, F. S., Cunningham, B. A., Crossin, K. L., and Edelman, G. M., 1990, Localization during development of alternatively-spliced forms of cytotactin by *in situ* hybridization, *J. Cell Biol.* **111**:685–698.

Prieto, A. L., Andersson-Fisone, C., and Crossin, K. L., 1992, Characterization of multiple adhesive and counteradhesive domains in the extracellular matrix protein cytotactin, *J. Cell Biol.* **119**:663–678.

Rakic, P., 1971, Neuron–glia relationship during granule cell migration in developing cerebellar cortex. A Golgi and electron microscopic study in *Macacus rhesus, J. Comp. Neurol.* **141**:283–312.

Raper, J. A., and Kapfhammer, J. P., 1990, The enrichment of a neuronal growth cone collapsing activity from embryonic chick brain, *Neuron* **2**:21–29.

Riou, J.-F., Shi, D.-L., Chiquet, M., and Boucaut, J.-C., 1990, Exogenous tenascin inhibits mesodermal cell migration during amphibian gastrulation, *Dev. Biol.* **137**:305–317.

Roberts, D. D., Rao, C. N., Liotta, L. A., Gralnick, H. R., and Ginsburg, V., 1986, Comparison of the specificities of laminin, thrombospondin, and von Willebrand factor for binding to sulfated glycolipids, *J. Biol. Chem.* **261**:6872–6877.

Spring, J., Beck, K., and Chiquet-Ehrismann, R., 1989, Two contrary functions of tenascin: Dissection of the active sites by recombinant tenascin fragments, *Cell* **59:**325–334.

Steindler, D. A., Cooper, N. G. F., Faissner, A., and Schachner, M., 1989, Boundaries defined by adhesion molecules during development of the cerebral cortex: The J1/tenascin glycoprotein in the mouse somatosensory cortical barrel field, *Dev. Biol.* **131:**243–260.

Stern, C. D., Norris, W. E., Bronner-Fraser, M., Carlson, G. J., Faissner, A., Keynes, R. J., and Schachner, M., 1989, J1/tenascin-related molecules are not responsible for the segmented pattern of neural crest cells or motor axons in the chick embryo, *Development* **107:**309–319.

Takeichi, M., 1990, Cadherins: A molecular family important in selective cell–cell adhesion, *Annu. Rev. Biochem.* **59:**237–252.

Tan, S.-S., Crossin, K. L., Hoffman, S., and Edelman, G. M., 1987, Asymmetric expression in somites of cytotactin and its proteoglycan ligand is correlated with neural crest cell distribution, *Proc. Natl. Acad. Sci. USA* **84:**7977–7981.

Tan, S.-S., Prieto, A. L., Newgreen, D. F., Crossin, K. L., and Edelman, G. M., 1991, Cytotactin expression in somites after dorsal neural tube and neural crest ablation in chicken embryos, *Proc. Natl. Acad. Sci. USA* **88:**6398–6402.

Van der Loos, H., and Woolsey, T. A., 1973, Somatosensory cortex: Structural alterations following early injury to sense organs, *Science* **179:**395–398.

Wehrle, B., and Chiquet, M., 1990, Tenascin is accumulated along developing peripheral nerves and allows neurite outgrowth in vitro, *Development* **110:**401–415.

Weller, A., Beck, S., and Ekblom, P., 1991, Amino acid sequence of mouse tenascin and differential expression of two tenascin isoforms during embryogenesis, *J. Cell Biol.* **112:**355–362.

Woolsey, T. A., and Van der Loos, H., 1970, The structural organization of layer IV in the somatosensory region (SI) of mouse cerebral cortex, *Brain Res.* **17:**205–242.

7

Adhesion Molecules and Bone Remodeling

DONALD R. BERTOLINI AND K. B. TAN

1. Introduction

Adhesion molecule involvement in bone remodeling is a relatively new area of research for bone biology. In the last 4 years there has been a considerable amount of data reported for adhesion molecules on bone cells, but this information is limited compared with that on other cell types, such as immune cells or tumor cells.

Since bone biology is not a widely studied area, we will begin with a brief overview of the physiology of bone and the process of bone remodeling. This will provide a basis for the subsequent discussion of adhesion molecules on the different bone cell types, the osteoblast and osteoclast, and their involvement in the processes of bone formation and resorption.

2. Bone Biology

2.1. Metabolism

Skeletal tissue is composed of a mineralized matrix that is uniquely adapted to its mechanical functions: protecting internal organs, supporting extremities and joints, and serving as levers to facilitate locomotion. In addition, bone

DONALD R. BERTOLINI and K. B. TAN • Department of Cellular Biochemistry, SmithKline Beecham Pharmaceuticals, King of Prussia, Pennsylvania 19406.
Cellular Adhesion: Molecular Definition to Therapeutic Potential, edited by Brian W. Metcalf, Barbara J. Dalton, and George Poste. Plenum Press, New York, 1994.

provides a reservoir for calcium, phosphorus, and magnesium, which are essential for homeostasis. Skeletal metabolism is therefore responsive to calcium-regulating hormones, particularly parathyroid hormone (PTH) and 1,25-dihydroxyvitamin D_3 [1,25$(OH)_2D_3$]. The level of circulating PTH increases in response to a calcium deficit, and the hormone maintains plasma and tissue calcium levels by increasing bone resorption as well as kidney reabsorption. Circulating 1,25$(OH)_2D_3$ increases in response to decreases in both plasma phosphate and calcium to maintain a supply of both of these ions. This is usually achieved by increasing intestinal absorption of these ions, but when dietary supply is not available, only bone resorption can provide an adequate amount of calcium and phosphorus.

The dynamic nature of bone has often gone unappreciated. Bone formation and resorption are continuous processes throughout life, and are required for the strength and vitality of skeletal tissue. These opposing processes, which appear to be coordinated, are encompassed by the terms *bone modeling* and *remodeling*. The modeling process is the balance of formation and resorption that occurs throughout bone growth. Remodeling is the process as it continues throughout adulthood. This remodeling process was thought to be exclusively under the control of the osteotropic hormones, but recent evidence has demonstrated the importance of local factors such as IL-1, TNF, or TGFβ. These factors are probably produced by the major cells of the bone, the osteoblast and the osteoclast, as well as cells of the marrow. The remodeling process is mediated by the osteoblast and osteoclast.

2.2. Cell Biology

2.2.1. The Osteoblast

The main function of the osteoblast is the synthesis of bone matrix (osteoid). This includes deposition of the matrix in appropriate amounts and orientation and initial mineralization. Osteoblasts are derived from stromal cell precursors (osteoprogenitor cells) within the periostium and bone marrow (Fig. 1). The progenitors, which are probably mesenchymal in origin, are usually quiescent, and are activated during remodeling and in response to injury. Osteoblasts have the typical appearance of actively synthesizing and secreting cells, with an abundant rough endoplasmic reticulum and prominent Golgi apparatus. Osteoblasts are related to bone lining cells and osteocytes. The bone lining cells show a flat, thin, elongated morphology and cover most bone surfaces in the adult skeleton (Fig. 1). There is evidence that at the end of the remodeling, when osteoblasts stop matrix synthesis, they become lining cells (Martin *et al.*, 1988). These cells may act as a pool of osteoblast precursors and serve as a continuous barrier between bone and extracellular fluid. The osteocyte is probably an osteoblast

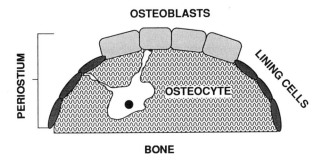

Figure 1. Diagrammatic representation of the relationship of osteoblasts, lining cells, and osteocytes.

trapped in the bone during elaboration of the matrix. These cells have long processes that extend through canals of the Haversian system to communicate with surface osteoblasts and bone lining cells (Fig. 1). It is thought that these contacts allow the osteocyte to communicate the extent of mechanical stress being placed on the bone and thus modulate remodeling to strengthen the bones (Skerry *et al.*, 1989).

2.2.2. The Osteoclast

Osteoclasts are large, multinucleated cells measuring several hundred microns in diameter. They are formed by the fusion of mononuclear cells, rather than by division. Osteoclasts are derived from hematopoietic precursors of monocytic lineage, possibly the granulocyte–macrophage colony-forming cells, and they may be the tissue macrophage of bone (Vaes, 1988). The osteoclast contains numerous mitochondria and lysosomes, and is especially rich in the lysosomal enzyme type V acid phosphatase, a tartrate-resistant enzyme, which is considered a marker for this cell type. There are still unanswered questions about the origin of the osteoclast: (1) Are the osteoclast and the macrophage derived from a common precursor or a subpopulation of cells that can be committed to the osteoclast lineage during develoment? (2) What is the sequence of events, both molecular and cellular, that leads to the development of the osteoclast? (3) What are the signals, humoral and physical, that control the process of osteoclast development? This last question has proven to be the most complex to address. As seen in Fig. 2, numerous factors, both systemic and local, affect osteoclast function and development. Adding to the complexity is the fact that some of these factors, such as PTH and interleukin 1 (IL-1), may act through the osteoblast in either of two ways. There may be a release of soluble factors from the

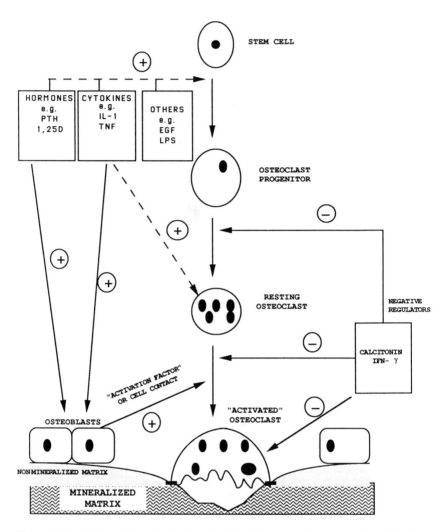

Figure 2. Possible interactions of molecules and cells in the control of bone remodeling. Stimulators of resorption (+) are shown in the boxes on the upper left. Arrows indicate where these molecules are thought to act in the process of stimulating the osteoclast to resorb. Dashed arrows indicate only a proposed site of action where no definitive data exists. Negative regulators (−) are shown in the box on the right. The interaction of osteoblasts and osteoclasts may be through a soluble factor "activation factor" or through direct cell–cell contact.

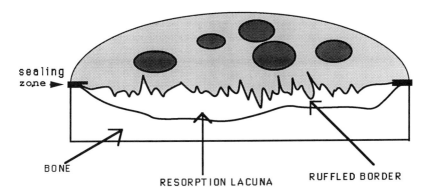

Figure 3. Diagrammatic representation of the resorbing osteoclast with the tight sealing zone, the resorption pit or lacuna, and the ruffled border.

osteoblasts, which affects the osteoclast or its precursor (Vaes, 1988), or there may be direct cell–cell contacts between osteoblasts and osteoclasts (Takahashi *et al.*, 1988). The osteoclast's function is to resorb the mineralized matrix. Resorption is accomplished using a highly specialized and complex region of the plasma membrane called the "ruffled border." This area consists of multiple membrane folds and fingerlike projections (Fig. 3). A region of the plasma membrane just adjacent to the ruffled border forms a tight bond with the bone and is referred to as the "sealing zone." The seal isolates the resorbing area and the ruffled border, forming the equivalent of an "extracellular lysosome." The sealing zone has been intensively investigated for adhesion receptors.

Bone resorption occurs through the secretion of hydrogen ions into the resorption area, acidifying the microenvironment beneath the sealed osteoclast and dissolving the mineral. This hydrogen ion secretion is accomplished with a specialized vacuolar-type proton pump, the exact nature of which has not been established (Chatterjee *et al.*, 1992). When lysosomal enzymes, such as cathepsin B or other serine proteases, are released into this region, they degrade the organic matrix, forming a resorption pit, or Howship's lacuna (Fig. 3).

2.3. Bone Matrix

The bone consists of inorganic mineral and organic matrix. The mineral, which constitutes 65% of bone, is mainly calcium and phosphorus in hydroxyapatite crystals. This mineral phase contains approximately 99% of the body calcium, 85% of the phosphorus, and 40–60% of the total body sodium and magnesium (Glimcher, 1990).

The matrix, which constitutes approximately 35% of the bone tissue, consists primarily of type I collagen (90%) and an assortment of noncollagenous proteins, which include osteopontin, bone sialoprotein, fibronectin, thrombospondin, osteonectin, osteocalcin, proteoglycans, and glycosaminoglycans (Robey *et al.*, 1990). Collagens are involved in cellular adhesion in a variety of tissues (Kleinman *et al.*, 1981) and this may also be true of bone, but little data are available. Several of the noncollagenous proteins found in the bone matrix, including osteopontin (Oldberg *et al.*, 1986), bone sialoprotein (Oldberg *et al.*, 1988), thrombospondin (Robey *et al.*, 1989), and fibronectin (Pierschbacher *et al.*, 1983), as well as collagen type I (Dedhar *et al.*, 1987), contain the Arg-Gly-Asp (RGD) sequences that can interact with integrins on various cell types. Osteopontin, bone sialoprotein, fibronectin, and thrombospondin have been demonstrated to act as ligands for RGD-binding integrins on bone cells (Flores *et al.*, 1992; Oldberg *et al.*, 1988; Puleo and Bizios, 1991; Robey *et al.*, 1989). It is not known whether other proteins in the bone matrix possess cell binding activity.

3. Adhesion and Bone Remodeling

3.1. The Remodeling Cycle

Bone remodeling occurs at discrete foci and involves resorption followed by new bone formation. This sequence has been referred to as a basic multicellular unit of bone turnover (BMU) by Frost (1984), and is illustrated in Fig. 4. A remodeling cycle begins, by a mechanism yet to be defined, when lining cells overlaying the bone surface retract and expose the mineralized matrix (activation phase). Osteoclast precursors are recruited from the circulation to this exposed matrix and fuse to form the activated osteoclasts (resorptive phase). Over a period of weeks the osteoclast will resorb a discrete amount of bone; this actively motile cell then moves on to another site. The resorptive phase is followed by an active reversal phase in which the cement line is deposited by the new osteoblasts recruited to the site. At this point the formation phase begins, and the resorption lacuna is filled in by the osteoblast over a period of months. As the lacuna fills, the osteoblasts become less cuboidal, and eventually become or are replaced by the flat lining cells.

The remodeling cycle has been referred to as a "coupled" process, with formation equaling resorption (Frost, 1964). One of the most intriguing questions for bone cell biology is how these cells are programmed to resorb a certain quantity of bone and then fill in with an equal quantity of new bone. This is likely to occur through the generation of local signals, possibly soluble factors, or through matrix interactions with specific receptors on the cells. We know that as an individual ages the regulation of resorption and formation is altered, and lacunae are not completely filled. This alteration results in a net loss of bone and

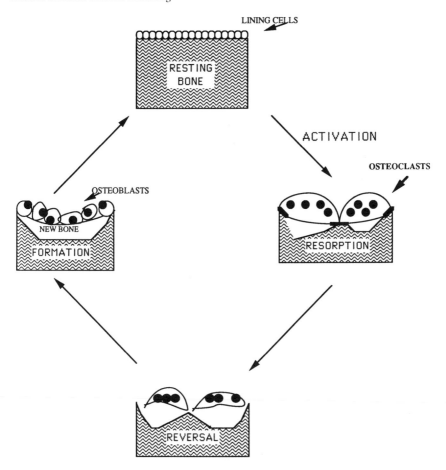

Figure 4. Diagrammatic representation of the bone remodeling cycle. Bone remodeling begins with the activation of bone resorption, in which osteoclasts are recruited to the bone surface. After resorbing a specific quantum of bone the osteoclasts leave the bone surface—the reversal phase. Osteoblasts are recruited to the resorption lacuna and fill in the resorption space with new osteoid—the formation phase. Finally, the bone surface is covered with lining cells and the site is again resting bone.

the gradual decrease in bone mass associated with aging. This process is sometimes accelerated, with a severe loss of bone resulting in osteoporosis and a concomitant increase in the frequency of fractures of the spine and hip.

3.2. Adhesion Molecules and the Osteoblast

Cellular adhesion probably plays an important role in both osteoblast development and function. The osteoblast develops from a committed precursor in an organized and apparently stepwise fashion (Fig. 5). At the periosteal surface of

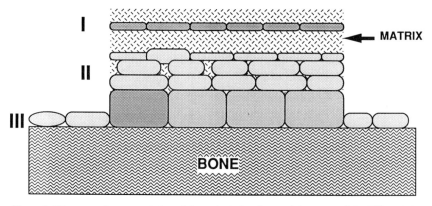

Figure 5. Diagrammatic representation of the periosteal surface and the stages of the differentiating osteoblasts. Osteoblasts appear to differentiate in an organized sequence. The mesenchymal stem cell (I) divides to supply progenitors for differentiation. Osteoblast precursors (II) then divide and differentiate until they reach the bone surface and begin to form bone as fully differentiated osteoblasts (III).

the bone there is a layer of cells that exhibit a spectrum of osteoblastic markers representing the precursor (I), the preosteoblast (II), and the fully differentiated osteoblast (III). All of these cells interact with each other and the matrix. Osteoblastic cells interacting with the matrix respond with shape changes, indicating that matrix binding to receptors transduces signals that modulate the cytoskeleton of the cell (Vukicevic *et al.*, 1990). It is also clear that osteoblasts on the surface of the bone are connected by gap junctions; these connections extend to the neighboring lining cells and probably to the osteocytes (Doty, 1981), allowing communication. Zimmermann *et al.*, (1988) have reported that osteoblastic cellular differentiation is increased by cell–cell interactions. The adhesion molecules and receptors involved in the signaling process remain to be elucidated. The majority of adhesion molecules identified in both osteoblasts and osteoclasts fall into the integrin family. Integrins are membrane glycoproteins consisting of two subunits, α and β. There have been 14 α and 8 β subunits identified so far. These subunits, in various combinations, can form at least 20 different integrins (Fig. 6). It is likely that more subunits will be discovered. Some, but not all, integrins recognize the RGD sequence in matrix proteins and mediate cell–substrate attachment; others interact with cell membrane proteins mediating cell–cell interactions (for review see Ruoslahti, 1991). Osteoblasts have been shown to express solely integrin molecules, whereas the osteoclast expresses other adhesion molecules.

3.2.1. Osteoblast Integrins

The expression of integrin molecules on the osteoblast has just begun to be reported (Horton and Davies, 1989; Clover *et al.*, 1992). These data indicate that

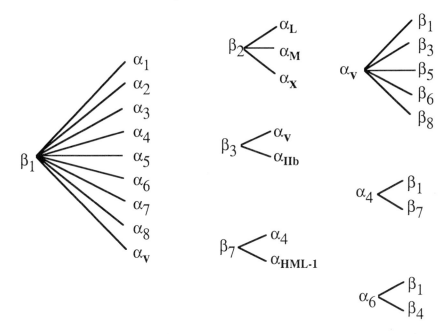

Figure 6. Integrin subunit associations.

osteoblasts express a different repertoire of integrin molecules than the osteo-clast. Human osteoblasts *in situ* primarily express α_1, α_3, and β_1, and weakly express α_2. There is no expression of α_4, α_5, α_6, α_I, α_M, and β_2 (Clover *et al.*, 1992). Others have reported the expression of α_5 on rat osteoblastic cells in culture (Horton and Davies, 1989; Majeska and Einhorn, 1992). Interestingly, an increased expression of α_1 and α_3 subunits was observed *in situ* when osteoblasts were actively synthesizing bone, and by lining cells when compared with pre-osteoblasts, while β_1 expression remained unchanged (Clover *et al.*, 1992). This expression pattern was also found when cultured osteoblasts were examined (Clover *et al.*, 1992).

Since osteoblasts express α_1, α_3, and α_5 in conjunction with β_1, it would be possible for these cells to contain $\alpha_1\beta_1$, $\alpha_2\beta_1$, $\alpha_3\beta_1$, and $\alpha_5\beta_1$ on their surface. $\alpha_1\beta_1$ and $\alpha_2\beta_1$ bind both collagen type I and laminin, $\alpha_5\beta_1$ binds fibronectin, and $\alpha_3\beta_1$ binds all three ligands. In addition, it is known that $\alpha_1\beta_1$, $\alpha_2\beta_1$, and $\alpha_3\beta_1$ can differentially bind to collagen or laminin, depending on the cell type in which they are expressed (Hynes, 1992). The natural ligands in bone that are recognized by these osteoblast receptors have not been identified. This is the present state of our knowledge on adhesion receptors expressed on osteoblasts. It is likely that these cells can express several more receptor types, and it will be important to determine how all of these receptors are involved in osteoblast development and function.

3.2.2. Osteoblast Adhesion to Matrix

Only a limited amount of data has been reported on osteoblast adhesion to matrix molecules. Osteoblasts have been shown to bind to RGDS peptides, which are also capable of inhibiting osteoblast adhesion to fibronectin (Puleo and Bizios, 1991). There appears to be some negative cooperativity in the adhesion or possibly binding sites on osteoblasts for RGDS peptides with differing affinity. The RGDS peptide has also been shown to disrupt mineralization in bone organ cultures. This disruption appears to result from a disorganization of the osteoblasts along the mineralization front (DeRome and Gronowicz, 1991). Similar effects are seen when these cultures are treated with glucocorticoids, which decrease fibronectin production by osteoblasts. Therefore, it seems likely that the RGDS peptides are blocking osteoblast adhesion to fibronectin, thereby inhibiting the mineralization of the newly produced matrix (DeRome and Gronowicz, 1991). IL-1β has been shown to elevate the expression of the β_1 subunit and its associated α subunits (Dedhar, 1989), indicating that this process of osteoblast adhesion to fibronectin can be regulated. Concomitant with the increased adhesion is an inhibition of cell proliferation, an altered cell morphology, and an increase in the expression of the osteoblastic phenotypic marker alkaline phosphatase (Dedhar, 1989).

Since several of the integrins reported on osteoblasts bind multiple ligands, it will be important to examine the interaction of these cells with multiple matrix molecules. To this end we have developed a simple adhesion assay to evaluate osteoblast binding to bone matrix molecules (Tan and Koncsics, 1992). We have demonstrated that osteoblastic cells (MC3T3-E1) bind to vitronectin, fibronectin, and collagen type I. The binding to collagen is slower than that for fibronectin or vitronectin: cell attachment requires 2 hr for collagen, whereas fibronectin or vitronectin attachment is complete in 1 hr (Fig. 7). The vitronectin binding is probably the result of use of cultured cell lines for the adhesion assay. The vitronectin receptor has not been demonstrated on osteoblasts *in situ*, but it is known that all established cell lines express a vitronectin receptor, whether or not it is normally expressed in primary cells of the same type (Horton, 1990). Adhesion to collagen does not depend on the collagen molecule being intact, as similar binding is seen with native or denatured collagen (Fig. 8). In order to adhere to collagen the osteoblast requires protein synthesis; cycloheximide partly inhibits adhesion (Fig. 8). This is not true for adhesion to fibronectin or vitronectin (Fig. 8). This difference may be related to the state of differentiation of the MC3T3 cells (Quarles *et al.*, 1992). As reported by Clover *et al.*, (1992), osteoblasts *in situ* only express $\alpha_1\beta_1$ and $\alpha_3\beta_1$ when actively synthesizing matrix. As the cells in culture may not be actively synthesizing matrix, the interaction with the collagen may induce the synthesis of integrin molecules or extracellular matrix.

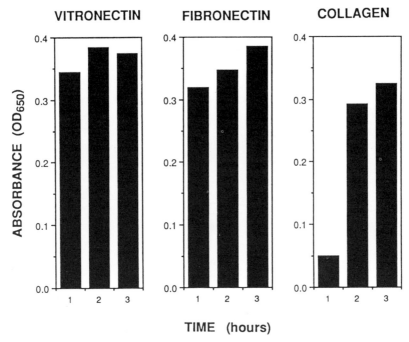

Figure 7. Adhesion of MC3T3 cells to vitronectin, fibronectin, and collagen. A 96-well microtiter plate was inoculated with 0.1 ml of vitronectin (1 μg/ml), fibronectin (10 μg/ml), or type I collagen (20 μg/ml), and incubated at 37°C for 2 hr. After blocking with BSA (3 mg/ml), each well was seeded with 20,000 cells and incubated at 37°C. At the indicated times, attached cells were fixed, stained, and quantitated as described by Tan and Koncsics (1992).

The binding of the osteoblast to fibronectin is blocked by peptides containing the RGD sequence (Fig. 9). The potency of these peptides is similar to that reported for other cell types (Ruoslahti, 1988). There is no activity of RGD peptides against the binding of osteoblasts to collagen-coated surfaces (Fig. 9). Further research into the adhesion molecules on the osteoblast, the way that they interact with the matrix molecules of bone, and the signals these interactions produce in the cell should allow us to better understand the processes of osteoblast development and bone formation.

3.3. Adhesion Molecules and the Osteoclast

Figure 10 (I–VI) shows that osteoclast adhesion is involved at different steps of development and function. Osteoclast precursors must leave the circulation to migrate to their site of action (I). This process appears to be similar to the

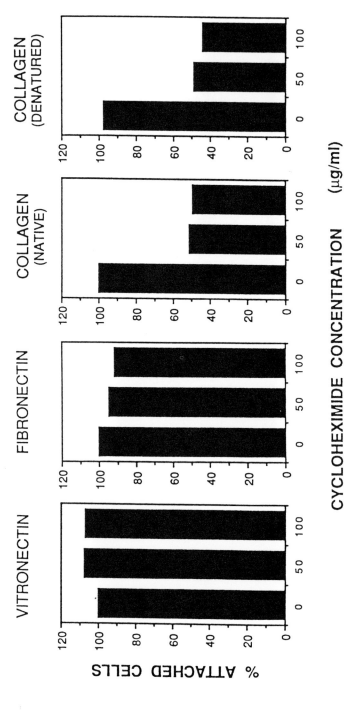

Figure 8. Effect of cycloheximide on adhesion. MC3T3 cells were seeded as described in Fig. 7 in the presence or absence of cycloheximide for 3 hr at 37°C. The values for vitronectin and fibronectin were normalized against their respective untreated controls. The values for both native and denatured collagen were normalized against the untreated native collagen control.

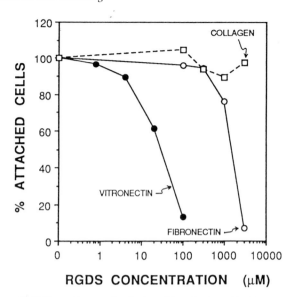

Figure 9. Effect of RGDS peptide on cell adhesion. The adhesion of MC3T3 cells to vitronectin, fibronectin, and collagen was determined in the presence of increasing concentrations of RGDS peptide. Adhesion to vitronectin and fibronectin was for 1 hr; 2 hr for collagen.

tissue infiltration of monocytes, a cell type to which the osteoclast is closely related. There is a large body of evidence for the involvement of several cell adhesion molecules in this process (Albelda and Buck, 1990). The osteoclast must then "home in" and attach to the bone (II and III). The precursor cells then fuse and differentiate to form mature multinucleated osteoclasts (IV and V). Finally, the unique process of bone resorption requires the tight seal of the osteoclast to the bone to form the "extracellular lysosome" (VI).

3.3.1. Osteoclast Adhesion Molecules

The first identification of an adhesion molecule on the osteoclast was achieved by Davies *et al.* (1989), who identified an osteoclast vitronectin receptor utilizing a monoclonal antibody to what was then called the osteoclast functional antigen (OFA). The OFA molecule had been shown to be expressed somewhat selectively in the osteoclast (Horton and Chambers, 1986), and to be developmentally regulated during embryogenesis (Simpson and Horton, 1989), appearing on the preosteoclast in the forming bone. Antibodies to OFA blocked bone resorption by isolated osteoclasts (Chambers *et al.*, 1986). All of these data suggested a functional role in osteoclast physiology. It was later demonstrated that OFA was not specific to the osteoclast, as it could be detected on other cell

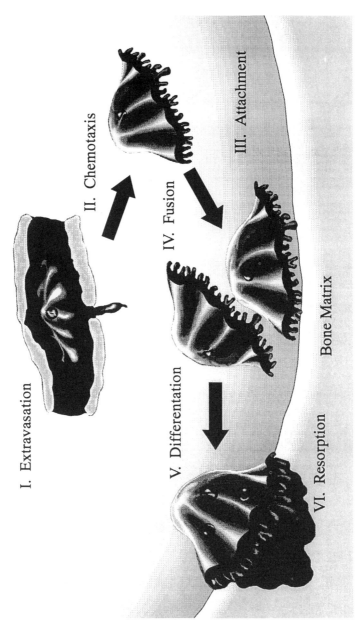

Figure 10. Role of adhesion in osteoclast differentiation/function.

Table I. Adhesion Molecules Identified on Osteoclasts

Integrin subunit	Human	Rat	Reference
α_v	+	−	Davies *et al.* (1989)
	+	NA	Quinn *et al.* (1991)
	+	NA	Clover *et al.* (1992)
$\alpha_v\beta_3$	+	−	Davies *et al.* (1989)
	+	+	Helfrich *et al.* (1992a)
β_3	+	NA	Clover *et al.* (1992)
	+	+	Helfrich *et al.* (1992a)
	+	NA	Quinn *et al.* (1991)
	+	NA	Teti *et al.* (1989)
β_1	+	NA	Quinn *et al.* (1991)
	+	NA	Clover *et al.* (1992)
	+	+	Helfrich *et al.* (1992a)
α_2	+	NA	Quinn *et al.* (1991)
α_4	+	NA	Quinn *et al.* (1991)
ICAM-1	+	NA	Athanasou and Quinn (1990)

types by immunohistochemical staining (Athanasou *et al.*, 1990). Since that first report, several other investigators have identified adhesion molecules on the osteoclast. Table I summarizes these data. The panel of adhesion molecules identified to date on the osteoclast is fairly restricted, and there is only limited overlap with those molecules expressed on the osteoblast. All of the possible combinations of the expressed molecules allow the osteoclast to express a limited number of receptors, including a vitronectin receptor ($\alpha_v\beta_3$), fibronectin receptors ($\alpha_v\beta_1,\alpha_4\beta_1$), a collagen/laminin receptor ($\alpha_2\beta_1$), and ICAM-1. The only receptor the osteoclast has in common with the osteoblast is $\alpha_2\beta_1$, the collagen/laminin receptor. This is not unexpected, since the major protein in the environment of these cells is type I collagen.

3.3.2. Osteoclast Adhesion

Figure 10 illustrates the steps in the development of the activated osteoclast in which adhesion may play a role. First, the mononuclear precursor cell must exit the circulation and get to the site of the resorption. The only adhesion molecule found on the osteoclast that is currently known to play a role in the extravasation of cells from the circulation is ICAM-1. This molecule has been shown to play an important role in the migration of monocytes, neutrophils, and lymphocytes into tissues in response to inflammatory signals (Albelda and Buck, 1990). It is known that ICAM-1 works in concert with several other adhesion molecules, including Mac-1, P150,95, and LFA-1, in the process of immune cell migration out of the circulation (Albelda and Buck, 1990). It is clear that the mature osteoclast does not express these other adhesion molecules (Athanasou and Quinn, 1990). It is not known whether the osteoclast precursor expresses these other molecules, but it is somewhat surprising that the mature osteoclast expresses only this one molecule. An interesting theory is that the ICAM-1 molecule may be involved in the cell–cell adhesion between the osteoclast and mononuclear precursors, for fusion leading to the multinucleation of the mature osteoclast. Upon fusion the other adhesion molecules may be lost from the cell surface, but ICAM-1 is maintained to facilitate further fusions.

One step in the process of osteoclastic bone resorption about which we have considerable information is the adhesion of the osteoclast to the bone surface. The osteoclast must adhere to the mineralized matrix and establish a seal around its perimeter to form the "extracellular lysosomal compartment" in which the resorptive process takes place. Several investigators have demonstrated that the osteoclast vitronectin receptor is involved in this adhesion process (Horton, 1990). The osteoclast vitronectin receptor has been shown to be required for bone resorption because specific monoclonal antibodies to this molecule disrupt the resorptive process (Chambers *et al.*, 1986). Localization of the β_3 subunit of the vitronectin receptor has been demonstrated in the osteoclast membrane in association with F-actin in the podosomes (adhesion plaques) of the osteoclast, as well as in a beltlike row of F-actin that encircles the entire circumference of the cell (Zambonin-Zallone *et al.*, 1989). Vinculin and talin are also associated with the actin and β_3 in these structures. All of these data have led to the suggestion that the vitronectin receptor of the osteoclast is involved in the tight sealing of the cell to the bone surface.

It has recently been reported that the osteoclast vitronectin receptor does not mediate the tight sealing zone attachment of the osteoclast (Lakkakorpi *et al.*, 1991). Immunolocalizaiton has demonstrated that the osteoclast vitronectin receptor is present on the basolateral membrane and on the "ruffled border" within the "extracellular lysosome," but not on the tight sealing zone.

With the discovery that the osteoclast vitronectin receptor played a role in the bone resorption process, a search for a ligand in the bone matrix began. At the time there was no evidence for vitronectin in bone matrix, so attention was focused on the other matrix proteins that contain an RGD sequence. Since that time vitronectin has been found in bone matrix, but the quantity that is present is not known (Mimura *et al.*, 1991). Most of the investigations for the ligands of the osteoclast vitronectin receptor have focused on osteopontin and bone sialoprotein, both of which contain an RGD sequence (Oldberg *et al.*, 1986, 1988). Reinholt *et al.*(1990) proposed osteopontin as a possible ligand for the osteoclast vitronectin receptor. They found colocalization of the protein with the receptor on the osteoclast membrane, especially in the clear zone of the osteoclast, which suggests that this is a natural ligand in bone matrix for this receptor.

It was recently found that bone sialoprotein is not associated with the osteoclast *in vivo* by immunolocalization, which suggests that it may not play a role in osteoclast binding to the bone surface (Flores *et al.*, 1992). Further evidence that osteopontin may play a role in the osteoclast adhesion to bone comes from a demonstration that peptides containing the RGD sequence found in this protein can inhibit bone resorption by isolated osteoclasts (Horton *et al.*, 1991; Lakkakorpi *et al.*, 1991; Flores *et al.*, 1992). The GRGDS sequence dose-dependently inhibits resorption of dentine slices by osteoclasts and blocked attachment of osteoclasts to bone. Osteoclasts have been shown to bind to other RGD-containing proteins, but osteopontin promotes the most and the strongest attachments (Helfrich *et al.*, 1992b; Flores *et al.*, 1992).

The first demonstration that the RGD sequence is important for osteoclastic bone resorption came from studies with the snake venom protein s-echistatin, a potent inhibitor of RGD-binding integrins. Echistatin was shown to be a potent inhibitor of bone resorption by isolated osteoclasts (Sato *et al.*, 1990). Echistatin was colocalized to the outer edge of the cell contact regions and was associated with vinculin, in a manner similar to the localization of the β_3 subunit of the osteoclast vitronectin receptor, seen by Zambonin-Zallone *et al.* (1989).

Our laboratory has extended this observation by demonstrating that RGD peptides can also block bone resorption in organ culture and *in vivo* (Bertolini *et al.*, 1991). A specific conformationally constrained cyclic RGD peptide, SK&F 106760, inhibited the release of radiolabeled calcium from fetal rat long bone rudiments in a dose-dependent manner (Fig. 11). To further demonstrate that RGD binding was essential for bone resorption, we infused this RGD peptide in an *in vivo* model of bone resorption, in which we infused the peptide subcutaneously by ALZA mini-osmotic pump over the skull of a mouse and simultaneously injected parathyroid hormone to induce the resorption (this protocol is diagrammed in Fig. 12). At the end of the experiment the skull bones were examined histologically to determine the extent of resorption; the results can be seen in Fig. 13. There was a mild inflammatory infiltrate at the dorsal aspect of the bone, resulting from the irritation caused by the catheter from the osmotic pump and the injection. This inflammation (arrow) was associated with moderate loss in bone mass, seen in Fig. 13B, compared with a normal untreated mouse, Fig. 13A. Figure 13D illustrates the effect of SK&F 106760 on this inflammation-mediated bone resorption. There was an almost complete preservation of the bone structure, with only a slight enlargement of the marrow space. Note that the peptide had no effect on the inflammation (closed arrowhead). When PTH was injected there was considerable loss of bone mass, with the typical "moth-eaten" appearance and conspicuous resorption surfaces associated with osteoclasts (curved arrow) (Fig. 13C). When SK&F 106760 was infused prior to and during the injection of PTH, there was a reduction in the loss of bone mass and there were few resorption surfaces and osteoclasts (Fig. 13E). There

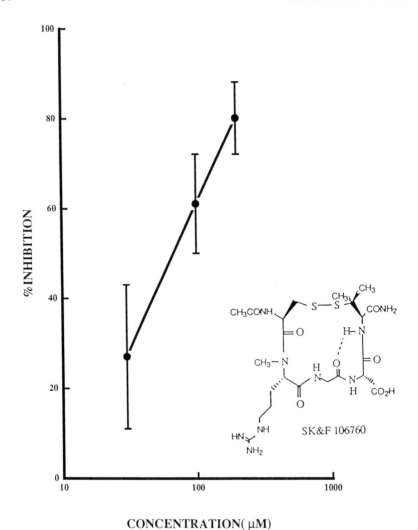

CONCENTRATION(μM)

Figure 11. Effect of SK&F 106760 on PTH-mediated bone resorption in fetal rat long bone cultures. Fetal rat long bone cultures were treated with PTH (12 nM) in the presence or absence of various concentrations of SK&F 106760 (200 μM) (•) for 48 hr and the release of $^{45}Ca^{2+}$ was monitored. Data are the mean ±SEM for three experiments.

were also prominent osteoprogenitor cells/tissue, and bony surfaces lined by conspicuous osteoblasts (open arrowheads) (Fig. 13F). These findings demonstrate that osteoclastic bone resorption *in vivo* was inhibited by SK&F 106760; this is best explained by the RGD inhibition of osteoclast adhesion. The most likely adhesion molecule is an $\alpha_v\beta_3$ vitronectin receptor, since antibodies that

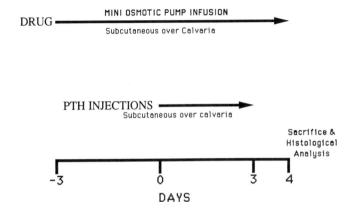

Figure 12. Protocol for *in vivo* PTH-induced bone resorption in mice.

recognize this protein also inhibit bone resorption (Chambers *et al.*, 1986). These data demonstrate that RGD peptides may also have utility as therapeutic agents to block bone resorption associated with diseases such as osteoporosis or Paget's disease.

3.3.3. Integrin Signaling in the Osteoclast

It has recently been demonstrated that integrin molecules not only mediate the adhesion of cells to matrix, but also transduce signals from the matrix to the cell (Kornberg and Juliano, 1992). The mechanism of signaling may involve changes in cytoplasmic pH (Ingber *et al.*, 1990), cytosolic calcium (Miyauchi *et al.*, 1991), and phosphorylation of proteins (Neff *et al.*, 1992; Hruska *et al.*, 1992; Rolnick *et al.*, 1992). These signals are thought to be transmitted by two possible mechanisms. The first functions through the cytoskeleton and regulation of cell shape. It is known that the integrin molecules can interact with specific cytoskeletal molecules at the site of cellular adhesion, which serves as a nucleation site for cytoskeletal organization (Burridge, 1988). The second mechanism is based on the idea that the integrin is a "true receptor" and gives rise to changes through biochemical signals (Kornberg and Juliano, 1992).

The transduction of signals by the osteoclast vitronectin receptor has recently been demonstrated. The binding of osteopontin and other RGD-containing molecules induces a transient reduction in cytosolic calcium (Miyauchi *et al.*, 1991). This change in calcium requires the RGD sequence, and is blocked by antibody directed against the vitronectin receptor (Miyauchi *et al.*, 1991). It has also been shown that the establishment of cell matrix attachment is characterized by changes in cytosolic calcium levels (Teti *et al.*, 1989), podosome assembly, and bone resorption (Miyauchi *et al.*, 1990). It is not clear whether this is a true

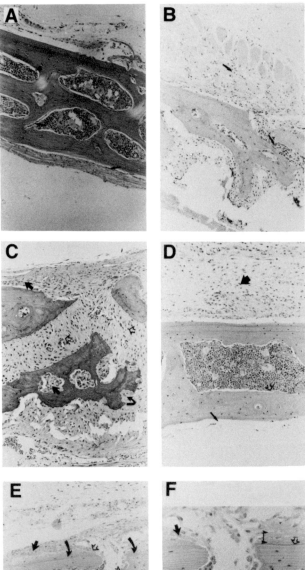
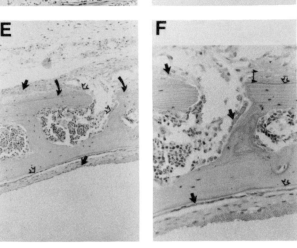

biochemical signal, or whether it functions through the cytoskeleton. The osteo-clast vitronectin receptor β_3 subunit has been shown to be associated with vin-culin and talin in the podosomes of the osteoclast (Zambonin-Zallone *et al.*, 1989). It is therefore possible that signal transduction occurs through this interac-tion. It is clear that blocking the interaction of the receptor with ligand causes a retraction of the cell without a decrease in the cytosolic calcium levels. A similar retraction of the cell is seen when the osteoclast is treated with calcitonin (Mal-garoli *et al.*, 1989; Zaidi *et al.*, 1990). Calcitonin has been shown to cause a transient rise in intracellular calcium; this is the direct opposite effect of what is seen when the osteoclast binds osteopontin, which may promote resorption. These data suggest that the integrin molecule is signaling by changing intracellu-lar calcium levels. Further evidence comes from the fact that osteoclast cytosolic calcium is dependent on extracellular calcium. There appears to be a receptor-operated mechanism stimulating calcium release from intracellular stores (Zaidi et al., 1991). In addition, osteoclasts transiently express voltage-gated calcium

---

←

Figure 13. Effect of SK&F 106760 on PTH-mediated bone resorption *in vivo* in the mouse. All panels are coronal sections, H&E. A–E, 300×; F, 600×. (A) Photomicrograph of the calvaria from a normal mouse. (The ventral periosteal surface is shown in the lower half of the frame; the brain is not shown.) Note the row of osteoblasts (arrow) intimately associated with osteoprogenitor cells/tissue at the ventral periosteal surface. Most trabecular bone surfaces are lined by osteoblasts (arrow) or flat "resting" cells (arrowhead). (B) Photomicrograph of the calvaria from a mouse given vehicle (0.1 mM acetic acid, 0.1% BSA in PBS) both by injection and by osmotic pump. The brain, not shown, is in the lower half of the frame. There is a mild inflammatory infiltrate at the dorsal aspect of the bone, involving the overlaying muscle and extending into the bony tissue (arrows). This inflammation is associated with moderate loss in bone mass (inflammatory-mediated osteopenia). (C) Photomicro-graph of the calvaria from a mouse given PTH (10 μg/bolus injection) s.c. for 3 consecutive days. There is a marked loss in bone mass. The residual bone has a "moth-eaten" appearance and conspicu-ous resorption surfaces associated with osteoclasts (curved arrow). There is also an increase in the amount of osteoprogenitor cells/tissue (closed arrowheads) intermingled with apparent (granulation) fibrous connective tissue (open arrowhead). (D) Photomicrograph of the calvaria from a mouse given SK&F 106760 (2 μg/hr, 48 μg/day) by pump for 6 consecutive days. Still present is the mild inflammatory infiltrate seen in all of the animals (closed arrowhead). There is only a mild loss in bone mass, few resorption surfaces and osteoclasts, and a reduction in/loss of osteoprogenitor cells/tissue and associated osteoblasts (arrow). Most bony surfaces are lined by flat "resting" cells (open arrow-head). (E) Photomicrograph of the calvaria from a mouse given SK&F 106760 by pump for 6 consecutive days and PTH three times a day (10 μg/injection) for 3 days. There are few resorption surfaces and osteoclasts. The osteoprogenitor cells/tissue (closed arrowheads) are markedly promi-nent, and bony surfaces are lined by conspicuous osteoblasts (open arrowheads). These changes in the bone cell population are associated with histologic evidence of apparent lamellar new bone formation (curved arrows). (F) Photomicrograph of the same section shown in E, but at higher magnification. Note that the newly deposited osteoid appears regular and lamellar in pattern and is associated with similar, linearly oriented new osteocytes (area between paired arrowheads). The new bone is separated from the original periosteal surface by prominent cement lines (open arrowheads). Both periosteal and endosteal surfaces are lined by conspicuous, plump, regularly arranged osteo-blasts (closed arrowheads).

channels. These channels, together with calcium-ATP, may regulate the intracellular calcium concentration in osteoclasts (Akisaka *et al.*, 1988). These data suggest that integrin receptors trigger biochemical signals in the osteoclasts in addition to reorganizing the cytoskeleton.

Does the vitronectin receptor participate directly in the tight seal that is required for bone resorption (Lakkakorpi *et al.*, 1991)? It seems unlikely that an integrin-type adhesion could establish a seal tight enough to retain hydrogen ions within the resorbing zone under the osteoclast. One likely way that this receptor may participate is through signaling. Recent signaling data suggest that this initial interaction between the osteoclast and the matrix may initiate the establishment of a different adhesion process, possibly through hemidesmosomes, which are more likely to produce a seal that would retain ions within the resorbing zone. Research into the expression of other adhesion molecules, in addition to the integrins, on the osteoclast may shed light on this prospect.

4. Summary

All of the data suggest that adhesion molecules participate in both the formation and resorption of bone during remodeling. Our knowledge of adhesion molecules on the osteoblast is limited; it must be expanded so that specific interactions with matrix molecules can be explored in more detail. Fibronectin appears to play a significant role in the function of this cell during mineralization, and this process appears to be under the regulation of locally produced molecules such as IL-1β.

There is considerably more information on the participation of integrin molecules in the process of bone resorption. There is clear evidence that a vitronectin receptor must participate in the adhesion process in order for bone resorption to occur. It is not known whether this is the only adhesion molecule that participates in this process; that will require further study. Recent data have begun to establish a connection between the interaction of the vitronectin receptor with matrix molecules and specific signals transduced within the cell. This area of research should shed considerable light on the mechanism of osteoclastic bone resorption. Another area of osteoclast biology that needs considerable exploration is the involvement of adhesion molecules in the fusion process required for the formation of multinucleated osteoclasts. There is considerable knowledge of cell fusion in other cell types, but no information is available for the osteoclast. This process is essential for active bone resorption, and may provide important information on the sequence of events, both molecular and cellular, in the development of the osteoclast. The area of adhesion molecules and bone biology promises to be an exciting area of research for many years to come.

References

Akisaka, T., Yamamoto, T., and Gay, C. V., 1988, Ultracytochemical investigation of calcium-activated adenosine triphosphatase (Ca^{++}-ATPase) in chick tibia, *J. Bone Miner. Res.* **3:**19–25.

Albelda, S. M., and Buck, C. A., 1990, Integrins and other cell adhesion molecules, *FASEB J.* **4:**2868–2880.

Athanasou, N. A., and Quinn, J., 1990, Immunophenotypic differences between osteoclasts and macrophage polykaryons—Immunohistological distinction and implications for osteoclast ontogeny and function, *J. Clin. Pathol.* **43:**997–1003.

Athanasou, N. A., Quinn, J., Horton, M. A., and McGee, J. O., 1990, New sites of cellular vitronectin receptor immunoreactivity detected with osteoclast-reacting monoclonal antibodies 13C2 and 23C6, *Bone Miner.* **8:**7–22.

Athanasou, N. A., Alvarez, J. I., Ross, F. P., Quinn, J. M., and Teitelbaum, S. L., 1992, Species differences in the immunophenotype of osteoclasts and mononuclear phagocytes, *Calcif. Tissue Int.* **50(5):**427–432.

Bertolini, D. R., Stadel, J., Samanen, J., Ali, F., Votta, B., Zembryki, D., and Greig, R., 1991, Inhibition of bone resorption by synthetic RGD containing peptides, *J. Bone Miner. Res.* **6:**S246.

Burridge, K., Fath, K., Kelley, T., Nuckolls, G., and Turner, C., 1988, Focal adhesions: Transmembrane junctions between the extracellular matrix and the cytoskeleton, *Annu. Rev. Cell Biol.* **4:**487–525.

Chambers, T. J., Fuller, K., Darby, J. A., Pringle, J. A. S., and Horton, M. A., 1986, Monoclonal antibodies against osteoclasts inhibit bone resorption in vitro, *Bone Miner.* **1:**127–135.

Chatterjee, D., Chakraborty, M., Leit, M., Neff, L., Jamsa-Kellokumpu, S., Fuchs, R., and Baron, R., 1992, Sensitivity to vanadate and isoforms of subunit-A and subunit-B distinguish the osteoclast proton pump from other vacuolar H^+ ATPases, *Proc. Natl. Acad. Sci. USA* **89:**6257–6261.

Clover, J., Dodds, R. A., and Gowen, M., 1992, Integrin subunit expression by human osteoblasts and osteoclasts in situ and in culture, *J. Cell Sci.* **103:**267–271.

Davies, J., Warwick, J., Totty, N., Philp, R., Helfrich, M., and Horton, M., 1989, The osteoclast functional antigen, implicated in the regulation of bone resorption, is biochemically related to the vitronectin receptor, *J. Cell Biol.* **109:**1817–1826.

Dedhar, S., 1989, Regulation of expression of the cell adhesion receptors, integrins, by recombinant human interleukin-1beta in human osteosarcoma cells: Inhibition of cell proliferation and stimulation of alkaline phosphatase activity, *J. Cell. Physiol.* **138:**291–299.

Dedhar, S., Ruoslahti, E., and Pierschbacher, M. D., 1987, A cell surface receptor complex for collagen type-I recognizes the Arg-Gly-Asp sequence, *J. Cell Biol.* **104:** 585–593.

DeRome, M., and Gronowicz, G., 1991, RGDS inhibits mineralization in vitro, *J. Bone Miner. Res.* **6:**S259.

Doty, S. B., 1981, Morphological evidence of gap junctions between bone cells, *Calcif. Tissue Int.* **33:**509–512.

Flores, M. E., Norgard, M., Heinegard, D., Reinholt, F. P., and Andersson, G., 1992, RGD-directed attachment of isolated rat osteoclasts to osteopontin, bone sialoprotein, and fibronectin, *Exp. Cell Res.* **201:**526–530.

Frost, H. M., 1964, Dynamics of bone remodeling, in: *Bone Dynamics* (H. M. Frost, ed), Little, Brown, Boston, pp. 315–333.

Frost, H. M., 1984, The origin and nature of transients in human bone remodeling, in: *Clinical Aspects of Metabolic Bone Disease* (B. Frame, A. M. Parfitt, and H. Duncan, eds), Excerpta Medica, Amsterdam, p. 124.

Glimcher, M. J., 1990, The nature of the mineral component of bone and the mechanism of calcification, in: *Metabolic Bone Disease* (L. V. Avioli and S. M. Krane, eds.), Saunders, Philadelphia, pp. 42–68.

Helfrich, M. H., Nesbitt, S. A., and Horton, M. A., 1992a, Integrins on rat osteoclasts: Characterization of two monoclonal antibodies (F4 and F11) to rat beta3, *J. Bone Miner. Res.* **7:**345–351.

Helfrich, M. H., Nesbitt, S. A., Dorey, E. L., and Horton, M. A., 1992b, Rat osteoclasts adhere to a wide range of RGD peptide-containing proteins, including the bone sialoprotein and fibronectin, via a beta3 integrin, *J. Bone Miner. Res.* **7:**335–343.

Horton, M., 1990, Vitronectin receptor: Tissue specific expression or adaptation to culture, *Int. J. Exp. Pathol.* **71:**741-759.

Horton, M. A., and Chambers, T. J., 1986, Human osteoclast-specific antigens are expressed by osteoclasts in a wide range of non-human species, *Br. J. Exp. Pathol.* **67:**95–104.

Horton, M. A., and Davies, J., 1989, Adhesion receptors in bone, *J. Bone Miner. Res.* **4:**803–807.

Horton, M. A., Taylor, M. L., Arnett, T. R., and Helfrich, M. H., 1991, Arg-Gly-Asp (RGD) peptides and the anti-vitronectin receptor antibody 23C6 inhibit dentine resorption and cell spreading by osteoclasts, *Exp. Cell Res.* **195:**368–375.

Hruska, K. A., Rolnick, F., and Huskey, M., 1992, Occupancy of the osteoclast $\alpha_v\beta_3$ integrin by osteopontin stimulates a novel SRC associated phosphatidylinositol 3 kinase resulting in phosphatidylinositol triphosphate formation, *J. Bone Miner. Res.* **7:**5106.

Hynes, R. O., 1992, Integrins—Versatility, modulation, and signaling in cell adhesion, *Cell* **69:**11–25.

Ingber, D. E., Prusty, D., Frangioni, J. V., Cragoe, E. J., Lechene, C., and Schwartz, M. A., 1990, Control of intracellular pH and growth by fibronectin in capillary endothelial cells, *J. Cell Biol.* **110:**1803–1811.

Kleinman, H. K., Klebe, R. J., and Martin, G. R., 1981, Role of collagenous matrices in the adhesion and growth of cells, *J. Cell Biol.* **88:**473–485.

Kornberg, L., and Juliano, L., 1992, Signal transduction from the extracellular matrix: The integrin–tyrosine kinase connection, *Trends Pharm. Sci.* **13:**93–95.

Lakkakorpi, P. T., Horton, M. A., Helfrich, M. H., Karhukorpi, E. K., and Vaananen, H. K., 1991, Vitronectin receptor has a role in bone resorption but does not mediate tight sealing zone attachment of osteoclasts to the bone surface, *J. Cell Biol.* **115:**1179–1186.

Majeska, R. J., and Einhorn, T. A., 1992, Expression of adhesion receptors by rat osteoblast-like cells in culture, *J. Bone and Miner. Res.* **7:**S209.

Malgaroli, A., Meldolesi, J., Zambonin-Zallone, A., and Teti, A., 1989, Control of cytosolic free calcium in rat and chicken osteoclasts, *J. Biol. Chem.* **264:**14342–14347.

Martin, T. J., Ng, K. W., and Nicholson, G. C., 1988, Cell biology of bone, *Balliere's Clinical Endocrinology and Metabolism* **2:**1–29.

Mimura, H., Chappel, J., Alverez, J., Farach-Carson, M., Butler, W., Gehron-Robey, P., Teitelbaum, S., and Ross, F., 1991, Antibodies to RGD-containing bone matrix proteins recognize them in whole bone and inhibit attachment to them by osteoclasts, *J. Bone Miner. Res.* **6:**s96.

Miyauchi, A., Hruska, K. A., Greenfield, E. M., Duncan, R., Alverez, J., Barattolo, R., Colucci, S., Zambonin-Zallone, A., Teitelbaum, S. L., and Teti, A., 1990, Osteoclast cytosolic calcium, regulated by voltage-gated calcium channels and extracellular calcium, controls podosome assembly and bone resorption, *J. Cell Biol.* **111:**2543–2552.

Miyauchi, A., Alverez, J., Greenfield, E. M., Teti, A., Grano, M., Colucci, S., Zambonin-Zallone, A., Ross, F. P., Teitelbaum, S. L., Cheresh, D., and Hruska, K. A., 1991, Recognition of osteopontin and related peptides by an alphavbeta3 integrin stimulates immediate cell signals in osteoclasts, *J. Biol. Chem.* **266:**20369–20374.

Neff, L., Horne, W. C., Male, P., Stadel, J. M., Samanen, J., Ali, F., Levy, J. B., and Baron, K., 1992, A cyclic RGD peptide induces a wave of tyrosine phosphorylation and the translocation of a C-SRC substrate in isolated rat osteoclasts, *J. Bone Miner. Res.* **7:**5106.

Oldberg, A., Franzen, A., and Heinegard, D., 1986, Cloning and sequence analysis of rat bone sialoprotein (osteopontin) cDNA reveals an Arg-Gly-Asp cell-binding sequence, *Proc. Natl. Acad. Sci. USA* **83:**8819–8823.

Oldberg, A., Franzen, A., and Heinegard, D., 1988, The primary structure of a cell-binding bone sialoprotein, *J. Biol. Chem.* **263:**19430–19432.

Pierschbacher, M. D., Hayman, E. G., and Ruoslahti, E., 1983, Synthetic peptide with cell attachment activity of fibronectin, *Proc. Natl. Acad. Sci. USA* **80:**1224–1227.

Puleo, D. A., and Bizios, R., 1991, RGDS tetrapeptide binds to osteoblasts and inhibits fibronectin-mediated adhesion, *Bone* **12:**271–276.

Quarles, L. D., Yohay, D. A., Lever, L. W., Caton, R., and Wenstrup, R. J., 1992, Distinct proliferative and differentiated stages of murine MC3T3-E1 cells in culture—An *in vitro* model of osteoblast development, *J. Bone Miner. Res.* **7:**683–692.

Quinn, J. M. W., Athanasou, N. A., and McGee, J. O., 1991, Extracellular matrix receptor and platelet antigens on osteoclasts and foreign body giant cells, *Histochemistry* **96:**169–176.

Reinholt, F. P., Hultenby, K., Oldberg, A., and Heinegard, D., 1990, Osteopontin—A possible anchor of osteoclasts to bone, *Proc. Natl. Acad. Sci. USA* **87:**4473–4475.

Robey, P. G., Young, M. F., Fisher, L. W., and McClain, T. D., 1989, Thrombospondin is an osteoblast-derived component of mineralized extracellular matrix, *J. Cell Biol.* **108:**719–727.

Robey, P. G., Bianco, P., and Termine, J. D., 1990, The cellular biology and molecular biochemistry of bone formation, in: *Disorders of Bone and Mineral Metabolism* (F. L. Coe and M. J. Favus, eds.), Raven Press, New York, pp. 241–263.

Rolnick, F., Huskey, M., Gupta, A., and Hruska, K. A., 1992, The signal generating complex of the occupied osteoblast $\alpha_v\beta_3$ integrin includes SRC, phosphatidyl inositol 3 kinase and phospholipase Cα, *J. Bone Miner. Res.* **7:**5105.

Ruoslahti, E., 1988, Fibronectin and its receptors, *Annu. Rev. Biochem.* **57:**375–413.

Ruoslahti, E., 1991, Integrins, *J. Clin. Invest.* **84:**1–5.

Sato, M., Sardana, M. K., Grasser, W. A., Garsky, V. M., Murray, J. M., and Gould, R. J., 1990, Echistatin is a potent inhibitor of bone resorption in culture, *J. Cell Biol.* **111:**1713–1723.

Simpson, A., and Horton, M. A., 1989, Expression of the vitronectin receptor during embryonic development: An immunohistological study of the ontogeny of the osteoclast in the rabbit, *Br. J. Exp. Pathol.* **70:**257–265.

Skerry, T. M., Bitensky, L. Chayen, J., and Lanyon, L. E., 1989, Early stain-related changes in enzyme activity in osteocytes following bone loading *in vivo*, *J. Bone Miner. Res.* **4:**783–788.

Takahashi, N., Akatsu, T., Sasaki, T., Nicholson, G. C., Moseley, J. M., Martin, T. J., and Suda, T., 1988, Induction of calcitonin receptors by 1,25-dihydroxyvitamin D3 in osteoclast-like multinucleated cells formed from mouse bone marrow cells, *Endocrinology* **123:**1504–1510.

Tan, K. B., and Koncsics, C., 1992, Assay for osteoblast adhesion mediated by RGD, *J. Bone Miner. Res.* **7:**S209.

Teti, A., Grano, M., Colucci, S., and Zambonin-Zallone, A., 1989, Immunolocalization of beta3 subunit of integrins in osteoclast membrane, *Boll. Soc. It. Biol. Sper.* **65:**1031–1037.

Vaes, G., 1988, Cellular biology and biochemical mechanisms of bone resorption, *Clin. Orthop.* **231:**239–271.

Vukicevic, S., Luyten, F. P., Kleinman, H. K., and Reddi, A. H., 1990, Differentiation of canalicular cell processes in bone cells by basement membrane matrix components: Regulation by discrete domains of laminin, *Cell* **63:**437–445.

Zaidi, M., Datta, H. K., Moonga, B. S., and Macintyre, I., 1990, Evidence that the action of calcitonin on rat osteoclasts is mediated by 2 G-proteins acting via separate post-receptor pathways, *J. Endocrinol.* **126:**473–481.

Zaidi, M., Kerby, J., Huang, C. L. H., Alam, A. S. M. T., Rathod, H., Chambers, T. J., and Moonga, B S., 1991, Divalent cations mimic the inhibitory effect of extracellular ionized calcium on bone resorption by isolated rat osteoclasts—Further evidence for a calcium receptor, *J. Cell. Physiol.* **149:**422–427.

Zambonin-Zallone, A., Teti, A., Grano, M., Rubinacci, A., Abbadini, M., Gaboli, M., and Marchisio, P. C., 1989, Immunocytochemical distribution of extracellular matrix receptors in human osteoclasts: A beta3 integrin is colocalized with vinculin and talin in the podosomes of osteoclastoma giant cells, *Exp. Cell Res.* **182:**645–652.

Zimmermann, B., Wachtel, H. C., Somogyi, H., Merker, H. J., and Bernimoulin, J. P., 1988, Bone formation by rat calvarial cells grown at high density in organoid culture, *Cell Differ. Dev.* **25:**145–154.

II

*BIOLOGICAL CONSEQUENCES
OF CELLULAR ADHESION*

8

Effects of Shear Stress on Leukocyte Adhesion

OMID ABBASSI, DAVID JONES, MICHELE MARISCALCO,
RODGER McEVER, L. V. McINTIRE, AND C. WAYNE SMITH

1. Neutrophil Adhesion under Conditions of Flow: CD18 Integrins are Insufficient

Neutrophils have been observed rolling along the luminal surface of endothelium in small venules (Atherton and Born, 1972, 1973; Fiebig *et al.*, 1991; Ley *et al.*, 1989, 1991a). In the early stages of acute inflammation this phenomenon is markedly increased (House and Lipowsky, 1987; Zimmerman and Granger, 1990; Kubes *et al.*, 1990; Hernandez *et al.*, 1987), and the rolling cells frequently stop, change shape, and emigrate into the surrounding tissue. Arfors *et al.* (1987) initially raised the possibility that the mechanisms accounting for the rolling phenomenon are different from those causing stationary adhesion and emigration. They observed, in a study on the effects of anti-CD18 monoclonal antibody 60.3 in a rabbit model of inflammation, that the systemic administration of this antibody prevented neutrophils from stopping and transmigrating, but the rolling behavior of these cells was apparently unaffected. Lawrence *et al.* (1990) used a parallel plate flow chamber *in vitro* to stimulate some of the forces

OMID ABBASSI and MICHELE MARISCALCO • Department of Pediatrics, Baylor College of Medicine, Houston, Texas 77030. *DAVID JONES and L. V. McINTIRE* • Department of Chemical Engineering, Rice University, Houston, Texas 77030. *RODGER McEVER* • Department of Medicine and Biochemistry, University of Oklahoma Health Sciences Center, Oklahoma City, Oklahoma 73104. *C. WAYNE SMITH* • Departments of Pediatrics and Microbiology and Immunology, Baylor College of Medicine, Houston, Texas 77030.
Cellular Adhesion: Molecular Definition to Therapeutic Potential, edited by Brian W. Metcalf, Barbara J. Dalton, and George Poste. Plenum Press, New York, 1994.

affecting neutrophil adhesion to endothelial cells, and assessed the ability of isolated neutrophils to adhere while flowing past a confluent monolayer of human umbilical vein endothelial cells (HUVEC). They demonstrated that in this flow system, few neutrophils interacted with the HUVEC monolayer even at wall shear stresses much lower than those predicted to occur *in vivo*. However, when the endothelial cells were stimulated with IL-1 for 3–4 hr prior to the flow experiment, a marked increase in adhesion was seen when the wall shear stresses were maintained below 3 dynes/cm^2. Much like the observations *in vivo*, two types of adhesion occurred: rolling and stationary. When the wall shear stress was approximately 2 dynes/cm^2, rolling neutrophils moved at less than 10 μm/sec. This slow rate was clearly an adhesive phenomenon, since cells that did not interact with the endothelial surface were flowing by at a rate of approximately 2000 μm/sec. As rolling cells stopped *in vitro*, they frequently changed shape and crawled beneath the monolayer, a well-known phenomenon under static culture conditions (Smith *et al.*, 1988). Abbassi *et al.* (1991) confirmed that these observations occur with canine neutrophils and canine jugular vein endothelial cell monolayers, and in studies by both Lawrence *et al.*, and Abbassi *et al.*, anti-CD18 monoclonal antibodies did not reduce the number of neutrophils rolling on the cytokine-stimulated monolayers at wall shear stresses of about 2 dynes/cm^2.

As Arfors *et al.* found *in vivo*, a pronounced effect of the anti-CD18 antibody can be demonstrated on the behavior of neutrophils *in vitro*. Transendothelial migration occurring under static conditions was almost completely blocked (Harlan, 1985; Smith *et al.*, 1988), and under conditions of flow, rolling neutrophils infrequently stopped on the apical surface of the endothelium (Lawrence *et al.*, 1990; Abbassi *et al.*, 1991), and the rate of rolling was significantly increased (Lawrence *et al.*, 1990; Abbassi *et al.*, 1991). Lawrence *et al.* (1990) also found that neutrophils from a patient with complete CD18 deficiency exhibited the same rolling phenomenon as normal cells, but the rate of rolling was greater than normal, rolling neutrophils did not stop, and transmigration failed to occur. Perry and Granger (1991) evaluated rolling velocities of neutrophils in cat mesenteric venules, and found that they increased following the systemic administration of anti-CD18 monoclonal antibody IB4.

Since neutrophil–endothelial adhesions under conditions of flow occurred *in vitro* only after cytokine activation of the endothelial monolayers, Lawrence *et al.* (1990) evaluated the possible role of intercellular adhesion molecule-1 (ICAM-1, CD54) (Rothlein *et al.*, 1986), a known ligand for CD18 integrins (Smith *et al.*, 1988, 1989b; Diamond *et al.*, 1990; Marlin and Springer, 1987) that is upregulated on endothelial cells following cytokine stimulation. In contrast to studies under static conditions *in vitro*, in which anti-ICAM-1 monoclonal antibody R6.5 inhibited neutrophil adhesion (Smith *et al.*, 1988), this anti-ICAM-1 antibody did not reduce the number of rolling neutrophils under conditions of flow. It did, however, inhibit transmigration under static or flow conditions. These studies indicate that both *in vitro* and *in vivo*, CD18 integrin–ICAM-1 interactions do not catch neutrophils as they flow past the endothelium. Rather, this adhesive mechanism is apparently necessary for events subsequent to the initial step of margination, when it influences the rate of rolling, the formation of stationary adhesions, and the process of transendothelial migration.

2. Selectins and Neutrophil Adhesion under Conditions of Flow

The first evidence that selectins (Bevilacqua *et al.*, 1991) could play an important role in neutrophil emigration came from the studies of Lewinsohn *et al.* (1987). They demonstrated that antibody MEL-14 (anti-L-selectin) inhibited the binding of neutrophils to high endothelial venules (HEV) *in vitro*, and when administered systemically significantly reduced (by 65%) the migration of fluorescence-labeled bone marrow neutrophils into endotoxin-treated Gelfoam sponges implanted subcutaneously in BALB/c mice. This antibody also reduced by 50% the accumulation of neutrophils in the inflamed peritoneal cavity. In both experiments, there was no evidence that MEL-14 produced a selective clearing of the treated neutrophils from the circulation.

Follow-up studies on the possible contributions of L-selectin to neutrophil emigration *in vivo* not only confirmed that MEL-14 had the ability to reduce peritoneal accumulation of neutrophils in mice, but also revealed that when neutrophils were activated *in vitro*, MEL-14 antigen was downregulated (Jutila *et al.*, 1989). The neutrophils with downregulated L-selectin exhibited a significantly reduced ability to localize in the inflamed peritoneal cavity, even though the activation process resulted in marked upregulation of CD11b/CD18 (Mac-1) to the cell surface. Kishimoto *et al.* (1989) proposed that the L-selectin was responsible for the initial margination of neutrophils at sites of inflammation, and that local chemotactic stimulation served to induce a transition from selectin-dependent adhesion to CD18-dependent adhesion by downregulating L-selectin and upregulating Mac-1.

2.1. Investigations of Neutrophil L-Selectin in Vitro

We have used two experimental settings *in vitro* to assess the contribution of L-selectin to neutrophil–endothelial adhesion under conditions of flow. The first utilizes human neutrophils and HUVEC monolayers in a parallel plate flow chamber (Lawrence *et al.*, 1990; Smith *et al.*, 1991); the second uses canine neutrophils and canine jugular vein endothelial cell (CJVEC) monolayers in a similar flow chamber (Abbassi *et al.*, 1991). In both experimental models, adhesion of neutrophils does not occur unless the endothelial cells have been stimulated. Then, rolling and stationary adhesions are observed to peak within 4 hr after exposing the endothelial cells to IL-1 or bacterial endotoxin (LPS), and can be seen most prominently at wall shear stresses below 3 dynes/cm². An important feature of this model is that optimum adhesion is seen only if the neutrophils are not activated before they contact the endothelial monolayer.

Monoclonal antibodies (MAbs) to L-selectin significantly reduced the number of neutrophils associating with the monolayer at wall shear stresses of about 2 dynes/cm² (Table I). MAbs against CD18 did not. The conclusion that L-selectin was necessary for optimum adhesion under flow was strengthened by the observation that chemotactic stimulation of neutrophils prior to their introduction in the flow chamber reduced adhesion to a level similar to that caused by anti-L-selectin MAbs. The stimulus conditions required attention to at least two important issues: first, L-selectin downregulation proceeds to a greater than 98% loss within 15 min following maximum stimulation at 37°C, and second, chemotactic stimulation of neutrophils induces a marked shape change in a high percentage of the neutrophils, characterized by extensive ruffling and the assumption of a bipolar configuration (Smith et al., 1979). This change would greatly alter the effects of shear forces on the cell's interaction with the endothelial monolayer, and thereby confound interpretation. However, when washed free of the stimulus, the shape change is reversible, and neutrophils assume a spherical shape much like that of unstimulated neutrophils. Stimulated and washed canine neutrophil adhesion was reduced by more than 66% following stimulation with zymosan-activated serum (Abbassi et al., 1991), and human neutrophil adhesion was reduced by more than 67% following stimulation with f-Met-Leu-Phe (fMLP) (Smith et al., 1991).

Additional studies on the contributions of L-selectin to neutrophil adhesion

Table I. L-Selectin and Neutrophil Adhesion under Conditions of Flow

Experimental conditions[a]	MAb[b]	Antigen	% inhibition
Canine PMN; CJVEC	CL2/6	L-selectin	61
	SL1	L-selectin	52–63
	R15.7	CD18	NS[c]
	SG10G7	Unknown	NS
Human PMN; HUVEC	DREG-56	L-selectin	54
	TS1/18	CD18	NS

[a]Unstimulated neutrophils (PMN) flowing past endothelial monolayers at a wall shear stress of around 2 dynes/cm². Canine jugular vein endothelial cell (CJVEC) monolayers were stimulated with bacterial endotoxin (LPS, 30 ng/ml) for 4 hr prior to being placed in the flow chamber; human umbilical vein endothelial cell (HUVEC) monolayers stimulated with IL-1β (10 U/ml) for 4 hr. PMN rolling at <50 μm/sec and stationary PMN were considered adherent, and the number of adherent PMN was determined after 10 min of flow.
[b]Monoclonal antibodies (MAbs) recognizing the cell surface antigens indicated were present throughout the flow experiment at 20 μg/ml. These MAbs saturate binding sites on neutrophils (10⁶/ml) at approximately 5 μg/ml. SG10G7 binds to canine neutrophils to an unknown antigen.
[c]NS indicates that the effect was not statistically significant.

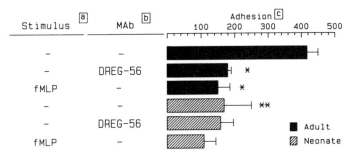

Figure 1. The contribution of L-selectin to the adhesion of neutrophils to cytokine-stimulated endothelial cell monolayers under conditions of flow *in vitro*. Human umbilical vein endothelial cell
(HUVEC) monolayers were prepared in parallel plate flow chambers as previously described (Smith
et al., 1991), following a 3-hr stimulation with IL-1β (10 U/ml, 37°C). (a) Isolated neutrophils from
adults or cord blood samples from healthy vaginal deliveries were suspended in either Dulbecco's
balanced salt solution with glucose (DPBS), or DPBS containing 10 nM f-Met-Leu-Phe (fMLP),
incubated at 37°C for 20 min, and then washed in DPBS to remove the fMLP. These cells were then
passed through the parallel plate flow chamber at a wall shear stress of approximately 2 dynes/cm^2.
(b) We have previously shown that monoclonal antibodies such as anti-CD18 do not significantly
reduce adhesion in this setting. Here we added either diluent (DPBS) or anti-L-selectin (DREG-56) at
a final concentration of 20 μg/ml prior to the infusion of neutrophils through the flow chamber.
(c) Adhesion was determined from the analysis of videotapes taken during the 10-min period of
infusion of neutrophils. The results indicate the number of neutrophils associated with the monolayer
during the last minute of observation; they include both stationary neutrophils and neutrophils rolling
at velocities of 50 μm/sec or less. The error bars are ± 1 S.D. *$p < 0.01$ compared with the control
adult samples (i.e., no fMLP or DREG-56), $n = 10$; **$p < 0.01$ compared with the control adult
samples, $n = 14$.

under flow were performed using human neonatal neutrophils. We found that
cord blood neutrophils and peripheral neutrophils from newborns within 48 hr of
delivery significantly reduced the levels of surface L-selectin, and their adhesion
to IL-1-stimulated HUVEC monolayers under conditions of flow was only 45%
of normal adult levels (Anderson *et al.*, 1991). The number of binding sites for
DREG-56 (anti-L-selectin) (Kishimoto *et al.*, 1990) on cord blood neutrophils
was $16 \pm 4 \times 10^3$; on adult neutrophils, the number was $31 \pm 2 \times 10^3$. As
shown in Fig. 1, the level of endothelial adhesion of cord blood neutrophils
under flow was significantly less than that of adult cells; in contrast to adult cell
adhesion, it was not significantly reduced by the anti-L-selectin MAb DREG-56
or by chemotactic stimulation.

2.2. Investigations of Neutrophil L-Selectin in Vivo

Our studies *in vitro* thus support the observations *in vivo* on the effects of
MAb MEL-14. A likely explanation for the anti-inflammatory effect of MEL-14

is that it inhibits the contribution of L-selectin to the initial phase of leukocyte localization. This concept has been directly addressed *in vivo* using intravital microscopy and antibodies against L-selectin. Ley *et al.* (1991b) and von Andrian *et al.* (1991, 1992) have shown directly that leukocyte rolling on mesenteric venules is inhibited by the administration of anti-L-selectin antibodies. Ley *et al.* (1991b) and Watson *et al.* (1991) extended these studies to include a soluble recombinant form of L-selectin that presumably inhibited leukocyte rolling and neutrophil influx into the inflamed peritoneum *in vivo* because it competed for a ligand on venular endothelial cells. von Andrian *et al.* (1992) extended their studies to show that activating neutrophils *in vitro* prior to injecting them into the mesenteric vessels markedly inhibited rolling; much like our studies *in vitro* (Smith *et al.*, 1991; Anderson *et al.*, 1991), they found that the degree of inhibition correlated quite well with the loss of L-selectin from the surface of the neutrophil following activation. This latter finding may have some bearing on a clinical observation in patients receiving GM-CSF (granulocyte–monocyte colony-stimulating factor), in which Peters *et al.* (1988) found that neutrophil migration into cutaneous inflammatory sites was dramatically diminished in patients receiving GM-CSF. Griffin *et al.* (1990) showed that GM-CSF stimulates neutrophils to downregulate L-selectin.

2.3. Investigations of E-Selectin in Vitro

Margination *in vitro* does not occur at venous wall shear stresses unless the endothelial cells have been stimulated (Lawrence *et al.*, 1990) under conditions that lead to upregulation of E-selectin (Bevilacqua *et al.*, 1989; Kishimoto *et al.*, 1991). Is it possible that E-selectin could play an important role in catching the neutrophil as it flows past, or that the need for endothelial stimulation is linked to the upregulation of a ligand for L-selectin? As discussed above, anti-L-selectin MAbs reduce neutrophil margination both *in vitro* (Smith *et al.*, 1991; Anderson *et al.*, 1991; Abbassi *et al.*, 1991) and *in vivo* (Ley *et al.*, 1991b; von Andrian *et al.*, 1991, 1992) by approximately 60%, leaving the impression that another unidentified adhesive mechanism contributes to the margination phenomenon.

In order to assess the potential contribution of E-selectin to human neutrophil adhesion under conditions of flow, we utilized murine L-cells transfected with human E-selectin (cDNA provided by Dr. B. Seed, Harvard University). E-selectin-expressing L-cells were sorted to obtain a cell line with a surface level of E-selectin comparable to that of IL-1-stimulated (4 hr, 37°C) HUVEC. These were grown to confluence and placed in flow chambers as previously described for HUVEC and canine endothelial cell monolayers (Smith *et al.*, 1991; Abbassi *et al.*, 1991). The adhesion of isolated human neutrophils at a wall shear stress of 1.85 dynes/cm^2 closely resembled that of IL-1-stimulated HUVEC monolayers (neutrophils adhering to E-selectin, 619 ± 61 cells/mm^2; to IL-1-stimulated

HUVEC, 657 ± 55 cells/mm², n = 16). Untransfected L-cell monolayers showed essentially no adhesiveness for isolated neutrophils (6.8 ± 1.8 cells/mm², n = 6). Pretreatment of the L-cells expressing E-selectin with monoclonal antibody CL2 [anti-E-selectin (Kishimoto *et al.*, 1991; Picker *et al.*, 1991)] reduced adhesion to a level close to that of untransfected L-cells (Fig. 2).

In additional experiments, L-cells expressing ICAM-1 at levels similar to those of HUVEC stimulated for 4 hr with IL-1 were grown in monolayers, and placed in the flow chambers. The adhesion of isolated human neutrophils at a wall shear stress of 1.85 dynes/cm² (20.2 cells/mm², n = 2) was only slightly higher than that to L-cells alone (Fig. 2). This finding supports previous work (Lawrence *et al.*, 1990; Smith *et al.*, 1989a) showing that ICAM-1 alone is insufficient to stop neutrophils that are flowing at this rate.

The behavior of neutrophils adhering to L-ELAM under flow was similar to that of neutrophils on IL-1-stimulated HUVEC, with some notable exceptions.

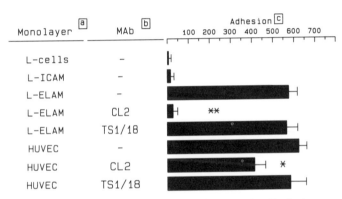

Figure 2. The adhesion of isolated neutrophils to L-cells expressing E-selectin compared with cytokine-stimulated HUVEC monolayers. (a) Monolayers of L-cells, L-cells transfected with human ICAM-1 cDNA (L-ICAM), L-cells transfected with E-selectin cDNA (L-ELAM), or HUVEC monolayers stimulated for 3 hr with IL-1β (10 U/ml) were mounted in parallel plate flow chambers as previously described (Smith *et al.*, 1991). The monolayers were washed with DPBS for 10 min at a wall shear stress of approximately 2 dynes/cm². (b) Isolated human neutrophils were suspended in DPBS with and without monoclonal antibodies CL2 [anti-human E-selectin, known to block the function of human E-selectin (Kishimoto *et al.*, 1991; Picker *et al.*, 1991)] or TS1/18 [anti-human CD18, known to block the functions of both CD11a/CD18 and CD11b/CD18 (Smith *et al.*, 1988, 1989b)]. The MAbs were maintained at a concentration of 20 μg/ml, a concentration shown to saturate all binding sites on the number of cells used in this experiment. The cells were then passed through the parallel plate flow chamber at a defined wall shear stress of approximately 2 dynes/cm². (c) Adhesion was determined from the analysis of videotapes taken during the 10-min period of infusion of neutrophils. The results indicate the number of neutrophils associated with the monolayer during the last minute of observation, and include both stationary neutrophils and neutrophils rolling at velocities of 50 μm/sec or less. **p < 0.01 compared with L-ELAM alone; *p < 0.01 compared with stimulated HUVEC alone.

As on IL-1-stimulated HUVEC, neutrophils were often rolling on the L-ELAM surface. When analyzed within a 5-sec time frame using video image subtraction techniques, 47.8 ± 6.0% of the cells were rolling at a mean rate of 10.6 ± 1.7 μm/sec, which was not significantly different from the results of using IL-1-activated HUVEC as the substratum. Thus, adhesive interactions of neutrophils with E-selectin expressed in a cell monolayer were of sufficient strength to partially withstand the drag force generated by the flowing fluid.

However, in clear contrast to the behavior of neutrophils on IL-1-stimulated HUVEC, stationary neutrophils on L-ELAM remained stationary for only a brief time, either resuming the rolling behavior or releasing into the flowing fluid. The shape change and transendothelial migration so prominent in neutrophils on cytokine-stimulated HUVEC were not seen. These observations indicate that simply binding to E-selectin is not sufficient to activate the motility needed for transmigration.

These results are consistent with the idea that E-selectin can support neu-trophil adhesion under flow, but they do not show that endothelial E-selectin does so. In fact, efforts to reduce neutrophil adhesion to IL-1-stimulated HUVEC by pretreating the monolayer with MAbs against E-selectin resulted in 32.2% inhi-bition ($n = 14$) with antibody CL2/6 (see Fig. 2) and 30.6% inhibition ($n = 2$) with antibody H18/7. Both CL2/6 (Kishimoto et al., 1991) and H18/7 (Lus-cinskas et al., 1991) have been shown to be effective in blocking E-selectin-dependent neutrophil adhesion.

2.4. Does E-Selectin Bind to Neutrophil L-selectin under Conditions of Flow?

Kishimoto et al. (1991) and Picker et al. (1991) have provided evidence that E-selectin binds neutrophil L-selectin. In contrast to lymphocyte L-selectin, neu-trophil L-selectin appears to be decorated with sLeX-like carbohydrates, and is predominantly displayed on projections from the surface of the neutrophil, where it would have some apparent advantage in the initial contacts between neu-trophils and stimulated endothelial cells. We addressed this question by attempt-ing to block neutrophil adhesion to L-ELAM in the parallel plate flow chamber by pretreating neutrophils with an anti-L-selectin antibody, DREG-56, previ-ously shown to inhibit neutrophil adhesion to cytokine-stimulated HUVEC monolayers. Adhesion was inhibited by approximately 70% (Fig. 3). The chemo-tactic stimulation of neutrophils under conditions that induced shedding of L-se-lectin also reduced the adhesion of neutrophils to L-ELAM monolayers. The residual adherence was not reduced by anti-CD18 monoclonal antibody R15.7, but was inhibited by anti-E-selectin, CL2/6. Thus, our current evidence does not contradict the conclusions of Kishimoto et al. (1991) and Picker et al. (1991) that neutrophil L-selectin interacts with E-selectin.

In order to carry this line of investigation further, we evaluated the ability of

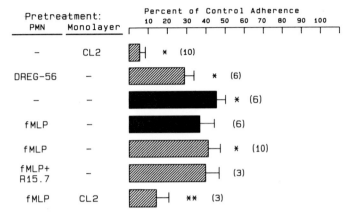

Figure 3. The effects of anti-L-selectin monoclonal antibody and chemotactic stimulation on the adhesion of neutrophils to E-selectin-expressing L-cell monolayers under conditions of flow. Monolayers of L-cells transfected with E-selectin cDNA were mounted in parallel plate flow chambers as previously described (Smith *et al.*, 1991) and washed with DPBS for 10 min at a wall shear stress of approximately 2 dynes/cm^2. Isolated neutrophils from adults or cord blood samples from healthy vaginal deliveries were suspended in either Dulbecco's balanced salt solution with glucose (DPBS) or DPBS containing 10 nM f-Met-Leu-Phe (fMLP), incubated at 37°C for 20 min, and washed in DPBS to remove the fMLP. The cells were then passed through the parallel plate flow chamber at a defined wall shear stress of approximately 2 dynes/cm^2. MAbs were present throughout the experiment at a concentration of 20 μg/ml. The results indicate the number of neutrophils associated with the monolayer during the last minute of observation and include both stationary neutrophils and neutrophils rolling at velocities of 50 μm/sec or less. DREG-56, anti-L-selectin; CL2, anti-human E-selectin; R15.7, anti-CD18 known to block the function of CD11a/CD18 and CD11b/CD18 (Entman *et al.*, 1990). $*p < 0.01$ compared with control cells without fMLP stimulation or MAbs; $**p < 0.01$ compared with the condition with fMLP stimulation only.

neonatal neutrophils to adhere to L-ELAM. These experiments were based on our previously published finding that neonatal neutrophils have significantly reduced levels of L-selectin. As shown in Fig. 3, isolated neonatal neutrophils adhere poorly under flow conditions to L-ELAM, and the adhesion that does occur is poorly blocked by anti-L-selectin, DREG-56. Figure 4 presents a plot of the levels of neutrophil L-selectin on the neonatal and adult neutrophils against levels of neutrophil adhesion to L-ELAM under flow. This correlation is highly significant: such results further support the hypothesis that E-selectin binds to neutrophil L-selectin and that this interaction is a significant determinant of neutrophil rolling under conditions of flow.

Our results are consistent with the concept that L- and E-selectin are both necessary for optimum rolling interactions of neutrophils with cytokine-stimulated endothelium. These interactions are apparently dependent on lectin–carbohydrate binding. It is very likely that L-selectin both presents sLeX for

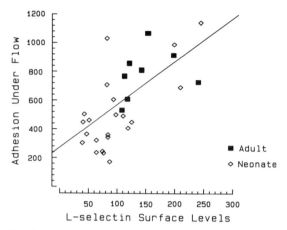

Figure 4. Correlation of neutrophil adhesion under flow to L cells expressing E-selectin with the surface levels of L-selectin on neutrophils. (The adhesion assay and analysis were the same as in Fig. 3.) Flow cytometry using FITC-labeled DREG-56 was employed to determine L-selectin surface levels on aliquots of neutrophils prior to being passed through the flow chamber. The correlation was highly significant, $p < 0.01$.

E-selectin binding (Picker *et al.*, 1991) and binds to an unidentified carbohydrate on the endothelial surface (Spertini *et al.*, 1991). In addition to binding to L-selectin, E-selectin appears to interact with other glycoproteins, such as CD66 (nonspecific cross-reacting antigens) (Kuijpers *et al.*, 1992a).

2.5. Investigations of E-Selectin in Vivo

Immunohistologic studies were performed by several investigators in a variety of inflammatory settings (Munro *et al.*, 1991; Mulligan *et al.*, 1991). There is now ample evidence to support the conclusion that E-selectin is expressed on endothelial cells located primarily on the venular side of the vascular bed in many acutely and chronically inflamed tissues. Current studies using monoclonal antibody CL3 (anti-human E-selectin that cross-reacts with rat E-selectin) were performed in several different inflammatory models. The results are entirely consistent with the expectation that blocking E-selectin function *in vivo* can produce a reduction in neutrophil emigration into tissue and a reduction in tissue injury. However, the presence of E-selectin in the inflamed tissue, as revealed by immunohistology, was not always an indication that anti-E-selectin would prevent tissue injury. In addition, $F(ab')_2$ fragments of the MAb were needed in order to produce a therapeutic effect, probably because Fc receptors on neutrophils provided a means for the attachment and activation of neutrophils on endothelial surfaces expressing E-selectin, thereby negating any beneficial effects of blocking E-selectin-dependent adherence.

2.6. P-Selectin Also Supports Neutrophil Rolling under Conditions of Flow

Lawrence and Springer (1991) have shown that isolated P-selectin inserted in a planar membrane *in vitro* supports neutrophil rolling at wall shear stresses below approximately 3 dynes/cm². We have evaluated the contribution of P-selectin to neutrophil rolling on thrombin-stimulated HUVEC monolayers. As shown in Fig. 5, the rolling adhesion of neutrophils was evident shortly after the introduction of thrombin into the flow chamber, and persisted for up to 30 min. The peak adhesion was seen within 10 min. Monoclonal antibody G1, anti-P-selectin, was almost completely inhibitory. Thus, it appears that both E- and P-selectin serve similar functions at the endothelial surface, with each adhesion molecule being upregulated by different stimuli. P-selectin is rapidly mobilized from Weibel–Palade bodies (Hattori *et al.*, 1989) to the endothelial cell surface following stimulation with histamine, bradykinin, or thrombin. This is in contrast to E-selectin, in which IL-1 and TNF induce synthesis and cell surface increases within 2–3 hr (Pober *et al.*, 1986; Bevilacqua *et al.*, 1989).

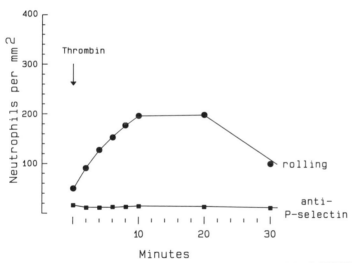

Figure 5. Rolling adhesion of neutrophils on human umbilical vein endothelial cell (HUVEC) monolayers following stimulation with thrombin. HUVEC monolayers were prepared in parallel plate flow chambers as previously described (Smith *et al.*, 1991). Isolated human neutrophils were suspended in DPBS and passed through the chamber at a wall shear stress of 1.85 dynes/cm². After 2 min of flow, thrombin (10 units/ml) was introduced into the chamber and the interaction of neutrophils with the monolayer was recorded on videotape. Representative data are shown as the number of rolling neutrophils per square millimeter of monolayer surface area. Only a small fraction of the neutrophils were stationary on the monolayer. The anti-P-selectin MAb was G1, which was previously shown to block neutrophil–endothelial adhesion under static conditions (Geng *et al.*, 1990; Patel *et al.*, 1991), and it was present throughout the observation period at a concentration of 2 μg/ml.

3. Three-Step Model of Neutrophil Extravasation

Our studies *in vitro* are consistent with the hypothesis that there are three distinct phases in the process of neutrophil emigration at sites of inflammation. The initial event involves the slowing of previously unstimulated neuterophils as they encounter activated endothelial cells. P-selectin mobilized to the endothelial surface appears to be the earliest adhesive mechanism, and is clearly evident within 30 sec to 1 min after stimulation with, for example, histamine. E-selectin expression is evident within 2 hr of cytokine stimulation, as is an unidentified ligand for L-selectin. The velocity of rolling is influenced by CD18 integrins, although these integrins are not sufficient to initiate rolling unaided.

As neutrophils stop on the endothelial surface, they become activated. There appears to be a coordinated loss of L-selectin, a reduced ability to adhere to E-selectin, an increased ability to adhere via CD18 integrins [primarily CD11a/CD18 and CD11b/CD18 (Smith *et al.*, 1989b)] and ICAM-1 (Smith *et al.*, 1988, 1989a), and a change in cell shape from a round to a motile configuration. The mechanisms accounting for this activation are poorly defined, but at least two distinct types of stimuli have some experimental support. First, the avidity of CD18 integrins may be increased by binding P-selectin (Lorant *et al.*, 1991) or E-selectin (Lo *et al.*, 1991; Kuijpers *et al.*, 1991) to surface glycoproteins [e.g., CD66 (Kuijpers *et al.*, 1992a)]. Our results do not indicate that E-selectin binding is sufficient to activate neutrophil motility, but results of others suggest that it may enhance CD11b/CD18 avidity. The second class of stimuli is chemokinetic. Stimulated endothelial cells present chemotactic factors such as platelet activating factor (PAF) (Lorant *et al.*, 1991) or IL-8 (Gimbrone *et al.*, 1989) on their surface. Chemokinetic stimulation increases the adhesion that is dependent on CD18 integrins (Smith, 1992), induces the shedding of L-selectin (Kishimoto *et al.*, 1989), and induces the shape change necessary for cell locomotion (Smith *et al.*, 1979). This phase in the sequence of events leading to neutrophil localization is particularly resistant to shear forces of flowing blood. Neutrophils adherent via CD18 integrins are quite shear resistant (Lawrence and Springer, 1991), and the activated neutrophils spread on the endothelial surface, thereby lessening the influence of fluid drag forces.

The third phase is transendothelial migration. Current evidence indicates that this process requires CD18 integrins CD11a/CD18 and CD11b/CD18 (Smith *et al.*, 1989b; Furie *et al.*, 1991b), utilizing ICAM-1 (Smith *et al.*, 1988; Furie *et al.*, 1991b) and another endothelial determinant yet to be defined (Furie *et al.*, 1991b). Conflicting data have been published regarding a direct role for L-selectin (Spertini *et al.*, 1991; Abbassi *et al.*, 1991; Smith *et al.*, 1991) and E-selectin (Luscinskas *et al.*, 1991; Kishimoto *et al.*, 1991; Furie *et al.*, 1991a). Our results fail to support a direct role for selectins in transmigration (Abbassi *et al.*, 1991; Smith *et al.*, 1991; Kishimoto *et al.*, 1991). The stimuli directly activating neutrophil locomotion are not completely defined, but current evidence supports a role for IL-8 (Huber *et al.*, 1991; Kuijpers *et al.*, 1992b), and possibly PAF (Kuijpers *et al.*, 1992b).

Thus, the hypothesis initially put forward by Kishimoto *et al.* (1989) that effective neutrophil extravasation involves a transition from L-selectin-dependent adhesion to CD18-dependent adhesion has been supported by much of the published data. Although many details of the mechanisms involved remain unsolved, the general concept appears sound.

4. Summary

Neutrophil–endothelial cell adhesion occurs *in vivo* in the context of fluid flow. A model for investigating the effects of wall shear stress on neutrophil adhesion *in vitro* has been developed using a parallel plate flow chamber containing monolayers of cultured endothelial cells over which isolated neutrophils are passed at defined shear rates. It has been determined, using various combinations of unstimulated or activated neutrophils, and unstimulated or activated endothelial monolayers, that stimulation of the endothelial cells is required for significant increases in adhesion at wall shear stresses of ≈ 2 dynes/cm^2. Abundant rolling-type adhesions occur when unstimulated neutrophils are passed over cytokine-stimulated endothelial cells, and this adhesion is most clearly dependent on neutrophil L-selectin and endothelial E-selectin. Rolling-type adhesions are also prevalent on histamine- or thrombin-stimulated endothelial cells; this adhesion is most clearly dependent on both L- and P-selectin. Neither CD18 integrins nor ICAM-1 appear to be capable of catching flowing neutrophils, but once neutrophils are rolling on the endothelial surface as a result of selectin-dependent mechanisms, CD18 integrins and ICAM-1 appear to function as accessory adhesion molecules capable of stopping the rolling cells. This latter function appears to require, or is greatly augmented by, a chemokinetic stimulus to the neutrophil at the endothelial surface. Thus, optimal neutrophil–endothelial adhesion under conditions of flow requires a coordinated sequence of events that apparently begins with the activation of the endothelial cell.

References

Abbassi, O., Lane, C. L., Krater, S. S., Kishimoto, T. K., Anderson, D. C., McIntire, L. V., and Smith, C. W., 1991, Canine neutrophil margination mediated by lectin adhesion molecule-1 (LECAM-1) *in vitro, J. Immunol.* **147:**2107–2115.

Anderson, D. C., Abbassi, O., Kishimoto, T. K., Koenig, J. M., McIntire, L. V., and Smith, C. W., 1991, Diminished lectin-, epidermal growth factor-, complement binding domain cell adhesion molecule-1 on neonatal neutrophils underlies their impaired CD18-independent adhesion to endothelial cells in vitro, *J. Immunol.* **146:**3372–3379.

Arfors, K. E., Lundberg, C., Lindbom, L., Lundberg, K., Beatty, P. G., and Harlan, J. M., 1987, A monoclonal antibody to the membrane glycoprotein complex CD18 inhibits polymorphonuclear leukocyte accumulation and plasma leakage *in vivo, Blood* **69**:338–340.

Atherton, A., and Born, G. V. R., 1972, Quantitative investigations of the adhesiveness of circulating polymorphonuclear leukocytes to blood vessel walls, *J. Physiol. (London)* **222**:447–474.

Atherton, A., and Born, G. V. R., 1973, Relationship between the velocity of rolling granulocytes and that of the blood flow in venules, *J. Physiol. (London)* **233**:157–165.

Bevilacqua, M. P., Stengelin, S., Gimbrone, M. A. Jr., and Seed, B., 1989, Endothelial leukocyte adhesion molecule 1: An inducible receptor for neutrophils related to complement regulatory proteins and lectins, *Science* **243**:1160–1165.

Bevilacqua, M. P., Butcher, E., Furie, B., Gallatin, M., Gimbrone, M. A., Harlan, J. M., Kishimoto, T. K., Lasky, L. A. McEver, R. P., Paulson, J. C., Rosen, S. D., Seed, B., Siegelman, M., Springer, T. A., Stoolman, L. M., Tedder, T. F., Varki, A., Wagner, D. D., Weissman, I. L., and Zimmerman, G. A., 1991, Selectins: A family of adhesion receptors, *Cell* **67**:233.

Diamond, M. S., Staunton, D. E., deFougerolles, A. R., Stacker, S. A., Garcia-Aguilar, J., Hibbs, M. L., and Springer, T. A., 1990, ICAM-1 (CD54): A counter-receptor for Mac-1 (CD11b/CD18), *J. Cell Biol.* **111**:3129–3139.

Entman, M. L., Youker, K., Shappell, S. B., Siegel, C., Rothlein, R., Dreyer, W. J., Schmalstieg, F. C., and Smith, C. W., 1990, Neutrophil adherence to isolated adult canine myocytes: Evidence for a CD18-dependent mechanism, *J. Clin. Invest.* **85**:1497–1506.

Fiebig, E., Ley, K., and Arfors, K.-E., 1991, Rapid leukocyte accumulation by "spontaneous" rolling and adhesion in the exteriorized rabbit mesentery, *Int. J. Microcirc. Clin. Exp.* **10**:127–144.

Furie, M. B., Burns, M. J., Tancinco, M. C. A., Benjamin, C. D., and Lobb, R. R., 1991a, Endothelial-leukocyte adhesion molecule-1 is not required for the migration of neutrophils across IL-1-stimulated endothelium *in vitro, J. Immunol.* **148**:2395–2404.

Furie, M. B., Tancinco, M. C. A., and Smith, C. W., 1991b, Monoclonal antibodies to leukocyte integrins CD11a/CD18 and CD11b/CD18 or intercellular adhesion molecule-1 (ICAM-1) inhibit chemoattractant-stimulated neutrophil transendothelial migration *in vitro, Blood* **78**:2089–2097.

Geng, J. G., Bevilacqua, M. P., Moore, K. L., McIntyre, T. M., Prescott, S. M., Kim, J. M., Bliss G. A., Zimmerman, G. A., and McEver, R. P., 1990, Rapid neutrophil adhesion to activated endothelium mediated by GMP-140, *Nature* **343**:757–760.

Gimbrone, M. A., Jr., Obin, M. S., Brock, A. F., Luis, E. A., Hass, P. E., Hebert, C. A., Yip, Y. K., Leung, D. W., Lowe, D. G., Kohr, W. J., Darbonne, W. C., Bechtol, K. B., and Baker, J. B., 1989, Endothelial interleukin-8: A novel inhibitor of leukocyte–endothelial interactions, *Science* **246**:1601–1603.

Griffin, J. D., Spertini, O., Ernst, T. J. Belvin, M. P., Levine, H. B., Kanakura, Y., and Tedder, T. F., 1990, Granulocyte–macrophage colony-stimulating factor and other cytokines regulate surface expression of the leukocyte adhesion molecule-1 on human neutrophils, monocytes, and their precursors, *J. Immunol.* **145**:576–584.

Harlan, J. M., 1985, Leukocyte–endothelial interactions, *Blood* **65**(3):513–525.

Hattori, R., Hamilton, K. K., Fugates, R. D., McEver, R. P., and Sims, P. J., 1989, Stimulated secretion of endothelial von Willebrand factor is accompanied by rapid redistribution to the cell surface of the intracellular granule membrane protein GMP-140, *J. Biol. Chem.* **264**:7768–7771.

Hernandez, L. A., Grisham, M. B., Twohig, B., Arfors, K. E., Harlan, J. M., and Granger D. N., 1987, Role of neutrophils in ischemia-reperfusion-induced microvascular injury, *Am. J. Physiol.* **238**:H699–H703.

House, S. D., and Lipowsky, H. H., 1987, Leukocyte-endothelium adhesion: Microhemodynamics in mesentery of the cat, *Microvasc. Res.* **34:**363–379.

Huber, A. R., Kunkel, S. L., Todd, R. F., III, and Weiss, S. J., 1991, Regulation of transendothelial neutrophil migration by endogenous interleukin-8, *Science* **254:**99–105.

Jutila, M. A., Rott, L., Berg, E. L., and Butcher, E. C., 1989, Function and regulation of the neutrophil MEL-14 antigen *in vivo:* Comparison with LFA-1 and MAC-1, *J. Immunol.* **143:**3318–3324.

Kishimoto, T. K., Jutila, M. A., Berg, E. L., and Butcher, E. C., 1989, Neutrophil Mac-1 and MEL-14 adhesion proteins inversely regulated by chemotactic factors, *Science* **245:**1238–1241.

Kishimoto, T. K., Jutila, M. A., and Butcher, E. C., 1990, Identification of a human peripheral lymph node homing receptor: A rapidly down-regulated adhesion molecule, *Proc. Natl. Acad. Sci. USA* **87:**2244–2248.

Kishimoto, T. K., Warnock, R. A., Jutila, M. A., Butcher, E. C., Lane, C. L., Anderson, D. C., and Smith, C. W., 1991, Antibodies against human neutrophil LECAM-1 (LAM-1/Leu-8/DREG-56 antigen) and endothelial cell ELAM-1 inhibit a common CD18-independent adhesion pathway *in vitro, Blood* **78:**805–811.

Kubes, P., Ibbotson, G., Russell, J., Wallace, J. L., and Granger, D. N., 1990, Role of platelet-activating factor in ischemia/reperfusion-induced leukocyte adherence, *Am. J. Physiol.* **259:**G300–G305.

Kuijpers, T. W., Hakkert, B. C., Hoogerwerf, M., Leeuwenberg, J. F. M., and Roos, D., 1991, Role of endothelial leukocyte adhesion molecule-1 and platelet-activating factor in neutrophil adherence to IL-1-prestimulated endothelial cells. Endothelial leukocyte adhesion molecule-1-mediated CD18 activation, *J. Immunol.* **147:**1369–1376.

Kuijpers, T., Hoogerwerf, M., van der Laan, L., Nagel, G., van der Schoot, C., Grunert, F., and Roos, D., 1992a, CD66 nonspecific cross-reacting antigens are involved in neutrophil adherence to cytokine-activated endothelial cells, *J. Cell Biol.* **118:**457–466.

Kuijpers, T. W., Hakkert, B. C., Hart, M. H. L., and Roos, D., 1992b, Neutrophil migration across monolayers of cytokine-prestimulated endothelial cells: A role for platelet-activating factor and IL-8, *J. Cell Biol.* **117:**565–572.

Lawrence, M. B., and Springer, T. A., 1991, Leukocytes roll on a selectin at physiologic flow rates: Distinction from and prerequisite for adhesion through integrins, *Cell* **65:**1–20.

Lawrence, M. B., Smith, C. W., Eskin, S. G., and McIntire, L. V., 1990, Effect of venous shear stress on CD18-mediated neutrophil adhesion to cultured endothelium, *Blood* **75:**227–237.

Lewinsohn, D. M., Bargatze, R. F., and Butcher, E. C., 1987, Leukocyte–endothelial cell recognition: Evidence of common molecular mechanism shared by neutrophils, lymphocytes, and other leukocytes, *J. Immunol.* **138:**4313–4321.

Ley, K., Lundgren, E., Berger, E., and Arfors, K. E., 1989, Shear-dependent inhibition of granulocyte adhesion to cultured endothelium by dextran sulfate, *Blood* **73:**1324–1330.

Ley, K., Cerrito, M., and Arfors, K. E., 1991a, Sulfated polysaccharides inhibit leukocyte rolling in rabbit mesentery venules, *Am. J. Physiol.* **260:**H1667–H1673.

Ley, K., Gaehtgens, P., Fennie, C., Singer, M. S., Lasky, L. A., and Rosen, S. D., 1991b, Lectin-like cell adhesion molecule 1 mediates leukocyte rolling in mesenteric venules *in vivo, Blood* **77:**2553–2555.

Lo, S. K., Lee, S., Ramos, R. A., Lobb, R., Rosa, M., Chi-Rosso, G., and Wright, S. D., 1991, Endothelial-leukocyte adhesion molecule 1 stimulates the adhesive activity of leukocyte integrin CR3 (CD11b/CD18, Mac-1, alpha m beta 2) on human neutrophils, *J. Exp. Med.* **173:**1493–1500.

Lorant, D. E., Patel, K. D., McIntyre, T. M., McEver, R. P., Prescott, S. M., and Zimmerman, G. A., 1991, Coexpression of GMP-140 and PAF by endothelium stimulated by histamine or thrombin: A juxtacrine system for adhesion and activation of neutrophils, *J. Cell Biol.* **115**:223–234.

Luscinskas, F. W., Cybulsky, M. I., Kiely, J.-M., Peckins, C. S., Davis, V. M., and Gimrone, M. A., 1991, Cytokine-activated human endothelial monolayers support enhanced neutrophil transmigration via a mechanism involving both endothelial-leukocyte adhesion molecule-1 and intercellular adhesion molecule-1, *J. Immunol.* **146**:1617–1625.

Marlin, S. D., and Springer, T. A., 1987, Purified intercellular adhesion molecule-1 (ICAM-1) is a ligand for lymphocyte function-associated antigen 1 (LFA-1), *Cell* **51**:813–819.

Mulligan, M. S., Varani, J., Dame, M. K., Lane, C. L., Smith, C. W., Anderson, D. C., and Ward, P. A., 1991, Role of endothelial-leukocyte adhesion molecule 1 (ELAM-1) in neutrophil-mediated lung injury in rats, *J. Clin. Invest.* **88**:1396–1406.

Munro, J. M., Pober, J. S., and Cotran, R. S., 1991, Recruitment of neutrophils in the local endotoxin response: Association with *de novo* endothelial expression of endothelial leukocyte adhesion molecule-1, *Lab. Invest.* **64**:295–299.

Patel, K. D., Zimmerman, G. A., Prescott, S. M., McEver, R. P., and McIntyre, T. M., 1991, Oxygen radicals induce human endothelial cells to express GMP-140 and bind neutrophils, *J. Cell Biol.* **112**:749–759.

Perry, M. A., and Granger, D. N., 1991, Role of CD11/CD18 in shear rate-dependent leukocyte–endothelial cell interactions in cat mesenteric venules, *J. Clin. Invest.* **87**:1798–1804.

Peters, W. P., Stuart, A., Affronti, M. L., Kim, C. S., and Coleman, R. E., 1988, Neutrophil migration is defective during recombinant human granulocyte–macrophage colony-stimulating factor infusion after autologous bone marrow transplantation in humans, *Blood* **72**:1310.

Picker, L. J., Warnock, R. A., Burns, A. R., Doerschuk, C. M., Berg, E. L., and Butcher, E. C., 1991, The neutrophil selectin LECAM-1 presents carbohydrate ligands to the vascular selectins ELAM-1 and GMP-140, *Cell* **66**:921–933.

Pober, J. S., Gimbrone, M. A., Jr., Lapierre, L. A., Mendrick, D. L., Fiers, W., Rothlein, R., and Springer, T. A., 1986, Overlapping patterns of antigenic modulation by interleukin 1, tumor necrosis factor and immune interferon, *J. Immunol.* **137**:1893–1896.

Rothlein, R., Dustin, M. L., Marlin, S. D., and Springer, T. A., 1986, A human intercellular adhesion molecule (ICAM-1) distinct from LFA-1, *J. Immunol.* **137**:1270–1274.

Smith, C. W., 1992, Transendothelial migration, in: *Adhesion: Its Role in Inflammatory Disease* (J. M. Harland and D. Y. Liu, eds.), Freeman, San Francisco, pp. 85–115.

Smith, C. W., Hollers, J. C., Patrick, R. A., and Hassett, C., 1979, Motility and adhesiveness in human neutrophils: Effects of chemotactic factors, *J. Clin. Invest.* **63**:221–229.

Smith, C. W., Rothlein, R., Hughes, B. J., Mariscalco, M. M., Schmalstieg, F. C., and Anderson, D. C., 1988, Recognition of an endothelial determinant for CD18-dependent human neutrophil adherence and transendothelial migration, *J. Clin. Invest.* **82**:1746–1756.

Smith, C. W., Marlin, S. D., Rothlein, R., Lawrence, M. B., McIntire, L. V., and Anderson, D. C., 1989a, Role of ICAM-1 in the adherence of human neutrophils to human endothelial cells *in vitro*, in: *Leukocyte Adhesion Molecules: Structure, Function, and Regulation* (T. A. Springer, D. C. Anderson, A. S. Rosenthal, and R. Rothlein, eds.), Springer-Verlag, Berlin, pp. 170–189.

Smith, C. W., Marlin,m S. C., Rothlein, R., Toman, C., and Anderson, D. C., 1989b, Cooperative interactions of LFA-1 and Mac-1 with intercellular adhesion molecule-1 in facilitating adherence and transendothelial migration of human neutrophils *in vitro*, *J. Clin. Invest.* **83**:2008–2017.

Smith, C. W., Kishimoto, T. K., Abbassi, O., Hughes, B. J., Rothlein, R., McIntire, L. V., Butcher, E., and Anderson, D. C., 1991, Chemotactic factors regulate lectin adhesion molecule 1 (LECAM-1)-dependent neutrophil adhesion to cytokine-stimulated endothelial cells *in vitro*, *J. Clin. Invest.* **87**:609–618.

Spertini, O., Luscinskas, F. W., Kansas, G. S., Munro, J. M., Griffin, J. D., Gimbrone, M. A., and

Tedder, T. F., 1991, Leukocyte adhesion molecule-1 (LAM-1, L-selectin) interacts with an inducible endothelial cell ligand to support leukocyte adhesion, *J. Immunol.* **147:**2565–2573.

von Andrian, U. H ., Chambers, J. D., McEvoy, L. M., Bargatze, R. F., Arfors, K.-E., and Butcher, E. C., 1991, Two step model of leukocyte–endothelial cell interaction in inflammation: Distinct roles for LECAM-1 and the leukocyte beta-2 integrins *in vivo, Proc. Natl. Acad. Sci. USA* **88:**7538–7542.

von Andrian, U. H., Hansell, P., Chambers, J. D., Berger, E. M., Filho, I. T., Butcher, E. C., and Arfors, K.-E., 1992, L-selectin function is required for beta-2 integrin-mediated neutrophil adhesion at physiologic shear rates *in vivo, Am. J. Physiol.* **263:**H1034–H1040.

Watson, S. R., Fennie, C., and Lasky, L. A., 1991, Neutrophil influx into an inflammatory site inhibited by a soluble homing receptor-IgG chimaera, *Nature* **349:**164–167.

Zimmerman, B. J., and Granger, D. N., 1990, Reperfusion-induced leukocyte infiltration: Role of elastase, *Am. J. Physiol.* **259:**H390–H394.

9

Blockade of Leukocyte Adhesion in in Vivo Models of Inflammation

LYNNE FLAHERTY, JOHN M. HARLAN, and ROBERT K. WINN

1. Introduction

The study of leukocyte–endothelial interaction has become one of the most active areas in biology in the past decade, as the molecules involved in leukocyte adherence to endothelium have been identified. Three families of adhesion molecules that participate in leukocyte–endothelial interactions have been identified: integrin receptors, selectin receptors, and members of the immunoglobulin supergene family. There are two subfamilies of the integrins that participate in leukocyte–endothelial cell adherence, the β_1 and β_2 integrin families. Only one member of the β_1 integrin subfamily, VLA-4 (CD49d/CD29), is involved in leukocyte–endothelial cell adherence. VLA-4 is found on mononuclear cells and eosinophils, but not on neutrophils. The β_2 subfamily consists of three receptors designated LFA-1 (CD11a/CD18), Mac-1 (CD11b/CD18), and p150,95 (CD11c/CD18). One or more of the β_2 integrin receptors are found on all leukocytes. The known counterstructures for these integrin receptors are members of the immunoglobulin supergene family. CD11a/CD18 recognizes intercellular adhesion molecules (ICAM) 1 and 2. ICAM-1 is also a ligand for CD11b/CD18. Little is known about the role that CD11c/CD18 plays in leuko-

LYNNE FLAHERTY • Department of Surgery, Harborview Medical Center, University of Washington, Seattle, Washington 98104. *JOHN M. HARLAN* • Department of Medicine, Harborview Medical Center, University of Washington, Seattle, Washington 98104. *ROBERT K. WINN* • Department of Surgery and Physiology–Biophysics, Harborview Medical Center, University of Washington, Seattle, Washington 98104.

Cellular Adhesion: Molecular Definition to Therapeutic Potential, edited by Brian W. Metcalf, Barbara J. Dalton, and George Poste. Plenum Press, New York, 1994.

cyte adherence to endothelium, and its counterstructure has not been identified. The selectin family consists of L-selectin (LAM-1, LECAM-1, Leu 8), P-selectin (GMP-140, PAPGEM, CD62), and E-selectin (ELAM-1). L-selectin is found on leukocytes, while P- and E-selectin are found on endothelial cells. The selectin receptors are lectinlike proteins that bind to carbohydrate counterstructures, particularly sialyl Lewis X (SLe^x).

Figure 1 shows a schematic representation of these adhesion molecules and their counterstructures. Each has been characterized by *in vitro* studies to deter-

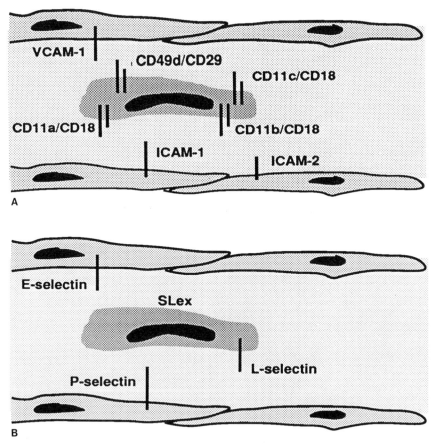

Figure 1. Schematic representation of leukocyte–endothelial cell adhesion molecules. Panel A shows the members of the integrin families (CD11/CD18 and CD49d/CD29) that are found on the leukocytes and their counterstructures in the immunoglobulin superfamily (ICAM-1 and VCAM-1). Panel B is a schematic representation of members of the selectin family and one known ligand, SLe^x. L-selectin is found on leukocytes; P- and E-selectin are found on endothelial cells.

mine their distribution, regulation, and response to stimuli such as endotoxin, complement, and cytokines. Monoclonal antibodies (MAbs) have been developed to each molecule, and they have been used to evaluate leukocyte interactions with cultured endothelial cells in response to various stimuli. However, the complexity of the inflammatory system and the multiplicity of interactions possible within the whole organism require that these *in vitro* findings be confirmed *in vivo*. This is especially true when therapeutic applications are contemplated. These *in vivo* experiments are often complicated, usually expensive, sometimes frustrating, and occasionally inconclusive. Nevertheless, these studies have yielded important information on the role of adhesion molecules in inflammatory and immune reactions in diverse models ranging from chemically produced inflammation to ischemia–reperfusion injury.

The normal function of inflammatory cells is to eliminate invading pathogens and to remove debris following tissue injury. Under some circumstances, however, this response may actually exacerbate the injury as phagocytes release their arsenal of microbicidal chemicals extracellularly. The recruitment of phagocytes is initiated by stimuli that elicit a cascade of events characteristic of all inflammatory reactions. Interruption of this cascade at any point may prevent phagocyte-induced vascular or tissue injury, but may also render the host incapable of an effective response to pathogens. For this reason, studies demonstrating the beneficial effects of the inhibition of phagocytes must be considered in the context of the potential deleterious effects on host defense.

The cascade of events leading to extravascular inflammation begins with an initial stimulus that causes phagocytes to slow and roll along the endothelium. This is followed by firm adherence to endothelial cells, migration over the surface of the endothelial cell to an intercellular junction, diapedesis through the junction, and then migration through the interstitium to the site of inflammation (Smith, 1992). The intercellular interactions in this process involve adhesion molecules. The initial slowing and rolling occurs as a result of the interaction of selectin receptors with their carbohydrate ligands, while the firm adherence and transendothelial migration are dependent on integrin receptors bonding to endothelial ligands (reviewed by Smith, 1992). The adherence interactions involved in this sequence of events are shown in schematic form in Fig. 2.

Inflammation is provoked by diverse exogenous or endogenous stimuli ranging from bacterial organisms to ultraviolet radiation to immune complex deposition. It has a characteristic temporal sequence of events, and each step exhibits distinct functions. It is beyond the scope of this chapter to consider all of the mechanisms involved in leukocyte recruitment in inflammation (e.g., lipid or peptide chemoattractants, cytokines or chemoattractant receptors, cytoskeleton and motility). Therefore, we will focus on the role of adhesion molecules in a number of *in vivo* models. Particular emphasis will be placed on those studies employing monoclonal antibodies that block adherence and aggregation.

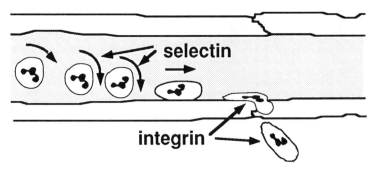

Figure 2. Stages of leukocyte emigration, from leukocyte rolling to leukocyte adherence and emigration. Leukocyte rolling is thought to be caused by selectin adhesion molecules, with adherence and emigration resulting from integrin molecules.

2. Chemical Inflammation

The intradermal or subcutaneous injection of inflammatory stimuli is a convenient method for eliciting neutrophil emigration *in vivo*. This approach has the advantages of allowing the rapid and simple administration of chemoattractants, as well as permitting multiple tests to be made on the same animal. Neutrophil emigration may be assayed by counting neutrophils in microscopic fields, determining the tissue radioactivity of labeled cells, or measuring myeloperoxidase concentration in tissue samples.

Arfors *et al.* (1987) examined neutrophil emigration in response to a variety of stimuli in rabbits pretreated with either saline or the adherence-blocking CD18 MAb designated 60.3. Following pretreatment, animals were given intradermal injections of fMLP, leukotriene (LTB_4), C5a, or histamine to provoke emigration. Neutrophil emigration was determined by myeloperoxidase activity in the tissue, and capillary leak was determined by the local accumulation of ^{125}I-labeled albumin that had been previously injected intravenously. PMN accumulation in response to fMLP, LTB_4, and C5a and plasma leakage was significantly reduced by MAb treatment compared with saline treatment. Plasma leakage resulting from histamine was unaffected by the CD18 MAb (Arfors *et al.*, 1987).

A similar set of experiments examined the recruitment of PMNs in response to intradermal injections of fMLP, C5a, LTB_4, and IL-1. In these experiments, indium-labeled neutrophils were incubated with MAb 60.3 prior to reinjection into experimental animals. A significant inhibition of neutrophil accumulation was seen at all doses of chemoattractant employed (Nourshargh *et al.*, 1989).

The inhibition of CD18-mediated adhesion with the CD18 MAb 60.3 virtually abolished neutrophil emigration into LPS-soaked sponges as long as high antibody concentrations were maintained (Price *et al.*, 1987). The MAb was given by intravenous injection 1 hr after insertion of the sponge, with the MAb dose varying from approximately 0.2 to 2 mg/kg. Sponges were removed 24 hr later; cells were mechanically expressed out of the sponges and counted electronically.

A similar set of experiments using LPS-impregnated Gelfoam demonstrated that a MAb directed to murine L-selectin (MEL-14) reduced PMN extravasation by 65% (Lewinshon *et al.*, 1987). In those experiments, fluorescence-labeled PMNs were injected 2 hr after sponge implantation, and sponges were harvested 1 hr later. The results from these studies are thus consistent with the model presented for leukocyte emigration: both the inhibition of rolling with L-selectin MAb and the inhibition of sticking with CD18 MAb prevented emigration.

2.1. Peritoneal Stimuli

Peritonitis induced by injecting 1% glycogen into rats resulted in an upregulation of CD11b/CD18 on the surface of neutrophils that emigrate into the peritoneum (Freyer *et al.*, 1989). Presumably, the upregulation resulted in increased emigration, but the study did not examine the function of this upregulation. However, the function of CD18 and CD49d was examined in rabbits when peritonitis was induced by protease peptone (Winn and Harlan, 1993). Neutrophil emigration in response to peritoneal inflammation was evaluated at 4 and 24 hr. The CD18 MAb 60.3 virtually abolished neutrophil influx at 4 hr. Notably, the CD18 MAb 60.3 did not block PMN accumulation when the inflammatory process was allowed to continue for 24 hr, which is consistent with prior studies showing a CD18-independent pathway of emigration in the peritoneum. Administration of the CD49d MAb HP1/2 alone did not inhibit neutrophil or mononuclear leukocyte emigration at 4 or 24 hr after induction of peritonitis. Concurrent administration of MAbs 60.3 and HP1/2 prevented most of the leukocyte emigration at both 4 and 24 hr, which is consistent with the necessity of blocking the β_1 and β_2 integrin function in order to prevent emigration. The combined treatment with CD18 and CD49d MAb also blocked PMN emigration. Since neutrophils do not express VLA-4 and do not bind the CD49d MAb, the reason for the blockade of neutrophil emigration at 24 hr is not apparent. Perhaps CD18-induced neutrophil emigration is dependent on mononuclear leukocytes, as suggested by Mileski *et al.* (1990a), and the inhibition of neutrophils reflects the inhibition of mononuclear leukocyte emigration.

The importance of L-selectin in chemical peritonitis has been examined in a number of ways. Jutila *et al.* (1989) induced PMN emigration in mice by the intraperitoneal instillation of thioglycollate. The emigration of fluorescence-labeled neutrophils was inhibited approximately 65% by pretreatment with MAb MEL-14, an antibody that recognizes a functional epitope on murine L-selectin. The administration of CD11b/CD18 MAbs resulted in a 70% inhibition of neu-

trophil emigration. Similarly, neutrophils treated with chymotrypsin were significantly inhibited in their ability to emigrate in response to thioglycollate-induced
peritonitis (Jutila et al., 1991). Chymotrypsin is thought to act specifically by
removing L-selectin, although an effect on other adhesion molecules has not been
excluded. The construction of an IgG–selectin chimera containing the L-selectin
extracellular domain provided Watson et al. with an alternative approach to
investigating this adhesion molecule. This L-selectin–Ig chimera inhibited neutrophil emigration in response to chemical peritonitis by the same amount as did
L-selectin MAbs in the mouse (Watson et al., 1991).

The participation of E-selectin in neutrophil emigration in response to
glycogen-induced peritonitis was examined by Mulligan et al. (1991). They
demonstrated that pretreating rats with a blocking antibody to E-selectin, designated CL-3, reduced neutrophil emigration into the glycogen-inflamed peritoneum by 70% as compared with a treated control.

The experiments described above examine the role of adhesion molecules in
the extravascular accumulation of neutrophils in response to inflammatory signals induced in the skin or peritoneum. They do not define where that blockade
occurs in the process of emigration. However, it is possible to observe the
behavior of neutrophils within the circulation by preparing mesentery, muscle, or
skin for intravital microscopy of the microcirculation. The effect of MAb on
leukocyte response to various stimuli can be observed directly in these preparations.

2.2. Intravital Microscopy

An inspection of postcapillary venules via intravital microscopy reveals
leukocytes in three states: (1) traveling with the flow of plasma and red cells, (2)
rolling slowly along endothelial surfaces, and (3) tightly adherent to endothelium
(Arfors et al., 1987). Treating rabbits with the CD18 MAb 60.3 prevented the
tight adhesion of leukocytes in response to LTB_4- or zymosan-activated plasma
(essentially C5a) (Arfors et al., 1987). This MAb had no effect on the rolling of
neutrophils. Subsequently, Argenbright et al. (1991) dissected the roles of
CD11a, CD11b, CD18, and ICAM-1 by using various MAbs. Intravital microscopy allowed the direct observation of rabbit mesentery superfused with
zymosan-activated plasma (ZAP). They confirmed that a CD18 MAb prevented
neutrophil adhesion, and in fact would displace cells that were already adherent.
Similar results were seen using a CD11a MAb. A CD11b MAb was less effective
in preventing adhesion, and did not displace adherent cells. ICAM-1 MAb was
very effective in preventing adhesion, but it also did not displace already adherent leukocytes. None of these MAbs affected leukocyte rolling.

The involvement of adhesion molecules in the rolling of leukocytes was
shown in experiments by von Andrian et al. (1991). L-selectin- and CD18-

mediated adherence were blocked with MAb DREG 200 and MAb IB4, respectively. Intravital microscopy was used to view rabbit mesentery superfused only with saline. In control animals, leukocyte rolling was observed along with a steady accumulation of tightly adherent leukocytes. The administration of the L-selectin MAb DREG 200 abolished rolling and greatly decreased the accumulation of adherent leukocytes. Similar results were found when the chimeric molecule made from the antigen of the MEL-14 mAb (the murine L-selectin) was used (Ley *et al.,* 1991). The CD18 MAb IB4 inhibited tight adhesion without affecting rolling, a finding similar to that seen with the other CD18 MAbs (Arfors *et al.,* 1987). These studies led von Andrian *et al.* to propose that leukocyte adhesion occurs in two steps following stimulation. First, leukocytes are slowed by a selectin-mediated interaction with the endothelium, which produces rolling. Second, this slowing and rolling permits the subsequent tight adhesion and emigration mediated by CD11/CD18 integrins (von Andrian *et al.,* 1991).

The complexity of the inflammatory system suggests that any response might be both stimulus and tissue specific. Doerschuk *et al.* showed that neutrophil infiltration into hydrochloric acid-soaked subcutaneous sponges was inhibited by CD18 MAb 60.3. However, emigration to this stimulus in the lungs was unaffected (Doerschuk *et al.,* 1990). Thus, at least in the lung, there is a CD18-independent pathway of neutrophil emigration. If von Andrian's conclusions are correct, then selectins may block emigration by blocking rolling, but a CD18-independent pathway of tight adherence must also exist.

3. Bacterial Inflammation

Chemically induced inflammation provides a mechanism for investigating the mechanisms of PMN emigration, but does not mimic human disease. Experiments using live bacteria may be somewhat more clinically relevant; they have been performed to evaluate the role of adhesion molecules in phagocyte emigration. Injection or instillation of bacteria can lead to the activation of other cells, leading to the production of various cytokines. This can potentially result in a more complex pathway of emigration than that observed with chemical stimuli. LPS from gram-negative bacteria and some cytokines (TNF, IL-1) are known to cause the expression of adhesion molecules on endothelial cells (Vadas *et al.,* 1992).

The peritoneum and lungs are accessible for bacterial inoculation, and allow the direct evaluation of cellular emigration by lavage. Mileski *et al.* (1990a) examined the effect of CD18 blockade on neutrophil emigration following the instillation of either *E. coli* or *S. pneumoniae* into the peritoneum. Cell counts were determined from lavage fluid taken 4 hr after instillation of bacteria. The

CD18 MAb 60.3 was very effective in preventing neutrophil emigration into the inflamed peritoneal cavity in response to both organisms in normal rabbits. A significant CD18-independent emigration occurred in response to intraperitoneal *S. pneumoniae* following an enhancement of the macrophage population in the peritoneum. The increased number of macrophages was accomplished by pretreating with protease peptone or transplanting macrophages from other animals. The response to *E. coli* was unaffected by the addition of macrophages. These results provide further evidence for the existence of CD18-independent mechanisms of PMN adhesion and emigration.

L-selectin is also involved in PMN emigration in response to bacterial peritonitis. Pretreatment of rabbits with the L-selectin MAb LAM1.3 was 90% effective in preventing neutrophil emigration resulting from *E. coli*-induced peritonitis. This MAb was compared with the nonblocking L-selectin MAb LAM1.14: both produced transient neutropenia of similar degree, but MAb LAM1.14 did not block neutrophil emigration. Thus, neutropenia alone cannot explain the blockade of emigration. The cause of the neutropenia is unknown, but it was not related to endotoxin contamination of the MAbs. Interestingly, if the induction of peritonitis was delayed for 3 hr after antibody administration, MAb LAM1.3 was no longer effective in blocking neutrophil emigration, even though L-selectin remained saturated with MAb. We do not know why MAb LAM1.3 was ineffective when peritonitis was delayed (Flaherty, unpublished).

MAbs were also used to examine the adhesion mechanism of neutrophil emigration in response to bacteria instilled into the lungs. Doerschuk *et al.* (1990) examined neutrophil emigration following subcutaneous and intratracheal *S. pneumoniae*. The CD18 MAb 60.3 inhibited neutrophil emigration to *S. pneumoniae* into subcutaneous sponges, but was ineffective in preventing PMN emigration to *S. pneumoniae* instilled into the lung. Similar results were found by Winn *et al.* (1993), whose studies showed that the CD18 MAb 60.3 did not significantly block PMN emigration into the lung in response to *S. pneumoniae*, but did in response to *E. coli*. In addition, the supernatant produced when macrophages were incubated *in vitro* with *S. pneumoniae* also produced a CD18-independent pathway of PMN emigration.

The experiments described above were conducted over a short time, and were designed to investigate the effects of adhesion blockade on neutrophil emigration. It is also clinically relevant to examine the effects of adhesion blockade on the outcome from bacterial challenge. Mileski *et al.* (1991) examined the effect of CD18 blockade in experimental appendicitis in the rabbit. The appendix was devascularized to induce appendicitis, and was removed 18 hr later. Animals were divided into four treatment groups: (1) control (no treatment), (2) antibiotic-treated, (3) CD18 MAb-treated, and (4) antibiotic plus CD18 MAb-treated. No difference in 10-day survival or infectious complications was seen between

MAb-treated and untreated animals. It should be noted that the MAb-treated animals did have some PMN emigration into the peritoneum at 18 hr, although in markedly reduced numbers.

Decreasing neutrophil emigration in response to bacterial infection may be beneficial in situations in which the edema produced by phagocyte-induced inflammation is itself injurious. In bacterial meningitis, two studies have indicated that the blockade of CD18-mediated adhesion can decrease mortality and morbidity. Tuomanen *et al.* (1989) demonstrated that MAb IB4 blocked the emigration of leukocytes into the CSF in response to bacteria. In addition, CD18 MAb-treated animals showed a delay in development of bacteremia, and IB4 prevented the development of brain edema and death in the presence of what should have been a lethal challenge with *S. pneumoniae*. More recently, S'aez-Llorens *et al.* (1991) have confirmed these findings in *H. influenza* meningitis in rabbits. They combined MAb administration with dexamethasone and demonstrated decreases in all indices of inflammation, including edema formation.

Much of the excitement associated with the field of leukocyte adhesion molecules relates to the potential clinical applications of antiadherence therapy in inflammatory diseases such as reperfusion injuries (see below). It is imperative that we examine the potential adverse effects of transiently blocking one portion of PMN function. Patients with leukocyte adherence deficiency (LAD) syndrome have a congenital deficiency of CD11/CD18, and suffer from recurrent, life-threatening soft tissue infections (Anderson and Springer, 1987). Soft tissue infections in animals treated with CD18 MAb were examined by Sharar *et al.* (1991) following intradermal *S. aureus* inoculation. Animals received inocula of up to 10^9 colony-forming units (CFU). An increase in the rate of abscess formation was seen at an inoculum of 10^8 CFU, and larger abscesses were seen at 10^9. At inocula of less than 10^7, no abscesses developed in either treated or untreated animals. These experiments confirm that the effectiveness of the immune response to staphylococci is reduced by transient CD18 blockade. The high inocula required to produce abscesses in these experiments leave their clinical significance in question.

4. Ischemia–Reperfusion

The role of the neutrophil and neutrophil adhesion in ischemia–reperfusion injury has been extensively investigated in a variety of organs. Ischemia–reperfusion injury may occur clinically in such common ailments as myocardial infarction, stroke, gut ischemia, organ transplantation, and peripheral vascular disease. In addition, hemorrhagic shock may be associated with ischemia–reperfusion injury that contributes to the susceptibility of trauma patients to subsequent multiple organ failure. Leukocyte adhesion blockade with MAbs has been examined in ischemia–reperfusion of soft tissue, central nervous system, heart, visceral organs, skeletal muscle, and the whole body. The length of observation following reperfusion has varied in these models from short-term observations of fluid fluxes and edema formation over hours to evaluations of organ function, tissue necrosis, and survival at several days.

The development of edema is an early and detectable change associated with ischemia–reperfusion injury, and results from increased microvascular permeability. Increased gut microvascular permeability following ischemia–reperfusion was reduced in cats treated with the CD18 MAb 60.3 (Hernandez *et al.*, 1987). Similarly, a morphologic evaluation of mesenteric venules after ischemia and reperfusion showed protection with CD18 MAbs (Oliver *et al.*, 1991). Ischemia–reperfusion of skeletal muscle resulted in increased microvascular permeability, and CD11/CD18 blockade prevented some of this increase (Carden *et al.*, 1990). CD18 blockade was protective in ischemia–reperfusion in the lung, as determined from the fluid filtration coefficient (Horgan *et al.*, 1990) and by improved perfusion (Bishop *et al.*, 1992). These experiments were short-term, involving 4 to 6 hr of reperfusion, and did not measure organ function or eventual tissue survival.

Edema formation and tissue viability were measured for 6 days by Vedder *et al.* (1990) following ischemia–reperfusion of soft tissue. Rabbit ears were amputated, reattached, rendered ischemic for 10 hr, and then allowed to reperfuse. The CD18 MAb 60.3 was administered either before or after ischemia, and the animals were followed for 6 days. Edema formation was markedly reduced in the antibody-treated animals compared with the saline-treated ones. These MAb-treated animals were not significantly different from control animals subjected to the same operation, but were not subjected to ischemia. In addition, tissue necrosis was minimal in antibody-treated animals, but was approximately 24% in saline-treated animals. Rabbits receiving MAb before ischemia were not different from those treated at the time of reperfusion, suggesting that the leukocyte-mediated injury occurred during the reperfusion phase.

Tissue viability after ischemia–reperfusion of the heart was improved with CD11b MAb in dogs. In these experiments, ischemia was for 90 min and reperfusion was for 6 hr. The size of the resulting infarction was reduced by 46% in dogs treated with the CD11b MAb 904 as compared with saline-treated controls. MAb was given by an intravenous injection halfway through the period of ischemia (Simpson *et al.*, 1988).

Although the ischemia-reperfusion injury of soft tissue and myocardium may be CD18-dependent in several species, it does not appear to be in the rabbit kidney. Thornton *et al.* (1989) found that CD18 MAb 60.3 produced no improvement in renal function after renal ischemia–reperfusion. No neutrophils were seen to infiltrate the kidneys in either control or MAb-treated rabbits, raising some question as to whether neutrophils are involved in this injury in the kidney at all. In addition, rats made neutropenic with antineutrophil serum had renal function similar to rats with normal neutrophil counts, again suggesting that neutrophils were not responsible for this ischemia–reperfusion injury.

The effect of a β_2-integrin blockade on the survival of liver after ischemia–reperfusion is not clear. The CD18 MAb 60.3 treatment produced no reduction in transaminase levels and no improvement in tissue necrosis in rabbits subjected to 45 min of hepatic ischemia followed by 24 hr of reperfusion (Langdale *et al.*, 1993). Neutropenia in this study did result in a less severe injury measured by tissue necrosis. In contrast, when Smith *et al.* (1992) used IgM CD11b MAb given before and after ischemia in rats, they showed a significant reduction in transaminases, tissue necrosis, and neutrophil accumulation in rat livers. The differences in these studies, apart from the MAbs, are not readily apparent.

Similar conflicting results have been reported in the central nervous system. Examining only cerebral reflow after balloon occlusion of the right middle cerebral artery in baboons, Mori *et al.* (1992) found significant injury measured by tissue necrosis. In contrast, Takeshima *et al.* (1992) examined the effect of CD18 MAb 60.3 on somatosensory evoked potentials and tissue injury in cat brains subjected to 90 min of left middle cerebral artery ischemia followed by 3 hr of reperfusion. They found no differences between treated and untreated animals. Likewise, the effects of a CD18 blockade with MAb 60.3 on neurologic function in rabbits following cardiac arrest and resuscitation were not different from those in saline control animals (Flaherty, unpublished). MAb was administered at the time of resuscitation, and animals were followed for 24 hr. Histologic specimens of selected brains showed no neutrophil infiltration in either control or antibody-treated animals. As in the kidney, this raises the question of neutrophil participation in this particular injury.

Clark *et al.* (1991a,b) also reported no protection when CD18 and anti-ICAM-1 MAbs were used to block the adhesion of leukocytes following a stroke produced by microsphere infusion. This might be expected, since microspheres produce an irreversible ischemia and reperfusion does not occur. However, they did detect some difference in the incidence of paraplegia in rabbits following treatment with these MAbs in spinal cord ischemia–reperfusion (Clark *et al.*, 1991a,b). Protection from spinal cord injury was not detected in a similar series of experiments examining CD18 blockade in rabbits (Forbes and Verrier, unpublished observations).

Vedder *et al.* (1988, 1989) reasoned that the inadequate tissue perfusion of hemorrhagic shock comprised, in effect, an ischemia–reperfusion injury to the entire body. The CD18 blockade that was shown to be effective in soft tissue ischemia–reperfusion might therefore be protective in hemorrhagic shock and resuscitation. Initial experiments were performed in rabbits, where histologic tissue damage and survival were improved by the administration of MAb 60.3. These findings were subsequently confirmed in nonhuman primates (Mileski *et al.*, 1990b). Monkeys were subjected to 90 min of severe shock (cardiac output = 30% of baseline) and treated with MAb 60.3 or saline. Cardiac output was maintained close to baseline by administering lactated Ringer's solution for 24 hr. The CD18 MAb-treated animals required less fluid for resuscitation compared with saline-treated animals. Hemorrhagic gastritis was observed in the saline- but not the CD18 MAb-treated monkeys. These results have exciting implications for the use of CD18 MAbs in treating patients suffering from severe shock as a result of hemorrhage, trauma, or other causes.

5. Conclusion

In vitro experiments have revealed much about the role of adhesion molecules in leukocyte–endothelial interactions. These *in vitro* studies have been confirmed and extended by *in vivo* experiments, initially relatively simple investigations involving neutrophil emigration in response to chemical irritants, and recently in more clinically relevant settings of bacterial inoculation or infection. In addition, the blockade of leukocyte integrin receptors has proved protective in several models of ischemia-reperfusion injury. It therefore appears that there are significant possibilities for the clinical application of antiadhesion therapy in disorders involving inflammation.

References

Anderson, D. C., and Springer, T. A., 1987, Leukocyte adhesion deficiency: An inherited defect in the Mac-1, LFA-1, and p150,95 glycoproteins, *Annu. Rev. Med.* **38**:175–194.

Arfors, K.E., Lundberg, C., Lindbom, L., Lundberg, K., Beatty, P. G., and Harlan, J. M., 1987, A monoclonal antibody to the membrane glycoprotein complex CD18 inhibits polymorphonuclear leukocyte accumulation and plasma leakage *in vivo*, *Blood* **69**:338–340.

Argenbright, L. W., Letts, L. G., and Rothlein, R., 1991, Monoclonal antibodies to the leukocyte membrane CD18 glycoprotein complex and to intercellular adhesion molecule-1 inhibit leukocyte–endothelial adhesion in rabbits, *J. Leukocyte Biol.* **49**(3):253–257.

Bishop, M. J., Kowalski, T. F., Guidotti, S. M., and Harlan, J. M., 1992, Antibody against neutrophil adhesion improves reperfusion and limits alveolar infiltrate following unilateral pulmonary artery occlusion, *J. Surg. Res.* **52**(3):199–204.

Carden, D. L., Smith, J. K., and Korthuis, R. J., 1990, Neutrophil-mediated microvascular dysfunction in postischemic canine skeletal muscle. Role of granulocyte adherence, *Circ. Res.* **66**(5):1436–1444.

Clark, W. M., Madden, K. P., Rothlein, R., and Zivin, J. A., 1991a, Reduction of central nervous system ischemic injury by monoclonal antibody to intercellular adhesion molecule, *J. Neurosurg.* **75**(4):623–627.

Clark, W. M., Madden, K. P., Rothlein, R., and Zivin, J. A., 1991b, Reduction of central nervous system ischemic injury in rabbits using leukocyte adhesion antibody treatment, *Stroke* **22**(7):877–883.

Doerschuk, C. M., Winn, R. K., Coxson, H. O., and Harlan, J. M., 1990, CD18-dependent and -independent mechanisms of neutrophil emigration in the pulmonary and systemic microcirculation of rabbits, *J. Immunol.* **144**(6):2327–2333.

Flaherty, L. C. (unpublished).

Freyer, D. R., Morganroth, M. L., and Todd, R. F., 1989, Surface Mol (CD11b/CD18) glycoprotein is up-modulated by neutrophils recruited to sites of inflammation *in vivo, Inflammation* **13**(5):495–505.

Hernandez, L. A., Grisham, M. B., Twohig, B., Arfors, K.-E., Harlan, J. M., and Granger, D. N., 1987, Role of neutrophils in ischemia-reperfusion induced microvascular injury, *Am. J. Physiol.* **253**:H699–H703.

Horgan, M. J., Wright, S. D., and Malik, A. B., 1990, Antibody against leukocyte integrin (CD18) prevents reperfusion-induced lung vascular injury, *Am. J. Physiol.* **259**:L315–L319.

Jutila, M. A., Rott, L., Berg, E. L., and Butcher, E. C., 1989, Function and regulation of the neutrophil MEL-14 antigen *in vivo:* Comparison with LFA-1 and MAC-1, *J. Immunol.* **143**(10);3318–3324.

Jutila, M. A., Kishimoto, T. K., and Finken, M., 1991, Low-dose chymotrypsin treatment inhibits neutrophil migration into sites of inflammation *in vivo:* Effects on Mac-1 and MEL-14 adhesion protein expression and function, *Cell. Immunol.* **132**(1):201–214.

Langdale, L. A., Flaherty, L. C., Liggitt, H. D., Harlan, J. M., Rice, C. L., and Winn, R. K., 1993, Neutrophils contribute to hepatic ischemia-reperfusion injury by a CD18-independent mechanism, *J. Leukoc. Biol.* **53**:511–517.

Lewinshon, D., Bargatze, R., and Butcher, E., 1987, Evidence of a common molecular mechanism shared by neutrophils, lymphocytes, and other leukocytes, *J. Immunol.* **138**:4313–4321.

Ley, K., Gaehtgens, P., Fennie, C., Singer, M. S., Lasky, L. A., and Rosen, S. D., 1991, Lectin-like cell adhesion molecule 1 mediates leukocyte rolling in mesenteric venules *in vivo, Blood* **77**:2553–2555.

Mileski, W., Harlan, J., Rice, C., and Winn, R., 1990a. Streptococcus pneumoniae-stimulated macrophages induce neutrophils to emigrate by a CD18-independent mechanism of adherence, *Circ. Shock* **31**(3):259–267.

Mileski, W. J., Winn, R. K., Vedder, N. V., Pohlman, T. H., Harlan, J. M., and Rice, C. L., 1990b, Inhibition of CD18-dependent neutrophil adherence reduces organ injury after hemorrhagic shock in primates, *Surgery* **108**:205–212.

Mileski, W. J., Winn, R. K., Harlan, J. M., and Rice, C. L., 1991, Transient inhibition of neutrophil adherence with the anti-CD18 monoclonal antibody 60.3 does not increase mortality rates in abdominal sepsis, *Surgery* **109**:497–501.

Mori, E., del Zoppo, G. J., Chambers, J. D., Copeland, B. R., and Arfors, K. E., 1992, Inhibition of polymorphonuclear leukocyte adherence suppresses no-reflow after focal cerebral ischemia in baboons, *Stroke* **23**(5):712–718.

Mulligan, M. S., Varani, J., Dame, M. K., Lane, C. L., Smith, C. W., Anderson, D. C., and Ward, P. A., 1991, Role of endothelial–leukocyte adhesion molecule 1 (ELAM-1) in neutrophil-mediated lung injury in rats, *J. Clin. Invest.* **88**(4):1396–1406.

Nourshargh, S., Rampart, M., Hellewell, P. G., Jose, P. J., Harlan, J. M., Edwards, A. J., and Williams, T. J., 1989, Accumulation of ^{111}In-neutrophils in rabbit skin in allergic and non-allergic inflammatory reactions *in vivo.* Inhibition by neutrophil pretreatment *in vitro* with a monoclonal antibody recognizing the CD18 antigen, *J. Immunol.* **142**(9):3193–3198.

Oliver, M. G., Specian, R. D., Perry, M. A., and Granger, D. N., 1991, Morphologic assessment of leukocyte–endothelial cell interactions in mesenteric venules subjected to ischemia and reperfusion, *Inflammation* **15**(5):331–346.

Price, T., Beatty, P. G., and Corpuz, S. R., 1987, *In vivo* inhibition of neutrophil function in the rabbit using monoclonal antibody to CD18, *J. Immunol.* **139:**4174–4177.

S'aez-Llorens, X., Jafari, H. S., Severien, C., Parras, F., Olsen, K. D., Hansen, E. J., Singer, I. I., and McCracken, G. H. J., 1991, Enhanced attenuation of meningeal inflammation and brain edema by concomitant administration of anti-CD18 monoclonal antibodies and dexamethasone in experimental Haemophilus meningitis, *J. Clin. Invest.* **88**(6):2003–2011.

Sharar, S., Winn, R., Murry, C., Harlan, J., and Rice, C., 1991, A CD18 monoclonal antibody increases the incidence and severity of subcutaneous abscess formation after high-dose *Staphylococcus aureus* injection in rabbits, *Surgery* **110:**213–220.

Simpson, P. J., Todd, R. F., III, Fantone, J. C., Mickelson, J. K., Griffin, J. D., and Lucchesi, B. R., 1988, Reduction of experimental canine myocardial reperfusion injury by a monoclonal antibody (Anti-Mo-1, Anti-CD11b) that inhibits leukocyte adhesion, *J. Clin. Invest.* **81:**624–629.

Smith, C. W., 1992, Transendothelial migration, in: *Adhesion: Its Role in Inflammatory Disease* (J. M. Harlan and D. Y. Liu, eds.), Freeman, San Francisco.

Smith, C. W., Farhood, A., and Jaeschke, H., 1992, Monoclonal antibody against CD11b/CD18 (Mac-1) integrin protects in a model of hepatic ischemia-reperfusion injury, *FASEB J.* **6:**A1303.

Takeshima, R., Kirsch, J. R., Koehler, R. C., Gomoll, A. W., and Traystman, R. J., 1992, Monoclonal leukocyte antibody does not decrease the injury of transient focal cerebral ischemia in cats, *Stroke* **23**(2):247–252.

Thornton, M. A., Winn, R., Alpers, C. E., and Zager, R. A., 1989, An evaluation of the neutrophil as a mediator of *in vivo* renal ischemic-reperfusion injury, *Am. J. Pathol.* **135**(3):509–515.

Tuomanen, E. I., Saukkonen, K., Sande, S., Cioffe, C., and Wright, S. D., 1989, Reduction of inflammation, tissue damage, and mortality in bacterial meningitis in rabbits treated with monoclonal antibodies against adhesion-promoting receptors of leukocytes, *J. Exp. Med.* **170:**959–968.

Vadas, M. A., Gamble, J. R., and Smith, W. B., 1992, Regulation of myeloid blood cell–endothelial interaction by cytokines, in: *Adhesion: Its Role in Inflammatory Disease,* Freeman, San Francisco, pp. 65–81.

Vedder, N. B., Winn, R. K., Rice, C. L., Chi, E., Arfors, K.-E., and Harlan, J. M., 1988, A monoclonal antibody to the adherence promoting leukocyte glycoprotein CD18 reduces organ injury and improves survival from hemorrhagic shock and resuscitation in rabbits, *J. Clin. Invest* **81:**939–944.

Vedder, N. B., Fouty, B. W., Winn, R. K., Harlan, J. M., and Rice, C. L., 1989, Role of neutrophils in generalized reperfusion injury associated with resuscitation from shock, *Surgery* **106:**509–516.

Vedder, N. B., Winn, R. K., Rice, C. L., Chi, E., Arfors, K.-E., and Harlan, J. M., 1990, Inhibition of leukocyte adherence by anti-CD18 monoclonal antibody attenuates reperfusion injury in the rabbit ear, *Proc. Natl. Acad. Sci. USA* **81:**939–944.

von Andrian, U. H., Chambers, J. D., McEvoy, L. M., Bargatze, R. F., Arfors, K.-E., and Butcher, E. C., 1991, Two-step model of leukocyte–endothelial cell interaction in inflammation: Distinct roles for LECAM-1 and the leukocyte beta 2 integrins *in vivo*, *Proc. Natl. Acad. Sci. USA* **88**(17):7538–7542.

Watson, S. R., Fennie, C., and Lasky, L. A., 1991, Neutrophil influx into an inflammatory site inhibited by a soluble homing receptor-IgG chimaera, *Nature* **349**(6305):164–167.

Winn, R. K. and Harlan, J. M., 1993, CD18-independent neutrophil and mononuclear leukocyte emigration into the peritoneum of rabbits, *J. Clin. Invest.* **92:**1168–1173.

Winn, R. K., Mileski, W. J., Kovach, N. L., Doerschuk, C. M., Rice, C. L., and Harlan, J. M., 1993, The role of protein synthesis and CD11/CD18 adhesion complex in neutrophil emigration into the lung, *Exp. Lung Res.* **19:**221–235.

10

Proadhesive Cytokine Immobilized on Endothelial Proteoglycan
A New Paradigm for Recruitment of T Cells

YOSHIYA TANAKA, DAVID H. ADAMS, and STEPHEN SHAW

1. Introduction

Inflammation and normal immune responses depend on the recruitment of specific leukocyte populations. Similarly, the maturation of lymphocytes requires their trafficking from blood into the appropriate lymphoid organ at the correct time. These processes require that leukocytes bind to endothelium in a specific and selective fashion, i.e., the right leukocyte subpopulation at the right time and place. In recent years, it has become apparent that numerous molecules are involved in this process. During the last year a consensus view has emerged, based on cumulative contributions from many laboratories, which proposes that leukocytes interact with endothelium by a multistep process involving several different classes of adhesion molecules. This model, which is the starting point for the present analysis, has been reviewed in detail elsewhere (Dustin and Springer, 1991; Shimizu *et al.*, 1992; Butcher, 1991; Zimmerman *et al.*, 1992; Pardi *et al.*, 1992; Schweighoffer and Shaw, 1992b), and will be restated here only briefly, as it pertains to T cells.

Three conceptually distinct steps in the interaction have been distinguished, each generally mediated by a different structural class of molecules. (1) *Tethering*

YOSHIYA TANAKA, DAVID H. ADAMS, and STEPHEN SHAW • The Experimental Immunology Branch, National Cancer Institute, National Institutes of Health, Bethesda, Maryland 20892.

Cellular Adhesion: Molecular Definition to Therapeutic Potential, edited by Brian W. Metcalf, Barbara J. Dalton, and George Poste. Plenum Press, New York, 1994.

or primary adhesion: This process is necessary to slow the movement of T cells (by at least an order of magnitude) and to provide contact between the T-cell glycocalyx and the endothelial glycocalyx. This interaction is mediated by a specialized class of molecules designated *selectins* (L-, E-, and P-selectin), which bind to specific carbohydrates on the apposing surface. (2) *Triggering or activation:* Circulating T cells are not fully "adhesion-competent"; they express the integrin adhesion molecules, but in a functionally inactive state. Therefore, strong integrin-mediated T-cell adhesion is dependent on T-cell encounter with "triggering" ligands present at or near the endothelial surface. Although a number of molecules have been found to trigger the activation of T-cell integrins (Schweighoffer and Shaw, 1992a; Dustin and Springer, 1989; Tanaka *et al.,* 1992; Shimizu *et al.,* 1990; Koopman *et al.,* 1990), none of them appear to explain T-cell integrin activation at endothelial surfaces. This brief review highlights one molecule, macrophage inflammatory protein-1β (MIP-1β), which we believe participates in that process and which uses a novel mode of action that can be used as a paradigm for other proadhesive cytokines. (3) *Strong adhesion:* The powerful T-cell integrins (when activated) mediate adhesion if the relevant ligands are expressed on the endothelial surface; the known ligands (ICAM-1, ICAM-2, and VCAM-1) are all members of the Ig supergene family.

What are the relevant triggering ligands for T cells? Based on a variety of lines of evidence, we view cytokines as excellent candidates. First, many are produced at inflammatory sites. Second, they appear to act locally, a prerequisite for explaining site-specific recruitment. Third, *in vivo* models using cytokines such as IL-8 demonstrate that they rapidly recruit leukocytes into the site of injection (Larsen *et al.,* 1989). Fourth, many cytokines are chemotactic for T cells (Oppenheim *et al.,* 1991; Schall, 1991; Bacon *et al.,* 1990); since chemotaxis involves regulated adhesion, chemotactic cytokines are good candidates for adhesion induction. Fifth, and of particular importance, cytokines are potent inducers of the adhesion of various myeloid cells, including granulocytes and monocytes (Butcher, 1991; Wolpe *et al.,* 1988; Rot, 1992). Finally, there are a large number of potentially relevant cytokines. If many of them were involved in adhesion induction for T cells, this diversity could contribute to the specific recruitment of different T-cell subsets under different inflammatory conditions.

2. Results and Discussion

We surveyed a variety of cytokines for their ability to induce T-cell adhesion and found that one, MIP-1β, could do so (Tanaka *et al.,* 1993a; 1993b). The system we used assayed the binding of purified resting T cells to the protein ligand VCAM-1, which had been immobilized on plastic wells. We used

VCAM-1 as ligand because it is induced on inflamed endothelium both *in vitro* and *in vivo*, and it is believed to be important in T-cell recruitment. Resting T cells have a modest level of basal binding to VCAM-1, which increases significantly after exposure to nanogram concentrations of MIP-1β. We have coined the term *proadhesive* cytokines to describe MIP-1β and other cytokines that have the capacity to induce adhesion. Not all resting T cells are equally influenced by MIP-1β; purified CD8 cells respond more strongly than do CD4 cells (Tanaka *et al.*, 1993a). This suggests that MIP-1β will contribute to the preferential recruitment of CD8 cells relative to CD4 cells.

MIP-1β belongs to a large and growing family of cytokines known as "chemokines" (previously known as intercrines) (Schall, 1991; Oppenheim *et al.*, 1991). Members of this family have structural homologies, including: a molecular weight of 8000–10,000, distinctive conserved cysteines (C-C in the α subfamily and C-X-C in the β subfamily), and a conserved site for binding glycosaminoglycans (GAG) such as heparin or GAGs on proteoglycans. This cytokine family is particularly interesting because several members have already been shown to be proadhesive for myeloid cells, chemotactic for leukocytes (including T cells), and rapidly induce recruitment of leukocytes (including T cells) *in vivo*.

Although cytokines are good candidates for adhesion triggering, there is one aspect of the conventional concept of cytokine-induced adhesion of leukocyte binding to endothelium which we have found problematic, namely, that cytokines act in solution in the vessel lumen. We (Tanaka *et al.*, 1993b), and Antal Rot (1992), have independently proposed an alternative paradigm, in which cytokines are immobilized on the endothelial surface. On theoretical grounds this solves two problems. First, immobilized cytokines will activate only those cells that are at a site favorable to binding; the activation of cells in midstream by soluble cytokine is undesirable, since they would probably first encounter endothelium in other tissues distant from the site of release. Second, if cytokines bind to endothelium they can reach an effective local concentration, whereas they would be continually washed away if they were secreted into the lumen.

We have derived evidence to support the concept of the endothelial immobilization of cytokines by (1) documenting the endothelial localization of cytokine *in vivo* and (2) developing an *in vitro* model with a molecular mechanism to account for that immobilization. The *in vivo* evidence comes from immunohistological studies of human lymphoid tissue (Tanaka *et al.*, 1993a). When tonsil and lymph node are stained with antibody specific for MIP-1β, there is strong staining on the endothelium of small vessels, particularly those with the morphology of high endothelial venules (HEV). The staining of HEV is diffuse, and apparently includes the luminal aspect.

What molecular mechanism is likely to mediate the endothelial immobilization of MIP-1β? We explored the possibility that immobilization occurs via binding to proteoglycan on the endothelial surface. Proteoglycans were attractive candidates for this role for several reasons: (1) MIP-1β has a GAG-binding site characteristic of the chemokine family. (2) Endothelial cells are rich in proteoglycans, and are known to bind other heparin-binding proteins such as antithrombin III at their luminal surface (Marcum and Rosenberg, 1989). (3) The length, stiffness, and hydrophilic nature of GAG side chains suggest that they would be at the tips of the glycocalyx, and therefore ideally suited for presenting cytokine early in a cell–cell interaction. (4) There is an exciting and growing literature regarding the importance of cytokine interactions with proteoglycans (Huang *et al.*, 1982; Ruoslahti and Yamaguchi, 1991).

We therefore expanded our *in vitro* model to test the hypothesis that MIP-1β could function when immobilized by proteoglycan (Tanaka *et al.*, 1993a). We used two different proteoglycans: a synthetic one (heparin coupled to albumin) and a natural, cell surface one (CD44 from peripheral blood mononuclear cells). The proteoglycan and the integrin ligand VCAM-1 were coimmobilized on plastic wells. MIP-1β was then added, the unbound MIP-1β was washed away, and the binding of resting T cells was assessed. The results demonstrated increased adhesion to VCAM-1 after free cytokine was washed out, but only in the presence of coimmobilized proteoglycan. These results confirmed the prediction that proteoglycan could retain and present MIP-1β to T cells. This was observed with each of the two proteoglycans tested, but not with the control protein albumin or glycoprotein CD45.

How general is this paradigm of chemotactic cytokine presentation on the endothelial surface? The theoretical issues outlined above lead us to conclude that it is a general rule. In addition, the data of Rot (1992) provide strong evidence that IL-8, which is proadhesive and chemotactic for granulocytes, is similarly immobilized on endothelium. We expect that leukocyte recruitment will involve many chemokines, and that they will generally be immobilized on endothelium via proteoglycan. Other cytokines, immobilized in other ways, may also be involved; this is suggested by the fact that TGFβ binds to the syndecan proteoglycan via its protein core (Ruoslahti and Yamaguchi, 1991).

What endothelial proteoglycans mediate the presentation of MIP-1β? We do not know. They might include CD44, which is present on endothelium. CD44 has also been reported to play a role in T-cell activation (Haynes *et al.*, 1989). Since there are many other proteoglycans on endothelium, we view CD44 as a model, not necessarily as a physiologically critical proteoglycan. Proteoglycans expressed by endothelial cells vary according to anatomic site and state of activation. Such variation would be expected to contribute to the specificity of adhesion cascades, since it could determine which cytokines could bind to a particular endothelium.

3. Conclusions

Our studies of MIP-1β demonstrate that the proadhesive role of cytokines extends to T cells. Although our experimental data currently implicate only MIP-1β, we anticipate that other chemokines will regulate the adhesion of T cells under other circumstances. Taken together, the diversity of chemokines and the preferential effect of MIP-1β on CD8 cells suggest that different chemokines may exert diverse and powerful effects on different T-cell subsets.

The model that we and Rot have proposed regarding endothelial localization of the proadhesive cytokines obviates many of the conceptual problems of cytokines acting in solution under conditions of flow. The class of immobilizable cytokines takes on a special significance. Since they are soluble products, they can diffuse from their source within tissue to the endothelium. At that site their immobilization endows them with some of the characteristics of cell surface ligands. The endothelial cell can thus be decorated by a variety of molecules that "report" on the status of many cells within the tissue. This report, which is "read" by the passing leukocyte, allows it to decide whether to undergo triggering, a required step for tissue entry. It is clear from our discussion that proteoglycans are excellent candidates for mediating this immobilization. Their structure, diversity, and regulation make them desirable components in an adhesion cascade.

References

Bacon, K. B., Gearing, A., and Camp, R., 1990, Induction of in vitro human lymphocyte migration by IL-3, IL-4 and IL-6, *Cytokine* **2**:100–105.

Butcher, E. C., 1991, Leukocyte–endothelial cell recognition: Three (or more) steps to specificity and diversity, *Cell* **67**:1033–1036.

Dustin, M. L., and Springer, T. A., 1989, T-cell receptor cross-linking transiently stimulates adhesiveness through LFA-1, *Nature* **341**:619–624.

Dustin, M. L., and Springer, T. A., 1991, Role of lymphocyte adhesion receptors in transient interactions and cell locomotion, *Annu. Rev. Immunol.* **9**:27–66.

Haynes, B. F., Telen, M. J., Hale, L. P., and Denning, S. M., 1989, CD44—A molecule involved in leukocyte adherence and T-cell activation, *Immunol. Today* **10**:423–428.

Huang, S. S., Huang, J. S., and Deuel, T. F., 1982, Proteoglycan carrier of human platelet factor 4: Isolation and characterization, *J. Biol. Chem.* **257**:11546–11550.

Koopman, G., van Kooyk, Y., De Graaff, M., Meyer, C. J. L. M., Figdor, C. G., and Pals, S. T., 1990, Triggering of the CD44 antigen on T lymphocytes promotes T cell adhesion through LFA-1 pathway, *J. Immunol.* **145**:3589–3593.

Larsen, C. G., Anderson, A. O., Appella, E., Oppenheim, J. J., and Matsushima, K., 1989, The neutrophil-activating protein (NAP-1) is also chemotactic for T lymphocytes, *Science* **243**:1464–1466.

Marcum, J. A., and Rosenberg, R. D., 1989, Role of endothelial surface heparin-like polysaccharides, *Ann. N.Y. Acad. Sci.* **556**:81–94.

Oppenheim, J. J., Zachariae, C. O. C., Mukaida, N., and Matsushima, K., 1991, Properties of the novel proinflammatory supergene "intercrine" cytokine family, *Annu. Rev. Immunol.* **9:**617–648.

Pardi, R., Inverardi, L., and Bender, J. R., 1992, Regulatory mechanisms in leukocyte adhesion: Flexible receptors for sophisticated travelers, *Immunol. Today* **13:**224–231.

Rot, A., 1992, Endothelial cell binding of NAP-1/IL-8: Role in neutrophil emigration, *Immunol. Today* **13:**291–294.

Ruoslahti, E., and Yamaguchi, Y., 1991, Proteoglycans as modulators of growth factor activities, *Cell* **64:**867–869.

Schall, T. J., 1991, Biology of the RANTES/SIS cytokine family, *Cytokine* **3:**165–183.

Schweighoffer, T., and Shaw, S., 1992a, Concepts in adhesion regulation, in: *Handbook of Immunopharmacology: Adhesion Molecules* (C. D. Wegner, ed.), Academic Press, New York.

Schweighoffer, T., and Shaw, S., 1992b, Adhesion cascades: Diversity through combinatorial strategies, *Curr. Opin. Cell Biol.* **4:**824–829.

Shimizu, Y., van Seventer, G. A., Horgan, K. J., and Shaw, S., 1990, Regulated expression and function of three VLA (β1) integrin receptors on T cells, *Nature* **345:**250–253.

Shimizu, Y., Newman, W., Tanaka, Y., and Shaw, S., 1992, Lymphocyte interactions with endothelial cells, *Immunol. Today* **13:**106–112.

Tanaka, Y., Albelda, S. M., Horgan, K. J., van Seventer, G. A., Shimizu, Y., Newman, W., Hallam, J., Newman, P. J., Buck, C. A., and Shaw, S., 1992, CD31 expressed on distinctive T cell subsets is a preferential amplifier of β1 integrin-mediated adhesion, *J. Exp. Med.* **176:**245–253.

Tanaka, Y., Adams, D. H., Hubscher, S., Hirano, H., Siebenlist, U., and Shaw, S., 1993a, Proteoglycan-immobilized MIP-1β induces adhesion of T cells, *Nature* **361:**79–82.

Tanaka, Y., Adams, D. H., and Shaw, S., 1993b, Proteoglycan on endothelial cells present adhesion-inducing cytokines to leukocytes, *Immunol. Today* **14:**111–114.

Wolpe, S. D., Davatelis, G., Sherry, B., Beutler, B., Hesse, D. G., Hguyen, H. T., Moldawer, L. L., Nathan, C. F., Lowry, S. F., and Cerami, A., 1988, Macrophages secrete a novel heparin-binding protein with inflammatory and neutrophil chemokinetic properties, *J. Exp. Med.* **167:**570–581.

Zimmerman, G. A., Prescott, S. M., and McIntyre, T. M., 1992, Endothelial cell interactions with granulocytes: Tethering and signaling molecules, *Immunol. Today* **13:**93–99.

11

Lung Inflammation and Adhesion Molecules

MICHAEL S. MULLIGAN and PETER A. WARD

1. Introduction

Adhesion molecules have been shown to be important in inducing inflammatory responses in experimental animals. The topic has recently been the subject of a fairly extensive review (Harlan and Liu, 1992). Adhesion molecules on endothelial cells and leukocytes are diverse, and can be divided into two groups: those that are constitutively expressed, and those that are inducible after endothelial cell contact with appropriate stimuli. Table I presents some of the adhesion molecules of endothelial cells and neutrophils that are most important in the inflammatory response. In this context, the chief difference between endothelial cells and leukocytes is that adhesion molecules are generally not constitutively expressed on the former, whereas they are on the latter. ICAM-1 and -2 are normally constitutively expressed in small amounts on endothelial cells. When endothelial cells are stimulated with TNFα, IL-1, or endotoxin, gene activation occurs, and ICAM expression is slowly but steadily increased over the next 12 hr. Also triggered is the gene controlling expression of E-selectin (ELAM-1), with maximal expression developing in about 4 hr. P-selectin is an exception to the requirement for protein synthesis; this glycoprotein is stored in the Weibel–Palade granules of endothelial cells (and in the alpha granules of platelets), and can be rapidly translocated (in 5–10 min) to the endothelial cell surface after the addition of histamine or thrombin. The most recently described adhesion-

MICHAEL S. MULLIGAN and PETER A. WARD • Department of Pathology, The University of Michigan, Ann Arbor, Michigan 48109.

Cellular Adhesion: Molecular Definition to Therapeutic Potential, edited by Brian W. Metcalf, Barbara J. Dalton, and George Poste. Plenum Press, New York, 1994.

Table I. Adhesion Molecules Involved in the Inflammatory Response

Source	Common term	Synonyms	"Counterreceptor"
Leukocytes	LFA-1	CD11a/CD18	ICAM-1,2
	Mac-1	CD11b/CD18, Mo-1	ICAM-1
		CD11c/CD18	Unknown
	L-selectin	MEL-14	?Gly-CAM-1
	VLA-4	α4β2	VCAM-1
Endothelial cell	ICAM-1		As above
	ICAM-2		As above
	VACM-1		As above
	E-selectin	ELAM-1	Oligosaccharide
	P-selection	GMP-140, PADGEM	Oligosaccharide

promoting molecule of the endothelial cell is Gly-CAM-1, a heavily glycosylated protein which, unlike the other adhesion molecules, has no transmembrane-spanning segment and appears to be entirely extracellular, embedded in the glycocalyx of the endothelial cells (Lasky et al., 1992). VCAM-1 is another adhesion-promoting molecule of endothelial cells. This glycoprotein is not normally expressed; it appears on the cell surface approximately 4 hr after stimulation, with expression being retained for the next 12–18 hr. "Counterreceptors" for these adhesion molecules are diverse (Table I). In the case of ICAM-1 and -2, the complementary reactive molecules on leukocytes are the β2 integrins (LFA-1 and Mac-1, see below). E- and P-selectin react with leukocytic lectins, which are oligosaccharides of the structure sialyl Lewis[x] and sialyl Lewis[a] (reviewed by Harlan and Liu, 1992). Gly-CAM-1 appears to be the lectin-containing molecule that is reactive with leukocytic L-selectin (see below).

Adhesion-promoting molecules on leukocytes include the β1 integrin heterodimer, α4β2 (also known as VLA-4). This molecule is not expressed on neutrophils, but it is found constitutively on T cells and monocytes and is reactive with endothelial VCAM-1. Its blocking results in reduced recruitment of T cells and monocytes in delayed-type hypersensitivity reactions in the skin (Issekutz and Wykretowicz, 1991). The β2 integrins are the major adhesion molecules constitutively expressed on leukocytes. These are subclassified as follows: CD11a/CD18 (LFA-1); CD11b/CD18 (Mac-1; Mo-1); and CD11c/CD18 (p150/95). Reserves of CD11b/CD18 and CD11c/CD18 in cytoplasmic granules of neutrophils permit substantial upregulation of the molecules on the surfaces of neutrophils after leukocyte activation by contact with chemotactic factors and other agonists. LFA-1 is the "counterreceptor" for ICAM-1 and -2. Mac-1 is reactive with ICAM-1, while the ligand for CD11c/CD18 is unknown. L-selectin, as the name implies, is reactive with an oligosaccharide on the endothelial cell, recently termed Gly-CAM-1 (Lasky et al., 1992). L-selectin is unique among the neutrophil-related adhesion molecules: when the neutrophil is activated, it is shed from the surface of the cell, resulting in a profound decrease in concentration on the cell surface (Smith *et al.*, 1991).

It has recently been suggested that at least two different adhesion-promoting pathways become sequentially engaged during the inflammatory response (Jutila *et al.*, 1989; Harlan and Liu, 1992). The engagement of selectin-dependent pathways appears to be the first step in events leading to the recruitment of leukocytes. The slowing of the rapid flow of leukocytes results in the "rolling" of leukocytes along the surface of the endothelium. This is now assumed to be a reflection of intermittent adhesive interactions between leukocytes and the endothelium, and is thought to result from the engagement of selectin molecules of both endothelial cells and leukocytes. The infusion of oligosaccharides bearing the sialyl Lewis[x] structure appears to interfere with these adhesive interactions. The next phase in adhesive interactions seems to involve the engagement of the β2 integrins, namely LFA-1 and Mac-1 integrins and endothelial ICAMs, which results in a cessation of movement along the endothelial surface and the transmigration of leukocytes beyond the vascular barrier.

2. *IgG Immune Complex-Induced Lung Injury*

One useful model of acute lung injury involves the intrapulmonary deposition of IgG immune complexes. Rabbit polyclonal IgG antibody to bovine serum albumin (anti-BSA) is instilled into the airways of rats, and BSA is injected intravenously. As IgG immune complexes deposit in the pulmonary parenchyma, complement activation occurs, and large numbers of neutrophils are recruited into the interstitial and intraalveolar compartments. Intense vascular endothelial and alveolar epithelial cell injury follows. Injury is quantitated by permeability change (leakage of ^{125}I-labeled albumin from the blood) and hemorrhage (extravasation of ^{51}Cr-labeled RBC), in which radioactivity counts in the lung are related to the amount of radioactivity remaining in the blood (1.0 ml) obtained at the time of sacrifice. It has been shown that injury in this model is complement- and neutrophil-dependent (Johnson and Ward, 1981). It would appear that toxic products of oxygen (H_2O_2, O_2^-, and the hydroxyl radical, HO·) from these cells are responsible for the injury. It has recently been shown that the full expression of injury requires the availability of L-arginine (Mulligan *et al.*, 1991a). In all likelihood, L-arginine undergoes metabolism in phagocytic cells to nitric oxide (·NO), then to peroxynitrite anion ($ONOO^-$), and ultimately to HO· (Beckman *et al.*, 1990). $ONOO^-$ and HO· are highly reactive molecules, and can chemically alter a large variety of target molecules. It should also be noted that HO· generated in this manner does not require the availability of a heavy metal such as iron.

The IgG immune complex model of lung injury has now been shown to be dependent on the formation of TNFα and IL-1, based on the blocking effects of antibodies to these cytokines or the use of the recombinant IL-1 receptor antagonist or recombinant soluble TNFα receptor-1 protein (Warren *et al.*, 1989; Mulligan and Ward, 1992). Both anti-TNFα and anti-IL-1 reduce injury by interfering with the influx of neutrophils. Although it is not known how this occurs, it is likely that the cytokines are released from immune complex-activated pulmonary macrophages and subsequently cause the upregulation of pulmonary vascular endothelial adhesion molecules, such as ICAM-1 and E-selectin.

In the IgG immune complex model of lung injury, a requirement for β2 integrins, as defined by the use of F(ab')$_2$ to CD18, has been demonstrated (Mulligan *et al.*, 1992b). The F(ab')$_2$ version of the anti-CD18 (CL-26) was used because infusion of the intact antibody into rats caused neutropenia. *In vitro* the anti-CD18 reduced adherence of PMA-stimulated human to TNFα-stimulated human umbilical vein endothelial cells by more than 79%, and by a similar amount when PMN-stimulated rat neutrophils were incubated with TNFα-stimulated rat pulmonary artery endothelial cells (Table II). When the F(ab')$_2$ anti-CD18 was employed *in vivo* (with intravenous injections of the antibody at 2.5, 3.0, and 3.5 hr) in the model of IgG immune complex-induced lung injury, the permeability and hemorrhage indices were reduced by 71 and 75%, respectively, at the lower dose (0.75 mg) of anti-BSA (Table III). Under these conditions, the accumulation of neutrophils in lung tissue, as assessed by myeloperoxidase (MPO), was reduced by 52%. Not surprisingly, as the amount of anti-BSA was increased, there was a diminishing ability of the anti-CD18 to protect against injury and reduce neutrophil influx (Table III). The participation of CD11 (a, b, c) in this type of inflammatory injury is not known because of the lack of available blocking antibodies.

It is useful to note that engagement of β2 integrins in a given inflammatory

Table II. *In Vitro* Protective Effects of Anti-β2 Integrins[a]

Materials employed[b]	Antibody target	Functional outcome (% of positive control)	
		Adherence	Cytotoxicity
PMA–human PMN	CD11a	87.5 ± 11	98.6 ± 7.9
	CD11b	7.8 ± 2.3[c]	65.2 ± 4.7[c]
	CD11c	64.1 ± 9.7[c]	47.8 ± 4.8[c]
	C18	<3	5 ± 2
PMA–rat PMN	CD11b	37.2 ± 2.9[c]	
	CD18	46 ± 2.3[c]	

[a]Taken from data Mulligan *et al.* (1992b).
[b]TNFα-activated rat pulmonary artery endothelial cells were employed.
[c]$p < 0.05$.

Table III. Dose–Response Relationship of Blocking Interventions in IgG Immune Complex-Induced Lung Injury[a]

Expt	Dose of anti-BSA (mg)	Intervention[b]	Protection (%)[c] Permeability	Protection (%)[c] Hemorrhage	MPO content Reduction (%)
A	0.75	Anti-CD18	71 (<0.05)	75 (<0.001)	52 (<0.001)
	2.5	"	55 (<0.003)	60 (<0.001)	18 (<0.05)
	3.3	"	34 (<0.05)	50 (<0.001)	65 (N.S.)
B	0.75	Anti-E-selectin	94 (<0.04)		(<0.04)
	1.75		97 (<0.001)		(0.008)
	2.5		52 (<0.003)	94 (<0.001)	(0.007)
	3.3		13 (N.S.)		(<0.001)

[a] Taken from data Mulligan *et al.* (1991a, 1991b).
[b] 200 μg F(ab′)$_2$ anti-CD18 or anti-E-selectin was injected intravenously in equally divided dosees at 2.5, 3.0, and 3.5 hr.
[c] Calculated on the basis of rats treated with irrelevant F(ab′)$_2$ mouse IgG1. Numbers in parentheses represent *p* values calculated by comparisons to positive controls treated with irrelevant MOPC-21.

response can be estimated by the use of TNFα-stimulated endothelial cell monolayers in the presence of PMA-activated neutrophils, using monoclonal antibodies to the subunits of β2 integrins, as shown in Table II. When PMA-activated human neutrophils were used with monolayers of rat pulmonary artery endothelial cells, the blocking activities of intact monoclonal antibodies to human CD11a, CD11b, and CD11c (and CD18, as described above) were used to define interference with adhesive interaction and with killing of endothelial cells by activated neutrophils. It is clear that antibody to CD11a was ineffective in either the adhesion of neutrophils or the killing of the endothelial cells. In contrast, both anti-CD11b and anti-CD11c significantly reduced the adhesion and cytotoxicity outcomes. As expected, anti-CD18 was the most effective blocking antibody. This would suggest that CD11a (LFA-1) plays a minor role, while Mac-1 and CD11c are important in adhesive interactions that result in damage to endothelial cells by the products of activated neutrophils. The problem with this type of data is that it does not determine whether the results with anti-CD11a preclude a role for CD11a, or whether the particular antibody is a relatively poor blocking agent. The effects of anti-CD11c were surprising, since the biologic role for this molecule is largely unknown. The results with PMA-activated rat neutrophils were limited by the availability of anti-human CD antibodies that cross-react with rat epitopes. To date, only the anti-CD18 has been used *in vivo* [as the F(ab′)$_2$ fragment]. The anti-CD11b was of the IgM class and has had unpredictable effects, perhaps because of its large size. The observations demonstrate that to a limited extent, antibody candidates for *in vivo* use can be screened by these *in vitro* evaluations. In the course of these studies, and for reasons that will be described below (see Section 3), it became apparent that anti-

CD18 has some direct effect on the effector function of phagocytic cells, suggesting that the protective effects of antibodies to β2 integrins may extend beyond an impact on the recruitment of blood leukocytes.

The ability of anti-CD18 and anti-E-selectin to protect in the model of IgG immune complex-induced lung injury is shown as a function of the dose of anti-BSA employed. As Table III shows, anti-CD18 and anti-E-selectin were maximally protective at the lowest dose of anti-BSA employed, and therefore under conditions in which the smallest amount of immune complex would be deposited in the lung. Increasing the dose of anti-BSA ultimately resulted in losing the protective effects of these interventions. This is to be expected, since numerous cases in the literature document that when agents that trigger acute inflammatory responses are administered at very high doses, the requirements for specific inflammatory mediators are often overcome.

E-selectin has recently been shown to be vitally involved in expressing injury in the IgG immune complex model of lung injury. Substantial protection (as much as 95%) has been achieved by the blocking of E-selectin when monoclonal antibody to human E-selectin with cross-reactivity to a rat epitope is used (Mulligan *et al.*, 1991b). The protective effects have been directly correlated with a reduced influx of neutrophils, suggesting that E-selectin expression allows for early adhesive interactions between endothelial cells and neutrophils, followed by β2 integrin-dependent transmigration of neutrophils.

It is important to exercise caution in this type of study, both in using monoclonal antibodies and in interpreting results, especially when the results are negative. We evaluated two monoclonal antibodies (CL-3 and CL-37), which were produced to human E-selectin and found to cross-react with a rat endothelial epitope. Neither antibody in its intact form (IgG1) suppressed injury *in vivo*, and *in vitro* neither one reduced the adhesion of neutrophils to TNFα-stimulated rat pulmonary artery endothelial cells. However, when F(ab')$_2$ fragments of CL-3 were employed *in vitro*, they reduced neutrophil adhesion to TNFα-stimulated endothelial cells; *in vivo*, they protected against IgG immune complex-induced lung injury, and this could be directly linked to reduced neutrophil influx. When F(ab')$_2$ fragments of CL-37 were used, they did not prevent *in vitro* adhesion of neutrophils to TNFα-stimulated endothelial cells or show protective effects *in vivo*. Caution is thus clearly called for in such studies.

The IgG immune complex model also demonstrates other features important in the pathophysiology of events leading inflammatory tissue to injury (Johnson and Ward, 1974). A lipid product of phagocytic cells, platelet activating factor, was named because of early evidence that this compound had a powerful biological effect on platelets, causing them to aggregate. It was also soon recognized that this lipid had pleiotropic functions and was a potent agonist for neutrophils and macrophages, causing them to undergo signal transduction, which resulted in functions such as O_2^- generation. PAF at less than micromolar concentrations could also "prime" phagocytic cells for greatly enhanced effector responses after the introduction of a second agonist. Since it has been shown that PAF is released from both macrophages and neutrophils, the role of PAF in IgG immune complex-induced vascular injury in rat skin and lung was evaluated by using antagonists for the PAF receptor. PAF receptor antagonists were highly protective against the development of vascular injury in IgG immune complex-induced dermal vasculitis and lung injury. The important observation was made that there was no reduction in the tissue accumulation of neutrophils in the presence of PAF receptor antagonists (Warren *et al.*, 1990). It was thus concluded that PAF does not facilitate the recruitment of neutrophils, but enhances their function once they have arrived at tissue sites containing deposits of immune complexes. Thus, lung injury that is induced by intrapulmonary deposits of IgG immune complexes is intensified by the local release of PAF.

3. *IgA Immune Complex Model of Lung Injury*

The intrapulmonary deposition of IgA immune complexes represents another model of acute lung injury, and is characterized by substantial injury to rat alveolar epithelial and vascular endothelial cells. The model was developed using a murine myeloma cell line that produces IgA with reactivity to dinitrophenol (DNP). The antibody is instilled into the airways of rats, and the antigen (linked to BSA) is injected intravenously. Immune complexes form in the alveolar walls and, in a complement-dependent manner, apparently activate intrapulmonary macrophages that release toxic products derived from O_2^- and L-arginine (as described above). These products appear to impact directly on alveolar epithelial cells and capillary endothelial cells, causing damage and destruction. Support for these conclusions came from the discovery that treatment with catalase or superoxide dismutase provide 93 and 100% reduction in the permeability increase, while treatment with the arginine analogue, L-NMA, reduces injury by 92%. The permeability increase was almost totally reduced by the copresence of L-arginine, but not by D-arginine (Johnson *et al.*, 1984, 1986; Mulligan *et al.*, 1992a). This model appears to be unique: few neutrophils are recruited into the alveolar compartment, and even those that are recruited into the lung are irrelevant to the injury, since neutrophil depletion is not protective. This model is important because it is one of the few in which IgA immune complexes are tissue-damaging; it can thus be used to assess pathogenic mechanisms that may be relevant to human diseases that feature tissue accumulation of IgA immune complexes (Berger's nephropathy, Henoch–Schönlein vasculitis). Although injury in this model appears to be macrophage-dependent, it is ironic that little, if any, cytokine presence can be detected in the bronchoalveolar lavage fluids, and pretreatment of rats with antibodies to TNFα or IL-1 is not protective.

Furthermore, in dramatic contradistinction to the requirement for E-selectin in IgG immune complex-induced lung injury, there is no such requirement for injury that follows the intrapulmonary deposition of IgA immune complexes (Mulligan et al., 1992a). The lack of a role for E-selectin is not altogether surprising, since the chief effector cells in this model of injury appear to be residential tissue macrophages. The treatment of rats undergoing intrapulmonary deposition of IgA immune complexes with CL-3 (F(ab')$_2$ anti-E-selectin is remarkable because of the absence of any protective effects, in striking contrast to the first model of lung injury described above. Although a faint expression of E-selectin in the pulmonary vasculature can be detected immunohistochemically approximately 4 hr after IgA immune complex deposition, this finding seems to be largely irrelevant to the outcome of injury, for the reasons cited above.

Although the recruitment of blood leukocytes into the lung in the IgA immune complex-induced model of injury does not seem to be significant, we have recently demonstrated that the systemic treatment of rats with F(ab')$_2$ in anti-CD18 (using the same protocol described for the IgG immune complex model of lung injury) reduces the 4-hr permeability and hemorrhage indices by 87 and 74%, respectively (Table IV). Treating rats with IgM anti-CD11b did not provide protection, but this must be viewed with caution because of the class of antibody used. It is not yet possible to determine the requirements for CD11 in this model of injury.

The discovery that CD18 was necessary in the IgA immune complex model was interesting since, for the reasons given above, recruitment of phagocytic

Table IV. Comparison of Antiadhesion Interventions in Lung Injury Models[a]

Lung model of injury	Requirement for neutrophils[b]	Time of intervention endpoint	% protection	
			Permeability	Hemorrhage
IgG immune complex[c]	Yes	4 hr anti-CD18	55.4 (<0.01)[e]	60 (<0.01)
		anti-E-selectin	52.1 (<0.003)	94 (<0.001)
		anti-P-selectin	<10 (N.S.)	<10 (N.S.)
IgA immune complex[d]	No	4 hr anti-CD18	87 (<0.005)	74 (<0.004)
		anti-E-selectin	<5 (N.S.)	<5 (N.S.)
CVF-induced complement activation	Yes	30 min anti-CD11b	63.4 (<0.001)	50.0 (<0.01)
		anti-CD18	69.1 (<0.02)	74.4 (<0.001)
		anti-P-selectin	50.8 (<0.001)	70.9 (<0.001)

[a]Taken from data in Mulligan et al. (1991a, 1991b).
[b]200-μg antibody preparations were used. In the case of anti-CD18 and anti-E-selectin, F(ab')$_2$ fragments were used. For anti-P-selectin and anti-CD11b, intact IgG and IgM, respectively, were used.
[c]2.5 mg anti-BSA was used.
[d]1.2 mg IgA anti-DNP was used.
[e]Numbers in parentheses are *p* values related to comparisons with positive controls treated with irrelevant MOPC-21.

cells from the bloodstream was considered unlikely, and was not required for the processes leading to tissue injury. It has already been shown that the adherence of phagocytic cells to solid surfaces results in a β2-integrin-dependent enhancement in the generation of H_2O_2 and O_2^- (Nathan, 1987, 1989; Shappell et al., 1990; Gresham et al., 1991). Accordingly, experiments were designed using rat alveolar macrophages and BSA–anti-BSA immune complexes presented in solid phase (attached to a plastic surface) (Mulligan et al., 1992b). The production of O_2^- as well as TNFα was then measured. As shown in Table V, in a 30-min period alveolar macrophages produced 4.4 ± 0.3 nmole O_2^- and 2.9 ± 0.3 units of TNFα. When the complexes were pretreated with fresh human serum the O_2^- and TNFα values rose to 6.4 ± 0.5 and 6.2 ± 1.1, respectively. It appears that this enhancement was related to a requirement for complement, since neither heated serum nor serum manipulated to achieve complement consumption (by adding cobra venom factor) was able to cause any increase in the O_2^- and TNFα responses of the macrophages. The presence of anti-CD18 abolished the increments in O_2^- and TNFα formation caused by the exposure of immune complexes to fresh serum. Furthermore, anti-CD18 did not alter the responses of the alveolar macrophages to PMA (Table V). These data suggest that in order to achieve optimal O_2^- and TNFα responses of rat alveolar macrophages to serum-opsonized immune complexes opsonized with unheated serum, a part must be played by CD18. The extent to which CD11a, CD11b, or CD11c may be involved is unclear. It is tempting to attribute the response to the participation of CD11b, since this, with CD18, represents the complement receptor 3 (CR3) reactive with the complement activation product (iC3b) that has become physically affixed to the immune complexes, but there is no proof for this. Whatever the mechanism, these data underscore the fact that CD18 has a variety of biologic functions, including direct effector responses of phagocytic cells and influence on the recruitment of leukocytes from the vasculature.

Table V. CD18-Dependent Effector Functions of Rat Alveolar Macrophages[a]

Material added to macrophages[b]	Blocking antibody	O_2^- response (nmole/30 min)	TNFα production (units/30 min)
IgG immune complexes[c]	None	4.4 ± 0.3	2.9 ± 0.3
Complement-opsonized	None	6.4 ± 0.5	6.2 ± 1.1
immune complexes	Anti-CD18	3.6 ± 0.4^d	4.9 ± 0.4^d
PMA	None	6.1 ± 0.8	5.3 ± 0.5
	Anti-CD18	5.7 ± 0.9	5.7 ± 0.3

[a] Taken from data in Mulligan et al. (1992a).
[b] Immune complexes were present in solid phase.
[c] Negative control O_2^- and TNF responses were 0.4 ± 0.2 nmole/O_2^-/30 min and 0.6 ± 0.5 unit/30 min, respectively.
[d] As compared with response to complement-opsonized IgG immune complexes, $p < .05$.

4. Lung Injury following Systemic Activation of Complement

Several studies have confirmed that intravascular activation of complement leads to neutropenia. It was suggested that this was the result of intravascular stimulation of neutrophils, which formed leukoaggregates that became sequestered within capillary beds. Craddock et al. (1977) showed that a similar event occurred in humans during hemodialysis or with extracorporeal blood flow during cardiopulmonary bypass. A decade ago, we developed a rat lung model of acute injury in which cobra venom factor (CVF) was infused intravascularly (Till et al., 1982). This resulted in neutropenia, intravascular leukoaggregation and entrapment of leukocytes in the pulmonary capillary vasculature, and lung injury as demonstrated by increased vascular permeability and intraalveolar hemorrhage. Morphologically, focal necrosis of capillary endothelial cells was demonstrated wherever neutrophils were in direct contact with the vascular endothelium. The endothelial damage was demonstrated to be complement- and neutrophil-dependent, and was associated with the appearance in lung parenchyma and plasma of lipid peroxidation (conjugated dienes, fluorescent products, lipid peroxides). All manifestations of injury and evidence of lipid peroxidation were reversed in the presence of catalase, iron chelator (deferroxamine), or hydroxyl-radical scavengers (Ward et al., 1983). Vascular injury in this model was very rapid in onset, being measurable by 10–15 min and appearing to peak by 30 min. Because of the requirement for complement and neutrophils, this model appeared to be a prime candidate for the role of rapidly expressible adhesion molecules. Antibody to CD18 reduced the permeability and hemorrhage indices by 69 and 74%, respectively (Table IV). In this model, treatment of rats with anti-CD11b did show substantial protective effects, reducing the permeability and hemorrhage parameters by 50 and 63%, respectively (Mulligan et al., 1993a). By morphological examination, treatment with anti-CD11b greatly reduced the physical contact of neutrophils with the vascular endothelial surfaces of the lung. Thus, a role for Mac-1 (CR3) has been confirmed and might be consistent with the in vitro data in Table IV. The possible involvement of CD11a (CD11c) in the CVF model of lung injury is as yet unknown.

As indicated above, the very rapid nature of the developing lung injury in the CVF model suggests that the best candidates for adhesion molecules are those that can be rapidly upregulated. The finding for the role of CD11b/CD18 is consistent with this hypothesis. Of the endothelial adhesion molecules, P-selectin is a prime candidate. Since histamine is known to cause the translocation of P-selectin from Weibel–Palade granules to endothelial cell membrane surfaces, and because the complement-derived anaphylatoxins, C3a and C5a, are known to cause histamine release from basophils and mast cells, we evaluated a linkage in the CVF model between complement activation products, histamine, and P-selectin. Fortunately, a monoclonal antibody (PB1.3) to human P-selectin was found to be cross-reactive with a product on activated rat platelets (which also express P-selectin on their surfaces). When the lung sections were examined immunohistochemically with PB1.3 after infusion of CVF, there was dramatic staining of the interstitial capillary and venular surfaces, in contrast to the lack of staining in normal rat lung (Mulligan *et al.*, 1992c). This antibody also interfered with the ability of thrombin-activated rat platelets to bind to rat neutrophils. Pretreatment of rats with PB1.3 and anti-P-selectin dramatically reduced, in a dose-dependent manner, the evidence of lung injury. An infusion of 200 μg of the intact antibody reduced the permeability and hemorrhage parameters by 51 and 71%, respectively (Table IV). This was associated with a 41% reduction in lung MPO content, and with substantially reduced physical contact of neutrophils with the pulmonary vascular endothelial surfaces after infusion of CVF (Mulligan *et al.*, 1992c). The use of a companion monoclonal antibody (PNB1.6), which failed to block the interaction of activated rat platelets with rat neutrophils but nevertheless reacted with a rat P-selectin epitope, demonstrated the total lack of protective effects *in vivo* in the CVF model, confirming the specificity of PB1.3 for a key epitope of rat P-selectin involved with binding interactions with neutrophils. The CVF model of lung injury in the rat appears to be the first experimental model of inflammatory injury in which a key role for P-selectin has been demonstrated.

Table IV compares protective interventions intended to block adhesion molecules; these are shown for the three models of acute lung injury in rats. In each of the models, anti-CD18 reduced injury by 50% or more, whether permeability or hemorrhage was used as the endpoint. Anti-E-selectin, which has been used to date only in the two immune complex models, significantly reduced injury in the IgG immune complex-induced model (reducing permeability and hemorrhage parameters by 52 and 94%, respectively), but had no effect on the IgA immune complex model. Anti-CD11b was highly protective in the CVF model of injury (reducing permeability by 63% and hemorrhage by 50%), as was anti-P-selectin. The differences can be explained by the time differences in the endpoints as well as the difference in the involvement of blood-derived leukocytes, as described above.

5. Differing Tissue Requirements for TNFα and IL-1 in IgG Immune Complex-Induced Injury

We came to several unexpected conclusions when we evaluated the cytokine requirements for the development of IgG immune complex-induced vascular injury in the skin and lung of rats (Warren, 1991; Warren *et al.*, 1989, 1991). It was previously shown that blocking TNFα or IL-1 significantly reduced the development of lung injury. In the case of TNFα, anti-TNFα-induced protection was associated with a greatly reduced influx of neutrophils. In subsequent studies we used three systemic interventions: anti-TNFα, anti-IL-1, and IL-1 receptor antagonist (IL-1ra) (Mulligan and Ward, 1992). The results are summarized in Table VI. Anti-TNFα consistently had no protective effect over a range of anti-BSA doses used in the skin, and there was no reduction in MPO content. When animals were treated with anti-IL-1 or IL-1ra, there was protection at all doses of anti-BSA employed, and protection was correlated with reduced MPO content in the skin. Shown as parallel studies in which animals were systemically treated with the same protocols, all three interventions were highly protective in lungs following the deposition of IgG immune complexes. The protective effects were again associated with reduced levels of MPO content in tissue. Using immunoperoxidase technology, we showed that BSA–anti-BSA immune complex deposition in lung was associated with the appearance in pulmonary macrophages of TNFα and IL-1. This is consistent with the earlier finding that BAL fluids from these animals contained a mixture of both cytokines (Warren *et al.*, 1989). No staining for TNFα was found on skin sites with BSA–anti-BSA immune complexes, but interstitial cells with large amounts of cytoplasm demonstrated a granular-staining pattern for IL-1. The appearance was consistent with that of mast cells. These results suggest that the deposition of IgG immune complexes triggers a selective elaboration of IL-1 in the skin. They also suggest that the cellular source of IL-1 may involve a different population from that of the lung.

Table VI. Differing Cytokine Requirements for Vascular Injury in Skin and Lung[a]

Vascular bed[b]	Amount of anti-BSA (mg)	Intervention[c]	% reduction[d]	
			Permeability	MPO content
Skin	0.21	Anti-TNFα	<5 (N.S.)	65 (N.S.)
	0.42	"	<5 (N.S.)	65 (N.S.)
	0.84	"	<5 (N.S.)	65 (N.S.)
Skin	0.21	Anti-IL-1	42 (0.001)	83 (<0.001)
	0.42	"	41 (0.002)	80 (<0.001)
	0.84	"	28 (0.13)	84 (<0.001)
Skin	0.21	IL-1ra	38 (0.008)	77 (<0.001)
	0.42	"	34 (0.005)	84 (<0.001)
	0.84	"	32 (0.006)	87 (<0.002)
Lung	2.5	Anti-TNFα	65 (<0.001)	50 (0.007)
	2.5	Anti-IL-1	69 (<0.001)	41 (0.01)
	2.5	IL-1ra	65 (<0.001)	45 (0.001)

[a] Taken from data in Mulligan and Ward (1992).
[b] All analyses were performed on tissue 4 hr after deposition of BSA–anti-BSA complexes.
[c] Animals were injected intravenously at time 0 with 75 μg goat or rabbit IgG polyclonal antibody to murine recombinant TNFα or IL-1β. Both antibodies were shown to be cross-reactive with the relevant rat cytokine. IL-1ra, IL-1 receptor antagonist.
[d] Reduction was calculated by comparison with rats treated with irrelevant goat IgG. Numbers in parentheses refer to *p* values, which were calculated by comparisons with positive controls treated with irrelevant IgG.

It was considered possible that the dermal vasculature might also be refractory to the effects of TNFα. Accordingly, dermal vascular sites of immune complex deposition were coinjected with either human recombinant IL-1 or TNFα. In both circumstances, the exogenous administration of these cytokines resulted in intensified vascular injury, as represented by either permeability change or hemorrhage. These findings suggest that the dermal vasculature is reactive to TNFα. In the case of immune complex deposition, there is simply no generation of TNFα. It remains to be determined whether this implies that the dermal interstitium is incapable of generating TNFα after stimulation, or whether the appropriate agonist has not yet been used.

References

Beckman, J. S., Beckman, T. W., Chen, J., Marshall, P. A., and Freeman, B. A., 1990, Apparent hydroxyl radical production by peroxynitrite: Implications for endothelial injury from nitric oxide and superoxide, *Proc. Natl. Acad. Sci. USA* **87**:1620–1624.

Craddock, P. R., Fehr, J., Dalmasso, A. P., Brigham, K. L., and Jacob, H. S., 1977, Hemodialysis leukopenia. Pulmonary vascular leukostasis resulting from complement activation by dialyzer cellophane membranes, *J. Clin. Invest.* **59**:879–888.

Gresham, H. D., Graham, I. L., Anderson, D. C., and Brown, E. J., 1991, Leukocyte adhesion-deficient neutrophils fail to amplify phagocytic function in response to stimulation: Evidence for CD11b/CD18-dependent and -independent mechanisms of phagocytosis, *J. Clin. Invest.* **88**:588–597.

Harlan, J. M., and Liu, D. Y. (eds.), 1992, *Adhesion: Its Role in Inflammatory Disease*, Freeman, San Francisco.

Issekutz, T. B., and Wykretowicz, A., 1991, Effect of a new monoclonal antibody, TA-2, that inhibits lymphocyte adherence to cytokine stimulated endothelium in the rat, *J. Immunol.* **147**:109–116.

Johnson, K. J., and Ward, P. A., 1974, Acute immunologic pulmonary alveolitis, *J. Clin. Invest.* **54**:349–357.

Johnson, K. J., and Ward, P. A., 1981, Role of oxygen metabolites in immune complex injury of lung, *J. Immunol.* **126**:2365–2369.

Johnson, K. J., Wilson, B. S., Till, G. O., and Ward, P. A., 1984, Acute lung injury in the rat caused by immunoglobulin A immune complexes, *J. Clin. Invest.* **74**:358–359.

Johnson, K. J., Ward, P. A., Kunkel, R. G., and Wilson, B. S., 1986, Mediation of IgA induced lung injury in the rat: Role of macrophages and reactive oxygen products, *Lab. Invest.* **54**:499–506.

Jutila, M. A., Rott, L., Berg, E. L., and Butcher, E. C., 1989, Function and regulation of the neutrophil MEL-14 antigen in vivo: Comparison with LFA-1 and MAC-1, *J. Immunol.* **143**:3318–3324.

Lasky, L. A., Singer, M. S., Dowbenko, D., Imai, Y., Henzel, W. J., Grimley, C., Fennie, C., Gillett, N., Watson, S. W., and Rosen, S. D., 1992, An endothelial ligand for L-selectin is a novel mucin-like molecule, *Cell* **69**:927–938.

Mulligan, M. S., and Ward, P. A., 1992, Immune complex-induced lung and dermal vascular injury: Differing requirements for TNFα and IL-1, *J. Immunol.* **145**:331–339.

Mulligan, M. S., Hevel, J. M., Marletta, M. A., and Ward, P. A., 1991a, Tissue injury caused by deposition of immune complexes is L-arginine dependent, *Proc. Natl. Acad. Sci. USA* **88**:6338–6342.

Mulligan, M. S., Varani, J., Dame, M. K., Lane, C. L., Smith, C. W., Anderson, D. C., and Ward, P. A., 1991b, Role of ELAM-1 in neutrophil-mediated lung injury in rats, *J. Clin. Invest.* **88**:1396–1406.

Mulligan, M. S., Warren, J. S., Smith, C. W., Anderson, D. C., Yeh, C. G., Rudolph, A. R., and Ward, P. A., 1992a, Lung injury after deposition of IgA immune complexes: Requirements for CD18 and L-arginine, *J. Immunol.* **148**:3086–3092.

Mulligan, M. S., Varani, J., Warren, J. S., Till, G. O., Smith, C. W., Anderson, D. C., Todd, R. F., III, and Ward, P. A., 1992b, Roles of β2 integrins of rat neutrophils in complement- and oxygen radical-mediated acute inflammatory injury, *J. Immunol.* **148**:1847–1857.

Mulligan, M. S., Polley, M. J., Bayer, R. J., Nunn, M. F., Paulson, J. C., and Ward, P. A., 1992c, Neutrophil-dependent acute lung injury: Requirement for P-selectin (GMP-140), *J. Clin. Invest.* **90**:1600–1607.

Mulligan, M. S., Smith, C. W., Anderson, D. C., Todd, R. F. III, Miyasaka, M., Tamatani, T., Issekutz, T. B., and Ward, P. A., 1993a, Role of leukocyte adhesion molecules in complement-induced lung injury, *J. Immunol.* **150**:2401–2406.

Mulligan, M. S., Wilson, G. P., Todd, R. F. III, Smith, C. W., Anderson, D. C., Varani, J., Issekutz, T. B., Miyasaka, M., Tamatani, T., Rusche, J. R., Vaporciyan, A. A., and Ward, P. A., 1993b, Role of β1, β2 integrins and ICAM-1 in lung injury following deposition of IsG and IsA immune complexes, *J. Immunol.* **150**:2407–2417.

Nathan, C. F., 1987, Neutrophil activation on biological surfaces: Massive secretion of hydrogen peroxide in response to products of macrophages and lymphocytes, *J. Clin. Invest.* **80**:1550–1560.

Nathan, C. F., 1989, Respiratory burst in adherent neutrophils: Triggering by colony-stimulating factors CSF-GM and CSF-G, *Blood* **73**:301–306.

Shappell, S. B., Toman, C., Anderson, D. C., Taylor, A. A., Entman, M. L., and Smith, C. W., 1990, Mac-1 (CD11b/CD18) mediates adherence-dependent hydrogen peroxide production by human and canine neutrophils, *J. Immunol* **144**:2702–2711.

Smith, C. W., Kishimoto, T. K., Abbassi, O., Hughes, B., Rothlein, R., McIntire, L. V., Butcher, E., and Anderson, D. C., 1991, Chemotactic factors regulate lectin adhesion molecule 1 (LECAM-1)-dependent neutrophil adhesion to cytokine-stimulated endothelial cells *in vitro, J. Clin. Invest.* **87**:609–618.

Till, G. O., Johnson, K. J., Kunkel, R., and Ward, P. A., 1982, Intravascular activation of complement and acute lung injury. Dependency on neutrophils and toxic oxygen metabolites. *J. Clin. Invest.* **69**:1126–1135.

Ward, P. A., Till, G. O., Kunkel, R., and Beauchamp, C., 1983, Evidence for role of hydroxyl radical in complement and neutrophil-dependent tissue injury, *J. Clin. Invest.* **72**:789–801.

Warren, J. S., 1991, Intrapulmonary interleukin-1 mediates acute immune complex alveolitis in the rat, *Biochem. Biophys. Res. Commun.* **175**:604–610.

Warren, J. S., Yabroff, K. B., Remick, D. G., Kunkel, S. L., Chensue, S. W., Kunkel, R. G., Johnson, K. J., and Ward, P. A., 1989, Tumor necrosis factor participates in the pathogenesis of acute immune complex alveolitis in the rat, *J. Clin. Invest.* **84**:1873–1882.

Warren, J. S., Barton, P. A., Mandel, M., and Matrosic, K., 1990, Intrapulmonary tumor necrosis factor triggers local platelet-activating factor production in rat immune complex alveolitis, *Lab. Invest.* **63**:746–754.

Warren, J. S., Barton, P. A., and Jones, M. L., 1991, Contrasting roles for tumor necrosis factor in the pathogenesis of IgG and IgA immune complex lung injury, *Am. J. Pathol.* **138**:581–590.

12

Adhesion Molecule-Dependent Cardiovascular Injury

GILBERT L. KUKIELKA and MARK L. ENTMAN

1. Introduction: Reaction to Injury in Ischemia and Reperfusion

The purpose of this chapter is to discuss the potential mechanisms by which inflammatory injury may complicate ischemic heart disease. It is to be emphasized that no one seriously proposes that the initial injury associated with ischemic heart disease is inflammatory in nature; rather we are describing the mechanism of a reaction to injury, and will examine evidence that this secondary reaction might extend and complicate cardiac injury associated with ischemia and reperfusion. The concept that a reaction to injury extends the primary injury is classic in pathology, but its application to this disease entity is relatively recent (Hillis and Braunwald, 1977).

The concept that inflammation accompanies myocardial infarction is an old one. The infiltration of leukocytes into an experimental myocardial infarction was probably first described by Baumgarten in 1899. In their seminal study of human autopsy material, Mallory and colleagues concluded that in myocardial infarction "necrosis of muscle and infiltration by polymorphonuclear leukocytes are the important features of the first week" (Mallory *et al.*, 1939). However,

GILBERT L. KUKIELKA • Section of Cardiovascular Sciences, The Methodist Hospital, The DeBakey Heart Center, Department of Medicine, and Speros P. Martel Laboratory of Leukocyte Biology, Department of Pediatrics, Texas Children's Hospital, Baylor College of Medicine, Houston, Texas 77030. *MARK L. ENTMAN* • Section of Cardiovascular Sciences, The Methodist Hospital, The DeBakey Heart Center, Department of Medicine, Baylor College of Medicine, Houston, Texas 77030.

Cellular Adhesion: Molecular Definition to Therapeutic Potential, edited by Brian W. Metcalf, Barbara J. Dalton, and George Poste. Plenum Press, New York, 1994.

early descriptions paid little attention to the possibility that the inflammatory reaction might function in a deleterious way to complicate the myocardial infarction process. In the last two decades, more attention has been paid to this latter possibility, and a large body of literature has arisen utilizing multiple techniques to assess the effects of inflammation and inflammatory injury in ischemic models. The initial studies with anti-inflammatory drugs, which will be briefly described below, gave promising results, and clinical trials were undertaken utilizing high-dose steroids to treat myocardial infarction. The failure of these clinical trials (which resulted in an increased incidence of ventricular aneurysm and rupture of the ventricular wall) reemphasized the original concept that inflammation is important in the healing phase of myocardial infarction (Roberts *et al.*, 1976). It also emphasized the need for a greater understanding of the secondary inflammatory reaction and its role in both injury and healing so that more rational and specific interventions could be developed and evaluated.

The compelling need for strategies that will expose the mechanisms of inflammatory injury and allow us to carefully characterize them has come from our ability to reperfuse previously ischemic myocardium. For patients coming to medical attention early in the course of a myocardial infarction, attempted reperfusion of the ischemic, yet potentially viable, myocardium is the standard of care.

Reinstitution of coronary blood flow to the ischemic zone, whether accomplished with thrombolytic therapy or coronary angioplasty, can now be achieved rapidly, thus minimizing myocardial damage. This clinical advance serves as an impetus to address secondary causes of injury that might only occur during, or become significant because of, reperfusion. There is evidence that reinstitution of the circulation potentiates the inflammatory reaction to injury (perhaps by increasing the number of neutrophils circulating through the ischemic area), and substantial evidence suggests that neutrophils participate in a major way in reperfusion injury to the ischemic heart. An early clue that events taking place at the time of reperfusion may lead to cell injury was observed by Bulkley and Hutchins in 1977. They observed paradoxical necrosis in areas of successful revascularization in hearts that failed immediately following bypass surgery. Areas of contraction band necrosis and myocardial edema occurred beyond successfully bypassed vessels, whereas there was no evidence of injury beyond adjacent obstructed vessels that had not undergone bypass.

The primary focus of this chapter will be the cellular and molecular mediation of the secondary inflammatory response occurring in reperfusion injury. We will begin with a brief general description of reperfusion injury as a concept and describe some of the pathophysiological consequences of this injury. The remainder of the chapter will be aimed at the cellular and molecular mediation of inflammatory injury during reperfusion.

2. Ischemia–Reperfusion Injury and Its Etiology

2.1. Pathophysiologic Paradigm

In both clinical and experimental models, the initial insult resulting in injury is, in all cases, ischemia. Occlusion of the coronary vessel reduces the blood flow to the portion of the myocardium to critical levels, markedly impairing its energy metabolism and rapidly resulting in irreversible death. The cells in the epicenter of an occluded vascular bed are irreversibly injured by 20 min and will die regardless of any further changes. Because reduction in blood flow resulting from the occlusion of any one coronary vessel creates a distribution of degrees of ischemia, other cells are not as rapidly damaged, and pathologically the presence of live undamaged cells adjoining dead cells and scars is commonly seen. In the absence of either spontaneous or induced reperfusion, therefore, the predominant mechanism of injury is ischemia itself, although other sources of extension of the injury may likewise occur in relatively well-perfused areas.

The concept of reperfusion injury is an intriguing one, and has much to recommend it. It is important to understand, however, that at the time of this writing the evidence that there is a component of ultimate cell death and injury resulting from reperfusion is controversial. While many of the abnormalities observed during reperfusion with respect to endothelial function, macromolecular efflux from the myocardial cell, leukocyte infiltration, and pathological appearance of myocardial cells are surely associated with special circumstances initiated by reperfusion, evidence that this actually extends the amount of cells that will ultimately be destroyed relies on relatively few recent studies. The bulk of existing data relates to the observation that certain agents, such as antioxidants (Mullane, 1988; Jolly *et al.*, 1984) and adenosine (Olafsson *et al.*, 1987), have been shown to reduce infarct size when administered only during the reperfusion period. With that reservation in mind, we wish to summarize some of the more prominent abnormalities associated with ischemia and reperfusion from a functional standpoint in preparation for a discussion of the potential role of inflammation in these abnormalities.

There is general agreement that the reperfusion of ischemic myocardium before the onset of irreversible myocardial damage is not associated with any permanent reperfusion-associated injury (Braunwald and Kloner, 1982). Occlusions of 5 to 20 min demonstrate significant functional abnormalities, such as ventricular arrhythmias, myocardial dysfunction for a period of 24 to 48 hr (stunned myocardium), and vasomotor dysfunction (Hearse and Bolli, 1991), upon the onset of reperfusion. Evidence suggests, however, that these functional abnormalities are not attended by any lethal injury to the myocardium, and the ischemic segment will ultimately recover. Because it is clear that neutrophils are not associated with the abnormalities seen during this time (Hearse and Bolli, 1991), this entity will not be discussed further in this chapter. However, the absence of any neutrophil response under these circumstances emphasizes that neutrophil-induced injury is only seen secondary to lethal injury initiated by a previous toxic stimulus (e.g., ischemia). This relative specificity is in contrast to the role of oxygen free radicals (from other sources), which appear to influence both the short-term dysfunction (Hearse and Bolli, 1991) and, in some studies, the longer-term evidence of permanent injury (Jolly et al., 1984).

The functional abnormalities seen during reperfusion consequent to lethal myocardial injury can be conveniently grouped into three general categories: myocardial dysfunction, endothelium-related vasomotor dysfunction, and increased microvascular permeability and flow abnormalities. In addition, it has been clearly demonstrated that neutrophil accumulation into previously ischemic areas is markedly augmented during early reperfusion, with the highest rates seen in the first hour of reperfusion (Dreyer et al., 1991b). This has led to attempts to evaluate the role of the neutrophil in these observed pathophysiologic dysfunctions as a method of studying their role in reperfusion-associated lethal injury to cardiac myocytes.

There have been two general approaches to this problem: (1) the use of antineutrophil therapy to mitigate both the functional and pathological changes associated with reperfusion, and (2) cellular and molecular investigations into the mechanism of neutrophil localization in previously ischemic areas and the mechanism of neutrophil-induced cellular injury. Both of these subjects will be discussed in this chapter, but because of the nature of this volume we will focus on the latter.

2.2. The Effect of Antineutrophil Therapy on Ischemia–Reperfusion Injury

As stated above, although irreversible myocardial cell injury begins at 20–30 min after occlusion of a coronary vessel, the extent of myocardial damage induced by this occlusion continues to increase for up to 6 hr after the onset of ischemia. This has led to the recommendation for thrombolytic therapy in patients who present with myocardial infarctions less than 6 hr old (GISSI Study Group, 1986). In fact, this ability to reperfuse the myocardium has created considerably more interest in the possibility of reducing the actual total extent of injury induced by a coronary occlusion, since the ischemia factor is being removed. While the association of complement-derived components in the areas of ischemic injury was described by Hill and Ward in 1971, the first major body of evidence suggesting a role for the neutrophil in secondary ischemic injury came in the latter part of that decade, when enormous resources were used to develop strategies for minimizing myocardial infarct size. Among the strategies examined were ones thought to interfere with a putative inflammatory mechanism occurring after myocardium injury. Strategies aimed at reducing the generation of chemotactic factors, such as complement depletion by cobra venom toxin (Maroko *et al.*, 1978), administration of lipoxygenase inhibitors (Shappell *et al.*, 1990a), and leukotriene B_4 receptor antagonists (Taylor *et al.*, 1989), successfully reduced infarct in some models. Experiments using prostacyclin analogues (Simpson *et al.*, 1987) and adenosine (Olafsson *et al.*, 1987) to alter neutrophil function were likewise successful in some experimental models. Strategies to reduce neutrophil number, such as the use of antineutrophil antibodies (Romson *et al.*, 1983), neutrophil-depleting chemicals (Mullane *et al.*, 1984), or neutrophil filters (Engler *et al.*, 1986a), were all successful in reducing ischemia-related injury in some models. Finally, agents aimed at altering the effects of neutrophil localization, such as cyclooxygenase inhibitors (Romson *et al.*, 1982) and free radical scavengers (Lucchesi and Mullane, 1986; Jolly *et al.*, 1984), were shown to reduce infarct size or sensitivity to ischemia in some models of experimental ischemia. This body of data certainly pointed to a potential role of inflammation in ischemic damage; the conclusions were complicated by the following considerations.

First, there was no universal success utilizing any of these strategies, and success depended, to a great degree, on the model used. It is important to point out, however, that positive results were found with antineutrophil strategies in many of these experiments even when reperfusion was not instituted (Mullane and Smith, 1990). This has led to a series of important (and not completely resolved) questions. The question of how neutrophils in any significant number would get into the ischemic myocardial area could (with some certainty) be addressed by the concept of areas of intermediate ischemia that are potentially salvageable within the infarct and around the ischemic edges. One could easily visualize a strategy to reduce a "border zone" that remains at risk. On the whole, however, the advent of thrombolytic therapy to restore coronary blood flow and alleviate myocardial damage certainly enhances the potential for mediation of secondary myocardial injury by the neutrophil. There is no question that neutrophil localization in ischemic or previously ischemic myocardial tissue is markedly accelerated by reperfusion and is greatest at the onset of reperfusion (Engler *et al.*, 1986b), is associated with the areas of greatest myocardial injury and lowest coronary blood flow (Rossen *et al.*, 1985; Dreyer *et al.*, 1991b), and can be minimized by strategies involving the administration of therapy only upon the onset of reperfusion (Mullane, 1988; Jolly *et al.*, 1984; Olafsson *et al.*, 1987; Mullane and Smith, 1990).

It is important to point out that even after the early experiments the information gained in the above studies was compelling enough that Hillis and Braunwald (1977) concluded that all patients with acute infarction, regardless of their hemodynamic state, might benefit from an anti-inflammatory agent. A trial of methylprednisolone for acute myocardial infarction produced extremely deleterious results in patients, with an increased incidence of ventricular aneurysm and cardiac rupture (Roberts *et al.*, 1976). Thus, anti-inflammatory therapy also appeared to interfere with myocardial healing. This had led to a second experimental approach to defining the role of inflammation in myocardial injury, which seeks to characterize the cellular and molecular events associated with this inflammatory reaction in order to develop more site-specific interventions that could conceivably reduce secondary inflammatory injury with less risk of the deleterious side effects.

3. Factors Contributing to Neutrophil Localization in Ischemic and Reperfused Hearts

3.1. Trapping of Leukocytes in the Microvessels

Because classic studies of inflammation have emphasized the role of neutrophil accumulation in capillaries and postcapillary venules, a similar examination was made in experimental myocardial infarction. In the experiments of Schmid-Schonbein and Engler (1987), the earliest inflammatory changes in the ischemia reperfusion process occurred primarily in small vessels and capillaries where the neutrophils appeared to be trapped. In many cases, it appeared that neutrophils actually may have obstructed the capillary. The accumulation of the neutrophils in microvessels probably occurred as a function of a number of definable pathological processes consequent to chemotactic stimulation in the area of ischemia and reperfusion. In the first place, it is known that chemotactic stimulation of neutrophils induces shape changes, making them less deformable. Additionally, these neutrophils undergo homotypic aggregation, which may further contribute to obstruction. Activated neutrophils also release a variety of vasoactive products that promote vasoconstriction, such as thromboxane A_2 (Michael *et al.*, 1990), and a variety of lipoxygenase products, such as leukotriene B_4 (Mullane *et al.*, 1988), all of which induce vasoconstriction and activate platelet adherence and aggregation, which can contribute to the obstructive process. Thus, it appears possible that neutrophils may alter microvascular circulation, and perhaps damage the microvascular endothelium during ischemia–reperfusion injury (Engler *et al.*, 1984, 1986b; Carden *et al.*, 1990; Hernandez *et al.*, 1987). It is also possible that the complex cellular events described above may involve a more specific factor altering microvascular permeability, which has consistently been shown to be increased during ischemia and reperfusion injury of relatively short duration (Svendsen *et al.*, 1991).

Perhaps the most dramatic and pathologically significant microvessel abnormality, however, is the "no reflow" phenomenon associated with ischemia–reperfusion injury, which has been linked to neutrophil localization (Ambrosio *et al.*, 1989). This phenomenon is not seen in short periods of ischemia, and is generally associated with ischemic periods of more than 3 hr which, upon reperfusion, are followed by reactive hyperemia in the ischemic bed. Over the next several hours, however, several areas that are initially reperfused become ischemic again as a result of microvascular obstruction, so there is a recurrence of severe ischemia (Ambrosio *et al.*, 1989). When this phenomenon occurs, it appears to constitute an essentially irreversible injury.

Thus, the accumulation of neutrophils in the microvessels is associated with a number of pathophysiologically important changes that occur after relatively short periods of ischemia and reperfusion, but which can result in an irreversible state if prolonged. While changes in cell shape or deformability and vasoconstriction may be important mechanisms for microvascular neutrophil accumulation, a preponderance of evidence suggests that specific interactions between adhesion molecules are the most critical factor in control of leukocyte-induced pathophysiologic changes in ischemia and reperfusion.

3.2. Intercellular Adhesion and Inflammatory Injury—CD18 and the Leukocyte β_2 Integrins

Recent evidence suggests that intercellular adhesion is a critical step in the accumulation of leukocytes at sites of inflammation (Smith, 1992). In a variety of cardiovascular-related inflammatory models, monoclonal antibodies made to the CD18 subunit of the leukocyte integrin CD11/CD18 glycoprotein complex have markedly reduced the pathophysiologic consequences. Administration of anti-CD18 antibodies has reduced myocardial infarct size in a rapid model of 1 hr ischemia–5 hr reperfusion when administered systemically before the coronary occlusion (Seewaldt-Becker *et al.*, 1989). In another study in rabbits, a different anti-CD18 monoclonal antibody (60.3) was applied to radiolabeled rabbit neutrophils before they were introduced into a rabbit undergoing a 30-min occlusion and 3-hr reperfusion protocol; the accumulation of neutrophils was markedly reduced (Williams *et al.*, 1988). Our laboratory has performed similar studies in canine models of cardiac ischemia reperfusion; these demonstrated that the anti-CD18 monoclonal antibody R15.7 reduced neutrophil accumulation in the previously ischemic area (Dreyer *et al.*, 1991b). Anti-CD18 antibodies have also been shown to reduce myocardial infarct size (Ma *et al.*, 1991).

In addition to myocardial injury, there is evidence that adhesion-dependent injury to the endothelium is part of the pathophysiology of ischemia–reperfusion injury. While the mechanism of this dysfunction is probably complex, it involves other factors such as reactive oxygen generation (Tsao and Lefer, 1990; Suzuki *et al.*, 1989). However, the depression of endothelium derived relaxation in postischemic vessels has also been mitigated by anti-inflammatory strategies. Some evidence suggests that large vessel injury is associated with neutrophil migration

into the subendothelial layer of the vascular wall, and can be obviated by monoclonal antibodies to CD18 (Ma *et al.*, 1991).

Hernandez *et al.* (1987) have studied ischemia and reperfusion in the small
intestine, demonstrating a marked reduction in the increased microvascular permeability attributed to ischemia and reperfusion that could be affected by either
antineutrophil antiserum or by the monoclonal antibody 60.3. Korthuis and co-
workers have demonstrated similar phenomena after skeletal muscle ischemia
and reperfusion was blocked by IB_4, another anti-CD18 monoclonal antibody
(Carden *et al.*, 1990).

These studies, and many more which have followed, suggest a critical role
for the leukocyte "β_2 (CD18)" integrins in neutrophil-induced inflammatory
injury.

Finally, recent experiments from our laboratory have demonstrated a CD18-
dependent neutrophil adhesion to cardiac myocytes from adult dogs (Entman *et
al.*, 1990) that modulates a neutrophil-specific injury to the myocardial cell; both
the adhesion and the toxic effects of the neutrophil on the myocyte are completely blocked by anti-CD18 (Entman *et al.*, 1992). Thus, the discussion above
outlines two potential roles for CD18 in the inflammatory injury associated with
ischemia–reperfusion: (1) CD18 integrins are obligate for transmigration across
the postcapillary venule and tissue localization, and (2) CD18 integrins modulate
tissue injury associated with neutrophil–cell adhesion.

3.3. Adhesion-Dependent Cytotoxicity—Mac-1 Adhesion
to Parenchymal–Myocyte ICAM-1

ICAM-1 (CD-54) is a counterreceptor for LFA-1 and Mac-1 (Smith *et al.*,
1989; Dustin *et al.*, 1986; Marlin and Springer, 1987; Diamond *et al.*, 1990). In
striking contrast to the restricted cellular distribution of the β_2 integrins, ICAM-1
is widely distributed in nonhematopoietic cells. Whereas this molecule is expressed at low levels in unstimulated endothelial cells, it is markedly upregulated
by cytokine stimulation.

Recent work from our laboratory examined the potential mechanisms of
neutrophil adhesion to isolated adult canine cardiac myocytes *in vitro* (Entman *et
al.*, 1990). Intercellular adhesion occurred only if the myocytes were stimulated
with cytokines that induced expression of ICAM-1 (Smith *et al.*, 1991a), and
when the neutrophils were stimulated to enhance CD18-dependent adhesion.
Stimulation of the myocytes with cytokines such as IL-1, TNFα, and IL-6 induced the expression of ICAM-1 (Smith *et al.*, 1991a; Youker *et al.*, 1992), and
stimulation of the neutrophils with zymosan-activated serum (C5a source) or
platelet-activating factor (PAF) promoted CD18-dependent adhesion that was
blockable by both anti-CD18 (R15.7) and anti-ICAM-1 (CL18/6) monoclonal
antibodies (Entman *et al.*, 1990; Smith *et al.*, 1991a). As will be discussed

further in this chapter, cardiac lymph collected during the early reperfusion period after experimental myocardial ischemia (Michael *et al.*, 1979) appeared to contain both a neutrophil-activating factor (C5a) (Dreyer *et al.*, 1991a) and potent cytokine activity capable of stimulating ICAM-1 transcription and expression (Youker *et al.*, 1992). We thus considered the hypothesis that neutrophils might induce direct cytotoxicity on cardiac myocytes under circumstances of ischemia–reperfusion injury. We found that adhering neutrophils were apparently cytotoxic to cardiac myocytes within a few minutes, and we examined the role of neutrophil–myocyte adhesion and various adhesion ligands in this process.

The binding of neutrophils to activated cardiac myocytes was found to be specific for Mac-1–ICAM-1 interaction, and was completely blocked by antibodies to ICAM-1, CD11b, and CD18. This interaction was unaffected by antibodies to CD11a, which are capable of blocking neutrophil binding to endothelial cell monolayer (Youker *et al.*, 1992; Entman *et al.*, 1992). In separate experiments, either neutrophils or cardiac myocytes were loaded with 2′,7′-dichlorofluorescein (DCFH), and the oxidation of this marker to DCF was monitored. Using zymosan-activated serum (1%) to activate neutrophils and IL-1, TNFα, or IL-6 to stimulate ICAM-1 formation on myocytes, we were able to show that fluorescence of either cell was dependent on neutrophil–myocyte adhesion and was likewise blocked by antibodies to ICAM-1 (CL18/6), CD11b (MY904), and CD18 (R15.7). Neutrophils appeared to fluoresce immediately upon adhesion, suggesting a rapid adhesion-dependent activation of the NADP oxidase of the neutrophils. In contrast, fluorescence of the cardiac myocytes appeared after several minutes and was rapidly followed by irreversible myocyte contracture. The iron chelator, desferroxamine, or the hydroxyl radical scavenger, dimethylthiourea, did not inhibit neutrophil adhesion, but completely prevented fluorescence and contracture of the cardiac myocyte. In contrast, neither extracellular oxygen radical scavengers such as superoxide dismutase and catalase, nor the extracellular ion chelator, starch-immobilized desferroxamine, reduced fluorescence, adhesion, or cytotoxicity. Under the experimental conditions utilized, no superoxide production could be detected in the extracellular medium during the neutrophil myocyte adhesion (Entman *et al.*, 1992). We have interpreted these data as suggesting that there is a compartmented transfer of reactive oxygen from neutrophil to myocyte, which is dependent upon this adhesion reaction. The specific role for Mac-1 and ICAM-1 in this process suggests that neutrophil-induced cytotoxicity is specific to events produced by chemotactic activation. The possibility that other neutrophil-derived products are transferred at the neutrophil–myocyte interface via a similar compartmented mechanism has not been carefully investigated. Obviously, neutrophils secrete a variety of enzymes that might similarly be toxic (Mullane and Smith, 1990). However, contracture or cell death was never seen in myocytes that had not previously fluoresced, and was completely inhibited by the intracellular reactive oxygen scavenger systems despite the fact that many neutrophils adhered to the surface of these myocytes (Entman et al., 1992). The experiments were followed for only 2 to 3 hr, and the possibility of a later toxicity mediated by other neutrophil components cannot be ruled out.

Unlike the neutrophil–ICAM-1 interaction in endothelial cells, the integrin counterreceptor for ICAM-1 on myocytes is exclusively Mac-1; anti-LFA-1 antibodies do not alter this process (Entman et al., 1992). This suggests that there is an additional layer of specificity to integrin–ICAM-1 interaction that varies by cell type. The mechanism of this specificity is not yet known.

Perhaps more significant than the specificity and affinity of the neutrophil binding to cytokine-stimulated cardiac myocytes are the biological consequences of this interaction. Although mediated by similar counterreceptors, the direct consequence of neutrophil–endothelial cell binding is transmigration across the wall of the postcapillary venule. On the other hand, upon adhesion to myocytes neutrophils undergo a respiratory burst, followed by a transfer of reactive oxygen species and irreversible myocyte contracture.

This differential behavior emphasizes the important role of the leukocyte integrins in cell signaling, and their ability to synergize with inflammatory mediators (C5a, TNFα) to modulate the functions of the NADPH oxidase. It also demonstrates a key feature of integrin function: regulation from within the cell and modulation of cellular behavior by extracellular environment (Hynes, 1992). There is ample evidence in the literature for an activation of integrins that mediate integrin–ligand affinity. With regard to LFA-1, there is substantial evidence for a qualitative change induced by chemotactic agents or phorbol ester, which mediates LFA-1 interaction with ICAM-1 on endothelium. Similarly, there is strong evidence that the chemotactic activation of Mac-1 adhesion begins with the qualitative activation of Mac-1 constitutively expressed on the neutrophil surface (Hughes et al., 1992), and that newly surfaced Mac-1 will be similarly activated as the cell proceeds to move in a leukotactic gradient. Dransfield et al. (1992) have produced a monoclonal antibody to a conformational state of Mac-1 that depends on Mg^{2+} and temperature and may represent the activated state.

The association of Mac-1 adhesion with NADP oxidase activation of neutrophils has been previously demonstrated in studies describing H_2O_2 production after neutrophil adhesion to protein-coated plastic (Shappell et al., 1990b; Nathan, 1987) or cardiac myocytes (Entman et al., 1990). This recent report demonstrates for the first time that ICAM-1 is a specific cellular ligand for Mac-1 that is necessary for the adhesion-dependent respiratory burst when neutrophils adhere to cardiac myocytes. It also suggests that neutrophil toxicity of parenchymal cells may not happen randomly: it requires a highly specific set of circumstances. In order for this reaction to occur, not only must neutrophils localize in the extracellular fluids in response to a chemotactic gradient, but also synthesis of ICAM-1 on the parenchymal cell must be initiated by some source of cytokine. It thus appears that the occurrence of chemotactic factors and the expression of ICAM-1 must relate temporally so that both activated neutrophils and ICAM-1-containing parenchymal cells coexist.

In summary, adhesion mechanisms involving LFA-1 and Mac-1 and their interaction with ICAM-1 may mediate both the rapid activation and transmigration of neutrophils and also neutrophil-derived parenchymal cell toxicity. It would appear that the expression of ICAM-1 on tissue cells is obligate for direct neutrophil-induced parenchymal injury. These mechanisms may play important roles in the overall process of inflammatory injury.

3.4. The Selectin Family of Adhesion Ligands—A Potential Role in Ischemia

The majority of the studies described above were done in systems where there was no flow. More recent studies have shown, however, that neutrophil adherence mechanisms are sensitive to shear stress. Lawrence *et al.* (1990) have shown *in vitro* that CD11/CD18–ICAM-1 adhesion occurs very infrequently at a wall shear stress of 0.5 dyne/cm^2, which is well below the calculated shear stress in the postcapillary venule. Arfors *et al.* (1987) demonstrated that anti-CD18 monoclonal antibody 60.3 did not inhibit neutrophil rolling on postcapillary venules in rabbits. This has added a complexity to the studies of neutrophil endothelial adhesion and has emphasized the importance of another class of adhesion molecules, the selectins (Bevilacqua *et al.*, 1991). Each member of this family of transmembrane glycoproteins has an N-terminal lectinlike domain followed by an epidermal growth factor-like domain, and multiple repeats of a complement-binding-like domain in the extracellular space with a transmembrane region and a short cytoplasmic domain. Two of the known selectins, P- and E-selectin, are found on endothelial surfaces, and one L-selectin is found on the neutrophil; P-selectin is also found on the surface of activated platelets. Evidence would suggest that these molecules are critical in allowing neutrophil adhesion under conditions of venular wall shear stress during physiologic blood flow; this is discussed in detail elsewhere in this volume by our colleague, Dr. C. Wayne Smith. Their potential role in ischemia is discussed briefly later in this chapter.

4. Chemotactic Factors—Signals That Initiate Inflammation in Ischemia–Reperfusion Injury

4.1. Complement Activation

The role of complement in this process was first suggested by experiments of Hill and Ward (1971) in a rat model of myocardial ischemia in which they demonstrated activation after 3 hr of ischemia. Pinckard *et al.* (1973, 1975) attempted to approach the mechanism of complement activation in studies demonstrating that subcellular fractions rich in mitochondria from heart muscle activate both classic and alternative complement pathways. Other mechanisms that have been suggested for complement activation are the activation of lysosomal proteases (Hill and Ward, 1971) and factors associated with the release of reactive oxygen species (Shingu and Nobunaga, 1984; Vogt et al., 1987). Rossen et al. (1988) have reported elevations of C1q binding molecules in the circulation of patients who have had acute myocardial infarction; these molecules were not found in patients with acute coronary syndromes in which there were no elevations of cardiac enzymes. The elevations seen in patients during the first few days after myocardial infarction were temporally associated with transient depressions of C3 and C4. Anticardiac antibodies were not found in these patients. The presence of C1q binding proteins in the serum of patients undergoing acute myocardial infarction was compatible with the reports of Pinckard et al. (1973, 1975) suggesting that this entity was associated with the activation of the classic complement pathway in both experimental and clinical myocardial infarction.

In experimental models of myocardial ischemia in the dog, we have demonstrated the localization of C1q in the ischemic segments following a 45-min coronary occlusion, and have correlated C1q localization with neutrophil accumulation in the same segments (Rossen et al., 1985). Interestingly, small elevations of C1q localization could be demonstrated in occlusions as brief as 15 min. We were thus able to demonstrate that C1q binding proteins appeared in cardiac lymph after coronary occlusions, and that these proteins were present and demonstrable during the first 4 hr of reperfusion (Rossen et al., 1988). The macromolecules binding C1q ranged in size from 23 to 67 kDa, were of mitochondrial origin, and were fully capable of activating the classic complement cascade (Rossen et al., 1988). Subsequently, we demonstrated that postischemic cardiac lymph was strongly chemotactic, with activity present within the first hour of reperfusion (Dreyer et al., 1989). Postischemic cardiac lymph was able to stimulate neutrophils and induce shape change, Mac-1 surface expression, chemotaxis, and increased adhesion to endothelial monolayers in neutrophils. Moreover, neutrophils found in the cardiac lymph during reperfusion demonstrated upregulation of surface Mac-1 (Dreyer et al., 1989). Chemotactic activity was found within the first 4 hr of reperfusion, correlating very well with the time course of the appearance of C1q binding proteins in the cardiac lymph (Dreyer et al., 1989). Recent studies with the dog model have demonstrated that this chemotactic activity, seen early during reperfusion, is almost entirely neutralized by anti-C5a antisera (Dreyer et al., 1991a). Finally, again in the dog model, Dreyer et al. (1991b) have demonstrated that the rate of neutrophil localization in the ischemic and reperfused myocardium is highest in the first hour and dissipates within the first 4 hr.

Thus, it would appear that myocardial injury during ischemia results in the egress of macromolecules that are capable of initiating the early inflammatory reaction ("reaction to injury") seen during reperfusion (Rossen *et al.*, 1988; Dreyer *et al.*, 1989). It is important to point out, however, that the data regarding the time course of this reaction have come from experiments performed primarily in the dog, because of the availability of the chronic cardiac lymph duct cannulation method. Unlike humans, the dog does not utilize a 5-lipoxygenase system for production of leukotrienes as a prominent feature in its inflammatory responses. The role of lipid-derived chemotactic factors will require study in models of ischemia and reperfusion done with other species.

4.2. Lipid-Derived Autacoids

In many species, activated neutrophils release leukotriene B_4 (LTB_4), which is a potent chemotactic agent. In rabbit, pig, and human neutrophils, it is produced by the activation of an enzymatic cascade that begins with phospholipase A_2–arachidonic acid entering into a pathway involving 5-lipoxygenase. The latter enzyme is translocated to membrane components of the cell, where it is enzymatically active in the production of the chemotactic mediators (Rowzer and Kargman, 1985). There are a variety of lipoxygenase enzymes found in neutrophils, and their relative quantity varies according to species. Thus, 12-hydroxy isomers (12-HETES) are the major product in dog neutrophils (Mullane *et al.*, 1987), and 15-hydroxy isomers (15-HETES) are present at higher arachidonic acid concentrations in rabbit neutrophils.

Because of the variation in the synthesis of lipoxygenase derivatives and the sensitivity of neutrophils to these products among species, one might expect studies of myocardial salvage postischemia of various inhibitors to give different results according to the species studied. A direct LTB_4 receptor antagonist, LY223982, was effective in inhibiting LTB_4-induced neutrophil shape change and upregulation of neutrophil Mac-1 adhesion glycoprotein complexes *in vitro*, and was capable of reducing infarct size in a rabbit model of ischemia (Taylor *et al.*, 1989). However, a similar LTB_4 antagonist, LY255283, had no effect on a dog model of ischemia reperfusion, which is consistent with the observations regarding the relative absence of 5-lipoxygenase products in dog and the insensitivity of dog neutrophils to LTB_4 (Hahn *et al.*, 1990). In the rabbit, where cardioprotective effects were observed, the cardioprotective effect was associated with a reduction in tissue myeloperoxidase activity, suggesting that the inhibition of LTB_4 action had reduced chemotactic activity in the infarcted area (Taylor *et al.*, 1989). Thus, in some species, lipoxygenase derivatives function as chemotactic agents that activate neutrophils and induce their migration to ischemic tissue; whether this occurs early in ischemia or only after neutrophils have been localized to the ischemic reperfused area (and thus produce LTB_4) has not yet been determined.

There are other lipid-derived autacoids that may be pertinent in ischemia, although there is very little direct evidence at this writing. The peptide leukotrienes, LTC_4 and LTD_4, are produced by endothelial and vascular smooth muscle cells from LTA_4 that is secreted by leukocytes after cell adhesion (Feinmark and Cannon, 1986, 1987). Since these leukotrienes are potent vasoconstrictors, they play a role in the injury or potentiate it in some way.

Another lipid-derived molecule that is known to have chemotactic activity is PAF. The enzymatic pathway leading to its formation is induced by thrombin (Zimmerman *et al.*, 1990) in endothelial cells, and recent evidence suggests that PAF species are produced nonenzymatically through the action of reactive oxygen species (see below). PAF is a potent chemotactic agent that induces increased surface expression of Mac-1 and promotes neutrophil adhesion and other Mac-1-dependent processes. As its name implies, it is also a potent activator of platelet function. The role of PAF in myocardial ischemia and reperfusion has not been carefully studied, but some interesting possibilities will be discussed in the next section, where reactive oxygen species are considered as chemotactic agents. Thus far, a single study in rats suggests that PAF antagonists appear to reduce myocardial infarction after a 6-hr permanent occlusion (Stahl *et al.*, 1988).

4.3. Reactive Oxygen Species as Chemotactic Activators

The generation and secretion of reactive oxygen species has been suggested as a major mechanism by which neutrophil-induced injury to other cells might occur; this will be discussed in an ensuing section of this chapter. However, it is important to point out that the relationship between reactive oxygen and neutrophil function may be more complex: there is significant evidence that reactive oxygen may be an important factor in neutrophil localization. The origin of this concept rests with the work of McCord (Petrone *et al.*, 1980), who presented data suggesting that increased reactive oxygen resulted in the formation of a "neutrophil-activating factor" present in serum. While this factor was not identified, the possibility of an enzymatic or nonenzymatic formation of a lipid-derived autacoid must certainly be considered. Obviously, a reactive oxygen will potentiate lipoxygenase and cyclooxygenase reactions, and nonenzymatic oxidation of unsaturated lipids is a well-known result of oxidative stress. In recent years, the work of Granger and his co-workers (Granger, 1988; Inauen *et al.*, 1990) has provided strong evidence for an involvement of reactive oxygen in chemotaxis; they have demonstrated that the presence of free radical scavengers or inhibitors of free radical formation during ischemia and reperfusion in the gut markedly reduces neutrophil infiltration (Suzuki *et al.*, 1989). Similar studies have not been done in myocardium. In addition to the activation of complement described above, two general mechanisms by which oxidative stress might induce neutrophil adhesion have been described that may be pertinent in reperfusion injury: (1) P-selectin surface expression (Patel *et al.*, 1991) and (2) the production of PAF analogues of oxidatively fragmenting phosphatidylcholine (Smiley *et al.*, 1991). These two mechanisms probably relate to the modulation of the two adhesion mechanisms described above, with the former relating to the postulated early shear stress-resistant selectin-mediated adhesion and the latter relating to the upregulation of Mac-1 and, perhaps, the activation of LFA-1 (Smith, 1992).

5. Potential Mechanisms of Neutrophil-Induced Parenchymal Injury

The major focus of the previous sections has been the mechanism by which neutrophils are attracted to and activated in an area of injury. With respect to ischemia reperfusion injury, indirect evidence was cited that agents that altered neutrophil function or the number of neutrophils attracted to an area might actually prevent extension to the initial ischemic injury.

In addition to the potential role of neutrophil-induced microvascular obstruction in extending tissue injury, there is also substantial evidence from intervention studies that neutrophils may directly injure parenchymal cells through the release of specific products. Under conditions found *in vivo*, both reactive oxygen species and proteolytic enzymes are almost exclusively secreted by adherent neutrophils (Shappell *et al.*, 1990b; Entman *et al.*, 1990; Inauen *et al.*, 1990; Suzuki *et al.*, 1989). Thus, we hypothesize that ligand-specific adherence of neutrophils to parenchymal cells (e.g., cardiac myocytes) or endothelial cells is an important part of the secondary inflammatory injury occurring in ischemia and reperfusion. There have been very few data *in vivo*, but recent *in vitro* evidence from our laboratories suggests that ligand-specific adhesion of neutrophils to myocardial cells is necessary for a direct toxic effect associated with neutrophil activation (reviewed in a previous section). Our experiments *in vitro* demonstrated that the compartmented transfer of neutrophil-derived toxic effects (reactive oxygen) to cardiac myocytes was mediated by Mac-1–ICAM-1 adhesion (Entman *et al.*, 1992).

The possible relevance of these observations made *in vitro* to the processes occurring during reperfusion injury is strengthened by a series of studies we have done analyzing cardiac lymph collected during the reperfusion. Our model of the awake unanesthetized dog with a chronically cannulated cardiac lymph duct has enabled us to examine inflammatory mediators under circumstances in which the effects of surgical trauma are no longer a significant factor (Rossen *et al.*, 1988; Dreyer *et al.*, 1989; Youker *et al.*, 1992). We have described, in a previous section, the use of this model to define the time course and mechanism of complement activation during ischemia and reperfusion. In a more recent set of studies, we demonstrated that high dilutions of cardiac lymph collected during reperfusion were capable of stimulating ICAM-1 expression on the surface of cardiac myocytes *in vitro* (Youker *et al.*, 1992). Preischemic lymph contained no such activity even when more concentrated. This cytokine activity was highest within the first hour of reperfusion, but persisted up to 72 hr. Postreperfusion lymph contained cytokine activity that could stimulate the expression of ICAM-1 on both canine endothelial cells and canine myocytes. Significantly, the cytokine activity of the postischemic cardiac lymph with respect to cardiac myocytes was completely inhibited by antibodies to human recombinant IL-6 at all time periods (Youker *et al.*, 1992). In contrast, the presence of anti-IL-6 antibodies did not inhibit the ability of cardiac lymph to induce ICAM-1 expression in endothelial cells, suggesting a more complex cytokine content of extracellular fluid during reperfusion. Myocytes stimulated with postischemic lymph are also vulnerable to injury by activated neutrophils (Entman *et al.*, 1992).

    In further investigations of the pertinence of this mechanism *in vivo*, myocardial biopsies taken during reperfusion have been examined for the presence of mRNA for ICAM-1. We have recently demonstrated that ICAM-1 mRNA is present in ischemic areas (less than 40% normal flow) within 1 hr of reperfusion, and continues to rise in concentration for 24 hr (Anderson *et al.*, 1991). There is no detectable mRNA for ICAM-1 in control during the first 3 hr of reperfusion, while in the ischemic areas, ICAM-1 mRNA is an inverse function of coronary blood flow. At 24 hr, however, ICAM-1 mRNA is found to some degree in all myocardial samples (although it is still highest in the ischemic area), suggesting that circulating cytokines (e.g., IL-6) are now affecting normal as well as ischemic areas (Anderson *et al.*, 1991). Thus, during the early reperfusion there appears to be both an activation of neutrophils and an induction of ICAM-1 mRNA occurring in the ischemic area that can be demonstrated *in vivo*. The presence of these factors suggests that it is possible, during reperfusion, that a neutrophil-derived cytotoxic effect on cardiac myocytes may occur *in vivo*, which is similar to that observed *in vitro*.

    Our observations have led us to formulate the following hypothetical scheme to describe the cell biology of secondary inflammatory injury associated with ischemia and reperfusion of the heart.

6. The Cell Biology of Inflammatory Injury in Ischemia Reperfusion: A Working Hypothesis

    We propose the following working hypothesis, which describes the events that mediate inflammatory reaction to ischemic reperfusion injury. While reperfusion injury is unquestionably a more complex phenomenon, this construct deals specifically with the inflammatory component and makes the assumption that neutrophils are the principal determinant of this injury. Our hypothesis for the critical steps in this process is as follows:

6.1. Step One: Chemotaxis

We hypothesize that the initial chemotactic event occurs when the injured myocardial cell releases macromolecules (C1q binding proteins), which activate the classic complement pathway as described. Thus, the extracellular release of complement-activating macromolecules of mitochondrial origin (Rossen *et al.*, 1988; Kagiyama *et al.*, 1989) initiates the production of C5a. *In vitro,* postischemic cardiac lymph promotes chemotaxis, neutrophil shape change, increased surface expression of Mac-1, and neutrophil adhesion to endothelial cell and cardiac myocytes; this activity is almost wholly inhibited in dog by anti-C5a. Mac-1-dependent adhesion potentiates the release of oxygen free radicals. In the experimental animal there is good evidence that macromolecular efflux from ischemic tissue occurs after only 15 min of ischemia and is extremely prominent in coronary occlusions of 45 min or more (Rossen *et al.*, 1985). It is likely that specific lipoxygenase products such as leukotriene B_4 are also important leukotactic agents in other species, but they are not prominent in the canine species. It is not yet clear whether these latter agents are secreted in response to the initial stimulus, or as a result of the chemotactic stimulation by complement of leukocytes, which then secrete LTB_4 as a secondary leukotactic mechanism. Endothelial cells also secrete peptido-leukotrienes, but the mechanism by which this would be induced in ischemia is unknown. Regardless of the mechanism, however, evidence suggests that leukotactic activity exists within the cardiac extracellular fluid within 15 min after a coronary occlusion, and that the leukotactic activity itself is not dependent on reperfusion, although the transmigration of neutrophils out of the intravascular space appears markedly augmented by reperfusion.

6.2. Step Two: Neutrophil Adhesion to Endothelium

Activated neutrophils must first adhere to the endothelium in order to exert their effects on microvessels, augment secretory functions of toxic or vasoactive products, and emigrate into the extravascular space. Both the secretory and migratory functions require adhesion through CD11a/CD18 (LFA-1) and/or CD11b/CD18 (Mac-1). While both are constitutively found on the neutrophil surface to some degree, evidence suggests that an additional activation step is required for LFA-1-mediated transmigration as well as Mac-1 adhesion and motility. It is important to point out, however, that it is not necessary to augment or activate the endothelial cell in order to produce transmigration. It is well known that neutrophils can follow a leukotactic gradient across unstimulated endothelial cells as long as the leukotactic stimulus activates the neutrophil (Harlan *et al.*, 1985; Furie *et al.*, 1987, 1991; Smith *et al.*, 1989). It is apparently not necessary for ICAM-1 to be upregulated on the endothelium during the early neutrophil transmigration associated with reperfusion; constitutively expressed ICAM-1 may be sufficient.

While the above system is absolutely necessary for neutrophil transmigration and neutrophil-derived toxicity, it does not appear to be sufficient to account for the entire adhesion transmigration process. As described previously, integrin–ICAM-1 adhesion is quite sensitive to shear stress and is unstable at shear levels above 0.5 dyne/cm². Calculated shear stresses in the postcapillary venule are several times that level, and it appears unlikely that these shear stress levels can be sufficiently reduced by the relatively minor amount of microvascular plugging seen in early ischemia and reperfusion. L-selectin on neutrophils allows neutrophil endothelial adhesion (Smith *et al.*, 1991b; Abbassi *et al.*, 1991) under higher shear stress, and its endothelial counterligand is, at least in part, the E-selectin, which must be synthesized *de novo* by the endothelial cell; this takes several hours (Picker *et al.*, 1991). Thus, its involvement in early margination is questionable (Bevilacqua *et al.*, 1989).

Therefore, we would postulate that the early margination event may occur through the surface expression of P-selectin on endothelial cells. P-selectin is stored in the Weibel–Palade bodies and is rapidly translocated to the endothelial surface in response to thrombin and/or oxidative stress reactions, both of which would be prominent during a reperfusion event associated with a myocardial infarction. It is possible that P-selectin may react with L-selectin to effect margination (Picker *et al.*, 1991).

6.3. Step Three: Cytokine Induction of Adhesion Molecule Expression

An important area in which there is relatively little progress concerns the stimuli that ultimately induce the synthesis of some of the adhesion molecules. While ICAM-1 is constitutively expressed on the endothelium, and additional synthesis may not be required, the induction of ICAM-1 on parenchymal cells appears to be necessary in order for neutrophil-induced injury to occur (Entman *et al.*, 1992). E-selectin is not constitutively expressed; it is expressed only after endothelial cells are stimulated by cytokines to induce its *de novo* synthesis (Bevilacqua *et al.*, 1989).

The data cited above suggest that IL-6 is a critical cytokine in the induction of the expression of ICAM-1 upon myocardial cells; endothelial stimulation by postischemic cardiac lymph appears to result in response to other cytokines (Youker *et al.*, 1992). The involvement of other cytokines is further supported by the observation that IL-6 secretion by cells is induced by other "primary" cytokines, such as TNFα and IL-1. We do not yet know the primary signal for cytokine production and the cell of origin of individual cytokines released locally in the ischemic area, but it is an area of significant interest. Mononuclear cells and endothelial cells are known to release cytokines capable of stimulating endothelial cells *in vitro*. In addition, recent studies have demonstrated that cytokines sufficient to activate expression of endothelial ICAM-1 and E-selectin can be supplied by activated platelets and degranulated tissue mast cells (Matis *et al.*, 1990). The signal that induces these cells to release cytokines is also un-known, although mononuclear cells are certainly activated by complement.

It is possible that the species of cytokines and growth factors secreted by, for example, the endothelial cell and the mononuclear cell may change as a function of time of reperfusion. For example, it is known that C5a levels (which stimulate monocytes) are very high in the injured tissue during the early reperfusion time, but are rapidly dissipated within the first 4 hr of reperfusion (Dreyer *et al.*, 1989, 1991a); likewise, thrombin (which stimulates endothelium) may accumulate during occlusion and be cleared during reperfusion. Thus, some of the known primary signals for endothelial cells and mononuclear cells may change, and one might speculate that the cytokines and growth factors secreted might also quali-tatively or quantitatively change under those circumstances. It would be attrac-tive to postulate that such a transition initiates a chronic healing phase of the inflammatory cascade. These mechanisms remain to be defined.

6.4. Step Four: Cytotoxicity of Neutrophils

Finally, once neutrophils have bound to the endothelium and transmigrated across the postcapillary venules, they can migrate into the extracellular space. Our work with isolated cardiac myocytes and isolated neutrophils suggests that parenchymal cells expressing ICAM-1 on their surface are highly vulnerable to neutrophil adhesion and neutrophil-induced injury. From the standpoint of acute injury, it would appear that reactive oxygen products are most prominent. The role of neutrophil-derived proteolytic and lipolytic enzymes cannot be easily defined with studies of this type.

As emphasized in this review, the toxic effects of neutrophils are specifi-cally dependent on adhesion molecule interactions. Evidence from our laboratory (Shappell *et al.*, 1990b; Entman *et al.*, 1990) suggests that Mac-1 is critical to the activation of NADP oxidase-related tissue injury. Our recent paper demon-strates that this is specifically a Mac-1–ICAM-1 interaction (Entman *et al.*, 1992), and that the activation of NADP oxidase involves diglyceride-dependent activation of protein kinase C (Shappell *et al.*, 1990b). This latter finding, in addition to its interest with regard to neutrophil-associated parenchymal injury, reinforces the potential modulatory role of adhesion molecule expression and interaction in control of specific enzymatic responses in inflammatory cells.

7. Therapeutic Intervention into the Reaction to Injury: A Cell Biological Approach

Insights into the cellular and molecular mechanisms that mediate the acute inflammatory reaction to injury associated with ischemia and reperfusion have targeted new approaches to specific therapeutic interventions designed to reduce this process. These more specific and sophisticated approaches are particularly topical at this time because: (1) the past history of clinical intervention into myocardial infarct size utilizing more general anti-inflammatory strategies was catastrophic, and (2) a very high percentage of patients are now undergoing thrombolytic therapy with successful reperfusion of the ischemic myocardium during a period of time in which myocardial salvage is highly practical.

7.1. Anticomplement Strategies

Early attempts at anticomplement therapy using complement depletion strategies involving cobra venom factor are obviously not clinically practical. In addition to depleting complement, this approach results in extraordinarily high levels of C5a for extended periods of time. Recombinant DNA technology has allowed the production of a soluble form of complement receptor 1 (CR1) (Weisman et al., 1990), which is capable of binding active complement and preventing complement-induced inflammatory reactions. This molecule was successful in markedly reducing myocardial infarction size in a rat model. Such an approach does appear to have some practical application to human disease, although it depends on the ability of CR1 to get into the extracellular fluid of ischemic areas rapidly enough to bind up activated complement immediately upon reperfusion, and on our ability to administer enough CR1 to bind the majority of C5a generated proteolytically in the extracellular fluid.

7.2. Antiadhesion Approaches

As previously described, monoclonal antibodies to CD18 have been used by a variety of laboratories to reduce ischemic reperfusion injury. Simpson et al. (1988, 1990) utilized monoclonal antibody to human CD11b, which appeared to minimize infarct size in dogs after ischemia and reperfusion despite the fact that it did not inhibit adherence of dog neutrophils to dog endothelial cells or cardiac myocytes. Interestingly, we found that this antibody also inhibited the production of reactive oxygen species by adherent dog neutrophils, despite the fact that it did not interfere with neutrophil adhesion reactions. This suggests that the integrin system may also have a secondary controlling effect (in addition to adhesion) on reactive oxygen production (Shappell et al., 1990b; Entman et al., 1992). Subsequent papers by Simpson and colleagues were unable to demonstrate that neutrophil infiltration was reduced by antibody 904 (Simpson et al., 1988; Simpson et al., 1990), in correlation with our observations that this antibody did not inhibit adhesion in vitro. Antibody 904 may be acting primarily by reducing the oxidative reaction of these neutrophils (Shappell et al., 1990b). In addition, a chimeric model of IgG and soluble L-selectin reduced neutrophil accumulation in a mouse peritonitis model (Watson et al., 1991), indicating that soluble forms of single types of adhesion molecules may have therapeutic effects.

In the future, site-specific approaches to both neutrophil adhesion and neutrophil activation will probably be carefully studied and considered. It is possible that these approaches will be used in combination. While monoclonal antibodies, bioengineered proteins, and peptide antagonists of ligand adhesion may be the initial candidates for intervention, there will also be a need for small, site-specific, inorganic molecules.

8. Summary

We have presented a classic cell biological scheme as a framework for examining the molecular processes that initiate reaction to injury occurring after reperfusion in ischemia–reperfusion models. The goal of these investigations is to analyze the mechanism by which an inflammatory cascade, as a reaction to injury, extends a primary injury. An understanding of the specific controlling steps in such a cascade as a function of time and degree of injury is critical to the rational design of specific site-directed interventions. The ultimate aim of such investigations is to identify specific molecular targets and devise practical methods of intervening in the biology of the process. It is critical to better understand the relative importance of the specific cell biological steps during ischemia and reperfusion, in order to attain a practical therapeutic approach aimed at the specific biological targets.

References

Abbassi, O., Lane, C. L., Krater, S. S., Kishimoto, T. K., Anderson, D. C., McIntire, L. V., and Smith, C. W., 1991, Canine neutrophil margination mediated by lectin adhesion molecule-1 (LECAM-1) *in vitro, J. Immunol.* **147**:2107–2115.

Ambrosio, G., Weisman, H. F., Mannisi, J. A., and Becker, L. C., 1989, Progressive impairment of regional myocardial perfusion after initial restoration of post ischemic blood flow, *Circulation* **80**:1846–1861.

Anderson, D. C., Smith, C. W., Michael, L. H., and Entman, M. L., 1991, Stimulation of ICAM-1 mRNA in ischemic and reperfused canine myocardium, *Circulation* **84**:II85 (Abstract).

Arfors, K. E., Lundberg, C., Lindbom, L., Lundberg, K., Beatty, P. G., and Harlan, J. M., 1987, A monoclonal antibody to the membrane glycoprotein complex CD18 inhibits polymorphonuclear leukocyte accumulation and plasma leakage *in vivo, Blood* **69**:338–340.

Baumgarten, W., 1899, Infarction of the heart, *Am. J. Physiol.* **2**:243–265.

Bevilacqua, M. P., Stengelin, S., Gimbrone, M. A., Jr., and Seed, B., 1989, Endothelial leukocyte adhesion molecule 1: An inducible receptor for neutrophils related to complement regulatory proteins and lectins, *Science* **243**:1160–1165.

Bevilacqua, M. P., Butcher, E., Furie, B., Gallatin, M., Gimbrone, M. A., Harlan, J. M., Kishimoto, T. K., Lasky, L. A., McEver, R. P., Paulson, J. C., Rosen, S. D., Seed, B., Siegelman, M., Springer, T. A., Stoolman, L. M., Tedder, T. F., Varki, A., Wagner, D. D., Weissman, I. L., and Zimmerman, G. A., 1991, Selectins: A family of adhesion receptors, *Cell* **67**:233.

Braunwald, E., and Kloner, R. A., 1982, The stunned myocardium: Prolonged post-ischemic ventricular dysfunction, *Circulation* **66:**1146–1149.

Bulkley, B. H., and Hutchins, G. M., 1977, Myocardial consequences of coronary artery bypass graft surgery: The paradox of necrosis in areas of revascularization, *Circulation* **56:**906–913.

Carden, D. L., Smith, J. K., and Korthuis, R. J., 1990, Neutrophil-mediated microvascular dysfunction in postischemic canine skeletal muscle. Role of granulocyte adherence, *Circ. Res.* **66:**1436–1444.

Diamond, M. S., Staunton, D. E., deFougerolles, A. R., Stacker, S. A., Garcia-Aguilar, J., Hibbs, M. L., and Springer, T. A., 1990, ICAM-1 (CD54): A counter-receptor for Mac-1 (CD11b/CD18), *J. Cell Biol.* **111:**3129–3139.

Dransfield, I., Cabanas, C., Craig, A., and Hogg, N., 1992, Divalent cation regulation of the function of the leukocyte integrin LFA-1, *J. Cell Biol.* **116:**219–226.

Dreyer, W. J., Smith, C. W., Michael, L. H., Rossen, R. D., Hughes, B. J., Entman, M. L., and Anderson, D. C., 1989, Canine neutrophil activation by cardiac lymph obtained during reperfusion of ischemic myocardium, *Circ. Res.* **65:**1751–1762.

Dreyer, W. J., Michael, L. H., Rossen, R. D., Nguyen, T., Anderson, D. C., Smith, C. W., and Entman, M. L., 1991a, Evidence for C5a in post-ischemic canine cardiac lymph, *Clin. Res.* **39:**271A (Abstract).

Dreyer, W. J., Michael, L. H., West, M. S., Smith, C. W., Rothlein, R., Rossen, R. D., Anderson, D. C., and Entman, M. L., 1991b, Neutrophil accumulation in ischemic canine myocardium: Insights into the time course, distribution, and mechanism of localization during early reperfusion, *Circulation* **84:**400–411.

Dustin, M. L., Rothlein, R., Bhan, A. K., Dinarello, C. A., and Springer, T. A., 1986, Induction by IL-1 and interferon-gamma: Tissue distribution, biochemistry, and function of a natural adherence molecule (ICAM-1), *J. Immunol.* **137:**245–254.

Engler, R., Dahlgren, M., Schmid-Schonbein, G. W., and Dobbs, A., 1984, Leukocyte depletion prevents progressive flow impairment to ischemic myocardium, *Circulation* **70:**II–228.

Engler, R. L., Dahlgren, M. D., Morris, D. D., Peterson, M. A., and Schmid-Schonbein, G. W., 1986a, Role of leukocytes in response to acute myocardial ischemia and reflow in dogs, *Am. J. Physiol.* **251:**H314–H323.

Engler, R. L., Dahlgren, M. D., Peterson, M. A., Dobbs, A., and Schmid-Schonbein, G. W., 1986b, Accumulation of polymorphonuclear leukocytes during 3h experimental myocardial ischemia, *Am. J. Physiol.* **251:**H93–H100.

Entman, M. L., Youker, K., Shappell, S. B., Siegel, C., Rothlein, R., Dreyer, W. J., Schmalstieg, F. C., and Smith, C. W., 1990, Neutrophil adherence to isolated adult canine myocytes: Evidence for a CD18-dependent mechanism, *J. Clin. Invest.* **85:**1497–1506.

Entman, M. L., Youker, K., Shoji, T., Kukielka, G., Shappell, S. B., Taylor, A. A., and Smith, C. W., 1992, Neutrophil induced oxidative injury of cardiac myocytes: A compartmented system requiring CD11b/CD18-ICAM-1 adherence, *J. Clin. Invest.* **90:**1335–1345.

Feinmark, S. J., and Cannon, P. J., 1986, Endothelial cell leukotriene C4 synthesis results from intercellular transfer of leukotriene A4 synthesized by polymorphonuclear leukocytes, *J. Biol. Chem.* **261:**16466–16472.

Feinmark, S. J., and Cannon, P. J., 1987, Vascular smooth muscle cell leukotriene C4 synthesis: Requirement for transcellular leukotriene A4 metabolism, *Biochim. Biophys. Acta* **922:**125–135.

Furie, M. B., Naprstek, B. L., and Silverstein, S. C., 1987, Migration of neutrophils across monolayers of cultured microvascular endothelial cells, *J. Cell Sci.* **88:**161–175.

Furie, M. B., Tancinco, M. C. A., and Smith, C. W., 1991, Monoclonal antibodies to leukocyte integrins CD11a/CD18 and CD11b/CD18 or intercellular adhesion molecule-1 (ICAM-1) inhibit chemoattractant-stimulated neutrophil transendothelial migration *in vitro, Blood* **78:**2089–2097.

GISSI Study Group, 1986, Effectiveness of intravenous thrombolytic treatment in acute myocardial infarction, *Lancet* **1:**397–402.

Granger, D. N., 1988, Role of xanthine oxidase and granulocytes in ischemia-reperfusion injury, *Am. J. Physiol.* **255:**H1269–H1275.

Hahn, R. A., MacDonald, B. R., Simpson, P. J., Potts, B. D., and Parli, C. J., 1990, Antagonism of leukotriene B4 receptors does not limit canine myocardial infarct size, *J. Pharmacol. Exp. Ther.* **253:**58–66.

Harlan, J. M., Killen, P. D., Senecal, F. M., Schwartz, B. R., Yee, E. K., Taylor, R. F., Beatty, P. G., Price, T. H., and Ochs, H. D., 1985, The role of neutrophil membrane glycoprotein GP-150 in neutrophil adherence to endothelium *in vitro, Blood* **66:**167–178.

Hearse, D. J., and Bolli, R., 1991, Reperfusion-induced injury: Manifestations, mechanisms and clinical relevance, *Trends Cardiovasc. Med.* **1:**233–240.

Hernandez, L. A., Grisham, M. B., Twohig, B., Arfors, K. E., Harlan, J. M., and Granger, D. N., 1987, Role of neutrophils in ischemia-reperfusion-induced microvascular injury, *Am. J. Physiol.* **238:**H699–H703.

Hill, J. H., and Ward, P. A., 1971, The phlogistic role of C3 leukotactic fragment in myocardial infarcts of rats, *J. Exp. Med.* **133:**885–900.

Hillis, L. D., and Braunwald, E., 1977, Myocardial ischemia, *N. Engl. J. Med.* **296:**1093–1096.

Hughes, B. J., Hollers, J. C., Crockett-Torabi, E., and Smith, C. W., 1992, Recruitment of CD11b/CD18 to the neutrophil surface and adherence-dependent cell locomotion, *J. Clin. Invest.* **90:**1687–1696.

Hynes, R. O., 1992, Integrins: Versatility, modulations, and signaling in cell adhesion, *Cell* **69:**11–25.

Inauen, W., Granger, D. N., Meininger, C. J., Schelling, M. E., Granger, H. J., and Kvietys, P. R., 1990, Anoxia/reoxygenation-induced, neutrophil-mediated endothelial cell injury: Role of elastase, *Am. J. Physiol.* **259:**H925–H931.

Jolly, S. R., Kane, W. J., Bailie, M. B., Abrams, G. D., and Lucchesi, B. R., 1984, Canine myocardial reperfusion injury: Its reduction by the combined administration of superoxide dismutase and catalase, *Circ. Res.* **54:**277–285.

Kagiyama, A., Savage, H. E., Michael, L. H., Hanson, G., Entman, M. L., and Rossen, R. D., 1989, Molecular basis of complement activation in ischemic myocardium: Identification of specific molecules of mitochondrial origin that bind C1q and fix complement, *Circ. Res.* **64:**604–615.

Lawrence, M. B., Smith, C. W., Eskin, S. G., and McIntire, L. V., 1990, Effect of venous shear stress on CD18-mediated neutrophil adhesion to cultured endothelium, *Blood* **75:**227–237.

Lucchesi, B. R., and Mullane, K. M., 1986, Leukocytes and ischemia induced myocardial injury, *Annu. Rev. Pharmacol. Toxicol.* **26:**201–224.

Ma, X.-L., Tsao, P. S., and Lefer, A. M., 1991, Antibody to CD18 exerts endothelial and cardiac protective effects in myocardial ischemia and reperfusion, *J. Clin. Invest.* **88:**1237–1243.

Mallory, G. K., White, P. D., and Salcedo-Salgar, J., 1939, The speed of healing of myocardial infarction. A study of the pathologic anatomy in seventy-two cases, *Am. Heart J.* **18:**647–671.

Marlin, S. D., and Springer, T. A., 1987, Purified intercellular adhesion molecule-1 (ICAM-1) is a ligand for lymphocyte function-associated antigen 1 (LFA-1), *Cell* **51:**813–819.

Maroko, P. R., Carpenter, C. D., Chiariello, M., Fishbein, M. C., Radvany, P., Knostman, J. D., and Hale, S. L., 1978, Reduction by cobra venom factor of myocardial necrosis after coronary artery occlusion, *J. Clin. Invest.* **61:**661–670.

Matis, W. L., Lavker, R. M., and Murphy, G. F., 1990, Substance P induces the expression of an endothelial–leukocyte adhesion molecule by microvascular endothelium, *J. Invest. Dermatol.* **94**:492–495.

Michael, L. H., Lewis, R. M., Brandon, T. A., and Entman, M. L., 1979, Cardiac lymph from conscious dogs, *Am. J. Physiol.* **237**:H311–H317.

Michael, L. H., Zhang, Z., Hartley, C. J., Bolli, R., Taylor, A. A., and Entman, M. L., 1990, Thromboxane B2 in cardiac lymph: Effect of superoxide dismutase and catalase during myocardial ischemia and reperfusion, *Circ. Res.* **66**:1040–1044.

Mullane, K. M., 1988, Myocardial ischemia-reperfusion injury: Role of neutrophils and neutrophil derived mediators, in: *Human Inflammatory Disease: Clinical Immunology* (G. Marone, L. M. Lichtenstein, M. Condorell, and A. S. Fauci, eds.), Dekker, New York, pp. 143–159.

Mullane, K. M., and Smith, C. W., 1990, The role of leukocytes in ischemic damage, reperfusion injury and repair of the myocardium, in: *Pathophysiology of Severe Ischemic Myocardial Injury* (H. M. Piper, ed.), Kluwer Academic Publishers, Dordrecht, pp. 239–267.

Mullane, K. M., Read, N., Salmon, J. A., and Moncada, S., 1984, Role of leukocytes in acute myocardial infarction in anesthetized dogs. Relationship to myocardial salvage by anti-inflammatory drugs, *J. Pharmacol. Exp. Ther.* **228**:510–522.

Mullane, K. M., Salmon, J. A., and Kraemer, R., 1987, Leukocyte-derived metabolites of arachidonic acid in ischemia-induced myocardial injury, *Fed. Proc.* **46**:2422–2438.

Mullane, K. M., Westlin, W., and Kraemer, R., 1988, Activated neutrophils release mediators that may contribute to myocardial dysfunction associated with ischemia and reperfusion, in: *Biology of the Leukotrienes,* New York Academy of Sciences, New York, pp. 103–121.

Nathan, C. F., 1987, Neutrophil activation on biological surfaces: Massive secretion of hydrogen peroxide in response to products of macrophages and lymphocytes, *J. Clin. Invest.* **80**:1550–1560.

Olafsson, B., Forman, M. B., Puett, D. W., Pou, A., Cates, C. V., and Friessinger, G. C., 1987, Reduction of reperfusion injury in the canine preparation by intracoronary perfluorochemical, *Circulation* **76**:1135–1145.

Patel, K. D., Zimmerman, G. A., Prescott, S. M., McEver, R. P., and McIntyre, T. M., 1991, Oxygen radicals induce human endothelial cells to express GMP-140 and bind neutrophils, *J. Cell Biol.* **112**:749–759.

Petrone, W. F., English, D. K., Wong, K., and McCord, J. M., 1980, Free radicals and inflammation: Superoxide-dependent activation of a neutrophil chemotactic factor in plasma, *Proc. Natl. Acad. Sci. USA* **77**:1159–1163.

Picker, L. J., Warnock, R. A., Burns, A. R., Doerschuk, C. M., Berg, E. L., and Butcher, E. C., 1991, The neutrophil selectin LECAM-1 presents carbohydrate ligands to the vascular selectins ELAM-1 and GMP-140, *Cell* **66**:921–933.

Pinckard, R. N., Olson, M. S., Kelley, R. E., Detter, D. H., Palmer, J. D., O'Rourke, R. A., and Goldfein, S., 1973, Antibody-independent activation of human C1 after interaction with heart subcellular membranes, *J. Immunol.* **110**:1376–1382.

Pinckard, R. N., Olson, M. S., Giclas, P. C., Terry, R., Boyer, J. T., and O'Rourke, R. A., 1975, Consumption of classical complement components by heart subcellular membranes *in vitro* and in patients after acute myocardial infarction, *J. Clin. Invest.* **56**:740–750.

Roberts, R., DeMello, V., and Sobel, B. E., 1976, Deleterious effects of methylprednisolone in patients with myocardial infarction, *Circulation* **53**(Suppl. I);204–206.

Romson, J. L., Hook, B. G., Rigot, V. H., Schork, M. A., Swanson, D. P., and Lucchesi, B. R., 1982, The effect of ibuprofen on accumulation of ¹¹¹indium labeled platelets and leukocytes in experimental myocardial infarction, *Circulation* **66**:1002–1011.

Romson, J. L., Hook, B. G., Kunkel, S. L., Abrams, G. D., Schork, M. A., and Lucchesi, B. R., 1983, Reduction of the extent of ischemic myocardial injury by neutrophil depletion in the dog, *Circulation* **67**:1016–1023.

Rossen, R. D., Swain, J. L., Michael, L. H., Weakley, S., Giannini, E., and Entman, M. L., 1985, Selective accumulation of the first component of complement and leukocytes in ischemic canine heart muscle: A possible initiator of an extra myocardial mechanism of ischemic injury, *Circ. Res.* **57**:119–130.

Rossen, R. D., Michael, L. H., Kagiyama, A., Savage, H. E., Hanson, G., Reisbery, J. N., Moake, J. N., Kim, S. H., Weakly, S., Giannini, E., and Entman, M. L., 1988, Mechanism of complement activation following coronary artery occlusion: Evidence that myocardial ischemia causes release of constituents of myocardial subcellular origin which complex with the first component of complement, *Circ. Res.* **62**:572–584.

Rowzer, C. A., and Kargman, S. A., 1985, Translocation of 5-lipoxygenase to the membrane in human leukocytes challenged with ionophore A23187, *J. Biol. Chem.* **263**:10980–10988.

Schmid-Schonbein, G. W., and Engler, R. L., 1987, Granulocytes as active participants in acute myocardial ischemia and infarction, *Am. J. Cardiovasc. Pathol.* **1**:15–30.

Seewaldt-Becker, E., Rothlein, R., and Dammgen, J. W., 1989, CDw18 dependent adhesion of leukocytes to endothelium and its relevance for cardiac reperfusion, in: *Leukocyte Adhesion Molecules: Structure, Function, and Regulation* (T. A. Springer, D. C. Anderson, A. S. Rosenthal, and R. Rothlein, eds.), Springer-Verlag, Berlin, pp. 138–148.

Shappell, S. B., Taylor, A. A., Hughes, H., Mitchell, J. R., Anderson, D. C., and Smith, C. W., 1990a, Comparison of antioxidant and nonantioxidant lipoxygenase inhibitors on neutrophil function. Implications for pathogenesis of myocardial reperfusion injury, *J. Pharmacol. Exp. Ther.* **252**:531–538.

Shappell, S. B., Toman, C., Anderson, D. C., Taylor, A. A., Entman, M. L., and Smith, C. W., 1990b, Mac-1 (CD11b/CD18) mediates adherence-dependent hydrogen peroxide production by human and canine neutrophils, *J. Immunol.* **144**:2702–2711.

Shingu, M., and Nobunaga, M., 1984, Chemotactic activity generated in human serum from the fifth component on hydrogen peroxide, *Am. J. Pathol.* **117**:210–216.

Simpson, P. J., Mickelson, J., Fantone, J. C., Gallagher, K. P., and Lucchesi, B. R., 1987, Iloprost inhibits neutrophil function *in vitro* and *in vivo* and limits experimental infarct size in canine heart, *Circ. Res.* **60**:666–673.

Simpson, P. J., Todd, R. F., III, Fantone, J. C., Mickelson, J. K., Griffin, J. D., and Lucchesi, B. R., 1988, Reduction of experimental canine myocardial reperfusion injury by a monoclonal antibody (anti-Mo1, anti-CD11b) that inhibits leukocyte adhesion, *J. Clin. Invest.* **81**:624–629.

Simpson, P. J., Todd, R. F., III, Mickelson, J. K., Fantone, J. C., Gallagher, K. P., Lee, K. A., Tamura, Y., Cronin, M., and Lucchesi, B. R., 1990, Sustained limitation of myocardial reperfusion injury by a monoclonal antibody that alters leukocyte function, *Circulation* **81**:226–237.

Smiley, P. L., Stremler, K. E., Prescott, S. M., Zimmerman, G. A., and McIntyre, T. M., 1991, Oxidatively fragmented phosphatidylcholines activate human neutrophils through the receptor for platelet-activating factor, *J. Biol. Chem.* **266**:11104–11110.

Smith, C. W., 1992, Transendothelial migration, in: *Adhesion: Its Role in Inflammatory Disease* (J. M. Harlan and D. Y. Liu, eds.), Freeman, San Francisco, pp. 85–115.

Smith, C. W., Marlin, S. D., Rothlein, R., Toman, C., and Anderson, D. C., 1989, Cooperative interactions of LFA-1 and Mac-1 with intercellular adhesion molecule-1 in facilitating adherence and transendothelial migration of human neutrophils *in vitro*, *J. Clin. Invest.* **83**:2008–2017.

Smith, C. W., Entman, M. L., Lane, C. L., Beaudet, A. L., Ty, T. I., Youker, K., Hawkins, H. K., and Anderson, D. C., 1991a, Adherence of neutrophils to canine cardiac myocytes *in vitro* is dependent on intercellular adhesion molecule-1, *J. Clin. Invest.* **88**:1216–1223.

Smith, C. W., Kishimoto, T. K., Abbassi, O., Hughes, B. J., Rothlein, R., McIntire, L. V., Butcher, E., and Anderson, D. C., 1991b, Chemotactic factors regulate lectin adhesion molecule 1 (LECAM-1)-dependent neutrophil adhesion to cytokine-stimulated endothelial cells *in vitro*, *J. Clin. Invest.* **87**:609–618.

Stahl, G. L., Terashita, Z.-I., and Lefer, A. M., 1988, Role of platelet activating factor in propagation of cardiac damage during myocardial ischemia, *J. Pharmacol. Exp. Ther.* **244**:898–904.

Suzuki, M., Onauen, W., Kiretys, P. R., Grisham, M. B., Meninger, C., Schelling, M. E., Granger, H. J., and Granger, D. N., 1989, Superoxide mediates reperfusion-induced leukocyte–endothelial cell interactions, *Am. J. Physiol.* **257**:H1740–H1745.

Svendsen, J. H., Bjerrum, P. J., and Hawnso, S., 1991, Myocardial capillary permeability after regional ischemia and reperfusion in *in vivo* canine heart, *Circ. Res.* **68**:174–184.

Taylor, A. A., Gasic, A. C., Kitt, T. M., Shappell, S. B., Rui, J., Lenz, M. L., Smith, C. W., and Mitchell, J. R., 1989, A specific leukotriene B[4] antagonist protects against myocardial ischemia-reflow injury, *Clin. Res.* **37**:528A (Abstract).

Tsao, P. S., and Lefer, A. M., 1990, Time course and mechanism of endothelial dysfunction in isolated ischemic and hypoxic rat hearts, *Am. J. Physiol.* **28**:H1660–H1666.

Vogt, W., von Zabern, I., Hesse, D., Nolte, R., and Haller, Y., 1987, Generation of an activated form of human C5 (C5b-like C5) by oxygen radical, *Immunol. Lett.* **14**:209–215.

Watson, S. R., Fennie, C., and Lasky, L. A., 1991, Neutrophil influx into an inflammatory site inhibited by a soluble homing receptor–OgG chimaera, *Nature* **349**:164–167.

Weisman, H. F., Barton, T., Leppo, M. K., Marsh, H. C., Jr., Carson, G. R., Concino, M. F., Boyle, M. P., Roux, K. H., Weisfeldt, M. L., and Fearon, D. T., 1990, Soluble human complement receptor type 1: *In vivo* inhibitor of complement suppressing post-ischemic myocardial inflammation and necrosis, *Science* **249**:146–151.

Williams, F. M., Collins, P. D., Nourshargh, S., and Williams, T. J., 1988, Suppression of ^{111}In-neutrophil accumulation in rabbit myocardium by MoA ischemic injury, *J. Mol. Cell. Cardiol.* **20**:S33.

Youker, K., Smith, C. W., Anderson, D. C., Miller, D., Michael, L. H., Rossen, R. D., and Entman, M. L., 1992, Neutrophil adherence to isolated adult cardiac myocytes: Induction by cardiac lymph collected during ischemia and reperfusion, *J. Clin. Invest.* **89**:602–609.

Zimmerman, G. A., McIntyre, T. M., Mehra, M., and Prescott, S. M., 1990, Endothelial cell-associated platelet-activating factor: A novel mechanism for signaling intercellular adhesion, *J. Cell Biol.* **110**:529–540.

13

GPIIb/IIIa Antagonists as Novel Antithrombotic Drugs
Potential Therapeutic Applications

ANDREW J. NICHOLS, JANICE A. VASKO, PAUL F. KOSTER, RICHARD E. VALOCIK, and JAMES M. SAMANEN

1. Introduction

Cardiovascular and cerebrovascular diseases are the leading cause of morbidity and mortality in the United States (Gunby, 1992). The major pathological event that gives rise to symptoms of these diseases is occlusion of a major artery by a thrombus or an embolus to the extent that blood flow is severely compromised, such that oxygen supply cannot meet demand. Thus, thrombosis represents a major target for the development of drugs to prevent and treat a variety of cardiovascular and cerebrovascular diseases. Recent advances in our understanding of the biochemical events that occur during thrombosis have led to the discovery and development of agents that inhibit thrombosis by selectively interfering with critical pathways in the initiation and maintenance of the thrombotic process. This review will concentrate on one particular aspect of thrombosis, namely platelet aggregation, and its inhibition by antagonists of the adhesion molecule, GPIIb/IIIa, which is part of the final common pathway for platelet aggregation.

ANDREW J. NICHOLS, JANICE A. VASKO, PAUL F. KOSTER, and RICHARD E. VALOCIK
• Department of Cardiovascular Pharmacology, SmithKline Beecham Pharmaceuticals, King of Prussia, Pennsylvania 19406.   *JAMES M. SAMANEN* • Department of Medicinal Chemistry, SmithKline Beecham Pharmaceuticals, King of Prussia, Pennsylvania 19406.
Cellular Adhesion: Molecular Definition to Therapeutic Potential, edited by Brian W. Metcalf, Barbara J. Dalton, and George Poste. Plenum Press, New York, 1994.

2. GPIIb/IIIa, Platelets and Thrombosis

2.1. Platelet Aggregation and Thrombosis

Thrombosis is the pathological extension of the normal hemostatic process that is required to prevent blood loss following damage to the vascular wall. Vascular thrombi are broadly classified into three types: (1) red thrombi, consisting mainly of erythrocytes enmeshed in a fibrin clot, (2) white thrombi, consisting almost exclusively of aggregated platelets, and (3) mixed thrombi, which consist of erythrocytes and platelets enmeshed in a fibrin clot. Red thrombi that have little or no platelet component are commonly found in areas of low blood flow and low shear stress, such as deep vein thrombi caused by immobility of the lower extremities. White and mixed thrombi, which have a significant platelet component, are most often associated with endothelial damage in regions of high shear stress, such as in the region of a ruptured atherosclerotic plaque in a coronary artery. In such cases, thrombosis is initiated by platelet adhesion to von Willebrand factor, collagen, and other matrix proteins in the exposed vascular subendothelium. The adhesive interaction of platelets with collagen and von Willebrand factor produces platelet activation and causes the release of thromboxane A_2 (TxA_2) and serotonin, and catalyzes the conversion of prothrombin to thrombin on the platelet surface, all of which act to produce platelet aggregation in the region of the adherent platelets.

Following their initial activation, platelets undergo a primary phase of reversible aggregation followed by a secondary, irreversible phase of aggregation. The relative roles of these two waves of platelet aggregation vary with the agonist that stimulates aggregation and with the agonist concentration. During the secondary phase of platelet aggregation the platelets release second messengers (e.g., TxA_2) and undergo secretion in which they release the contents of their dense granules (e.g., ATP and serotonin) and the α granules (e.g., PDGF, TGFβ, PF-4, β-thromboglobulin, and PAI-1). Obviously, the stimulus for platelet aggregation *in vivo* involves multiple agonists, including serotonin, TxA_2, PAF, thrombin, and epinephrine. As such, physiological or pathological platelet aggregation *in vivo* is associated with platelet granule release. Indeed, increased circulating plasma levels of platelet granule contents (e.g., PF-4 and β-thromboglobulin) are used as diagnostic indicators of *in vivo* platelet activation and aggregation.

Platelet aggregation in the region of stenosis formed by the atherosclerotic plaque will lead to the formation of a white thrombus that may break down spontaneously or may progress to become a mixed thrombus. Platelet activation exposes phosphatidylserine on the platelet surface, which serves as the phospholipid substrate for the assembly of both the intrinsic tenase complex and the prothrombinase complex. Thus, activated platelets have a functional procoagulant activity, serving as a site for thrombin generation and subsequent fibrin formation. If platelet aggregates are not resolved within a short period of time, they can act as sites for the formation of fibrin. Moreover, the platelet surface provides an anchor site for Factor XIII, which acts to cross-link the fibrin monomers into fibrin polymers that are required to stabilize the fibrin component of the thrombus. Thus, activation of the coagulation system and further aggregation of platelets leads to a mixed thrombus, with both erythrocytes and platelets entrapped in a fibrin mesh. Platelet aggregation is central to the initiation and propagation of this process and is, therefore, necessary to ensure a fully developed arterial thrombus. The critical role of platelets in the thrombotic process that underlies the acute coronary syndromes, unstable angina and acute myocardial infarction, has recently been confirmed by Mizuno *et al.* (1992), who have used intravascular angioscopy to demonstrate the presence of platelets in the coronary arterial thrombus of nearly all patients undergoing an acute coronary syndrome. Moreover, Mizuno *et al.* (1992) also demonstrated that the relatively transient coronary artery thrombus that forms during an episode of unstable angina is almost exclusively white, and thus rich in platelets, whereas the longer-lasting, more stable coronary artery thrombus that forms during an acute myocardial infarction is almost exclusively mixed, and is thus composed of a mixture of platelets, fibrin, and erythrocytes.

2.2. The Role of GPIIb/IIIa in Platelet Aggregation

GPIIb/IIIa, also known as α_{IIb}/β_3, CD41/CD61, or the fibrinogen receptor, is a member of the integrin family of adhesion molecules. It is located exclusively on platelets and cells of megakaryoblastic potential, and serves as the receptor for the macromolecular ligands, fibrinogen, and von Willebrand factor in the mediation of platelet aggregation. The α and β subunits form a structure that is believed to have a globular head that is exposed on the extracellular surface, with two tails that insert into the plasma membrane (see Fig. 1). The critical role of GPIIb/IIIa in platelet aggregation is exemplified in patients with Glanzmann's thrombasthenia, in which there is a congenital defect or deficiency in intact GPIIb/IIIa on the platelets (Phillips and Agin, 1977). Platelets from these patients fail to bind fibrinogen (Mustard *et al.*, 1979; Coller, 1980), and do not aggregate when activated by any platelet agonist, whether it is a weak agonist (e.g., ADP) or a strong agonist (e.g., thrombin). However, these platelets can agglutinate following the stabilization of the interaction of von Willebrand factor with the GPIb/IX complex by ristocetin, demonstrating that the loss of GPIIb/IIIa function produces a selective deficit in platelet function. Since solution-phase fibrinogen and other macromolecules bind to GPIIb/IIIa only after the activation of platelets by an agonist (Bennet and Vilaire, 1979), and the molecular defect in Glanzmann's thrombasthenia is a loss of functional GPIIb/IIIa, it can be concluded that a change in conformation in GPIIb/IIIa

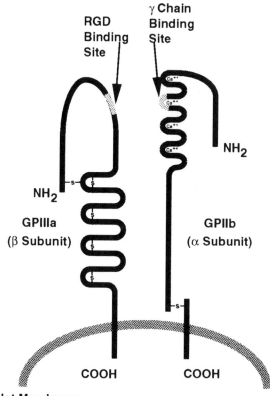

Platelet Membrane

Figure 1. The structure of GPIIb/IIIa (α_{IIb}/β_3), demonstrating the short cytoplasmic tails and globular extracellular heads of both subunits. The location of intramolecular disulfide bridges, the proposed Ca^{2+} binding domains, and the proposed binding sites for the RGD sequence and the γ-chain dodecapeptide are outlined to demonstrate the functional sites of the receptor.

following platelet activation allowing the interaction of macromolecules is an absolute requirement for platelet aggregation. This process is therefore the final common pathway for platelet aggregation, and represents a target for pharmacological intervention to inhibit platelet aggregation produced by all stimuli. The idea that GPIIb/IIIa is the final common pathway for platelet aggregation and represents a therapeutic target is further supported by studies with monoclonal antibodies against GPIIb/IIIa, one of which has been studied in preliminary clinical trials. 7E3 is a murine monoclonal antibody raised against human platelets that recognizes intact GPIIb/IIIa and binds preferentially to activated platelets (Coller, 1985). The F(ab')$_2$ fragments of this monoclonal antibody have been shown to produce a dose-related blockade of GPIIb/IIIa, and to inhibit *ex vivo* platelet aggregation in patients with unstable angina (Gold *et al.*, 1990).

3. RGD, from Adhesion Motif to GPIIb/IIIa Antagonists

3.1. RGD as an Adhesion Motif

The peptide sequence Arg-Gly-Asp (RGD) was originally recognized as the adhesion motif in fibronectin required for binding to fibronectin receptors (Pierschbacher and Ruoslahti, 1984). It was subsequently discovered that this sequence is also present in fibrinogen, von Willebrand factor, and vitronectin (Ruoslahti and Pierschbacher, 1984), the circulatory macromolecules that are believed to mediate platelet aggregation. Small RGD peptide derivatives will inhibit fibrinogen binding to activated platelets, and will inhibit platelet aggregation mediated by all stimuli (Plow *et al.*, 1985). Although fibrinogen is usually assumed to be the plasma macromolecule responsible for platelet aggregation (Hawiger *et al.*, 1989), there is evidence that von Willebrand factor (Weiss *et al.*, 1989) may also be involved, at least under certain conditions. Fibrinogen is a symmetrical multimeric protein that consists of two α chains, two β chains, and two γ chains (Fig. 2). The regions of fibrinogen that are responsible for binding to GPIIb/IIIa are residues 95–97 and 572–574, both representing RGD sequences in the α chain, and the 12 C-terminal residues, 400–411 (HHLGGAK-QAGDV), in the γ chain (Kloczewiak *et al.*, 1984). It appears that of the two pairs of RGD sequences, only those closer to the C-terminus (i.e., 572–574) are involved in the binding of solution-phase fibrinogen to GPIIb/IIIa (Cheresh *et al.*, 1989). Even if fibrinogen is not the major mediator of platelet aggregation *in vivo*, von Willebrand factor also has an RGD sequence that binds to GPIIb/IIIa. In receptor cross-linking experiments, RGD peptides have been found to bind to

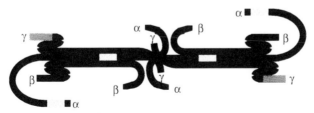

Figure 2. The structure of fibrinogen, demonstrating the interlinking of the α, β, and γ chains at their N-terminal regions, and the location of the terminal γ-chain dodecapeptide adhesive moiety (▬▬▬▬), and the location of the two RGD sequences in the α chain (▦▦▦). The RGD sequence toward the N-terminus is contained in an α helix, and thus hidden, whereas the RGD sequence toward the C-terminus is exposed and available to act as an adhesive moiety.

residues 109–171 of GPIIIa (D'Souza *et al.*, 1988), whereas γ-chain dodecapeptides have been found to bind to residues 294–314 of GPIIb (D'Souza *et al.*, 1990). The postulated sites of interaction of fibrinogen with GPIIb/IIIa are shown diagrammatically in Fig. 1. Interestingly, RGD peptides and γ-chain peptides (Timmons *et al.*, 1989) interact with GPIIb/IIIa in a mutually exclusive manner, suggesting some functional interaction between these two binding sites. The mechanism of this functional interaction is unknown, but it may be the result of the conformational change in GPIIb/IIIa induced by the binding of a small ligand (Parise *et al.*, 1987), such that when one binding site is occupied, the other site becomes inaccessible (Bennet *et al.*, 1988). Moreover, these data demonstrate that it is only necessary to inhibit one, but not both, fibrinogen binding sites in order to inhibit fibrinogen binding and platelet aggregation. Thus, regardless of which molecule is the mediator, or which sequence in fibrinogen, if it is the mediator, mediates platelet aggregation, RGD peptides will inhibit platelet aggregation.

3.2. RGD Analogues as GPIIb/IIIa Antagonists

The first GPIIb/IIIa antagonists to be developed were monoclonal antibodies, such as the 7E3 monoclonal antibody described briefly above. Indeed, such agents have paved the way for the development of other GPIIb/IIIa antagonists by demonstrating the critical role of platelet aggregation mediated by GPIIb/IIIa in a variety of cardiovascular disease models. However, the development of monoclonal antibodies as therapeutic agents may present many problems, not the least of which is the cost of production. Many groups have therefore concentrated on the development of small-molecule GPIIb/IIIa antagonists. Shebuski *et al.* (1989a) demonstrated that direct intracoronary infusion of Ac-RGDS-NH$_2$ could inhibit platelet-dependent coronary artery thrombosis, but high doses were required and the duration of action was short. The approach taken by most groups has thus been to develop analogues of RGDS in an attempt to increase affinity for GPIIb/IIIa, reduce affinity for other RGD-dependent integrins, and increase plasma stability. Limited data are available on the activity of the many compounds that have been synthesized. We will therefore concentrate on data obtained in our own laboratory, referring to compounds from other groups when available.

SK&F 106760 and SK&F 107260 (Table I) are RGD analogues with constrained conformations that impart higher affinity for GPIIb/IIIa and greater stability to degradation by plasma peptidases (Samanen *et al.*, 1991; Ali *et al.*, 1992). Table I shows the progressive increase in affinity for GPIIb/IIIa, and the increasing antiaggregatory potency, as Ac-Arg-Gly-Asp-Ser-NH$_2$ is modified by increasing the lipophilicity of the N-terminal region, constraining the conformation within a cyclic disulfide structure and the α-amino methylation of the ar-

Table I. Structural Modifications of Arg-Gly-Asp-Ser Leading to Potent, Selective GPIIb/IIIa Antagonists

Structure	Dog PRP/ADP[a] IC_{50} (μM)	L8 cell/VN[b] IC_{50} (μM)	GPIIb/IIIa[c] VNR
Ac-Arg-Gly-Asp-Ser-NH$_2$	91.3	150	1.64
Ac-Arg-Gly-Asp-Val-NH$_2$	55.5		
Ac-Cys-Arg-Gly-Asp-Cys-NH$_2$ ⌐S————S⌐	16.2		
Ac-Cys-Arg-Gly-Asp-Pen-NH$_2$ ⌐S————S⌐	4.1		
Ac-Cys-(NMe) Arg-Gly-Asp-Pen-NH$_2$ ⌐S————S⌐ (SK&F 106760)	0.35	67.2	192
Mba-(NMe) Arg-Gly-Asp-Man ⌐S————S⌐ (SK&F 107260)	0.09	2.0	22

[a]Concentration required to produce 50% inhibition of ADP-induced platelet aggregation in canine platelet-rich plasma.
[b]Concentration required to produce 50% inhibition of binding of L8 cells (rat skeletal muscle cell line) to vitronectin.
[c]Selectivity of GPIIb/IIIa relative to L8 cell vitronectin receptor.

ginine. In contrast to the increase in affinity for GPIIb/IIIa, these modifications do not increase affinity consistently for the vitronectin receptor on the L8 skeletal muscle cell line (Table I), thus leading to selectivity for GPIIb/IIIa. Figure 3 demonstrates the concentration-related inhibition of ADP-induced platelet aggregation in human platelet-rich plasma. In addition, it can be seen that SK&F 106760 does not inhibit GPIb/IX-mediated platelet agglutination induced by ristocetin in human platelet-rich plasma, indicating a lack of interaction with the adhesive von Willebrand factor receptor on platelets (Fig. 3B). The antiaggregatory activity of SK&F 106760 is independent of the agonist used, and is similar whether it is studied using light transmission in platelet-rich plasma or using impedance change in whole blood (Table II), indicating inhibition of a common pathway for platelet aggregation. SK&F 106760 and SK&F 107260 show significant species specificity in their ability to inhibit ADP-induced platelet aggregation. Table III shows that both compounds are equally effective at inhibiting aggregation of platelets from humans, cynomolgus monkeys, dogs, guinea pigs, and mice. However, neither compound can inhibit platelet aggregation in ferrets and gerbils. There are some species in which the compounds are weak, but equipotent (e.g., rabbits), and some species in which SK&F 107260 is significantly more potent than SK&F 106760 (e.g., Sprague–Dawley rats). Sur-

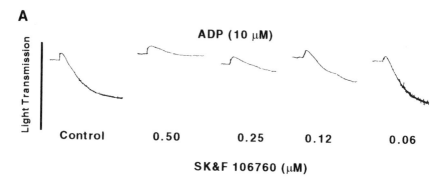

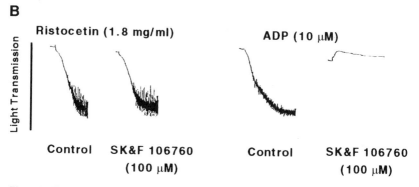

Figure 3. The concentration-related effect of SK&F 106760 in inhibiting ADP-induced platelet aggregation (A), in contrast to its inability to inhibit GPIb/IX-mediated platelet agglutination induced by ristocetin (B) in human platelet-rich plasma.

prisingly, SK&F 107260 is active in the mouse and the guinea pig. Thus, although there are significant species differences in the activity of peptide GPIIb/IIIa antagonists, these differences do not fit any obvious phylogenetic scheme.

Table II. The inhibition of Platelet Aggregation by SK&F 106760 *in vitro*

Method[a]	Agonist	IC$_{50}$ (μM)
Optical (PRP)	ADP	0.35 ± 0.04
	Collagen	0.26 ± 0.02
	U-46619	0.49 ± 0.09
Impedance (WB)	ADP	0.30 ± 0.06

[a]PRP, platelet-rich plasma; WB, whole blood.

Table III. Relative Potencies of SK&F 106760 and SK&F 107260 against ADP-Induced Platelet Aggregation in PRP from Different Species

Species	SK&F 106760	SK&F 107260	Potency ratio
	IC$_{50}$ (μM)		
Human	0.23 ± 0.06	0.06 ± 0.01	3.8
Cynomolgus monkey	0.14 ± 0.01	0.08 ± 0.01	1.8
Mongrel dog	0.35 ± 0.04	0.09 ± 0.01	3.9
Yucatán minipig	40.7 ± 4.4	28.2 ± 0.4	1.4
Sprague–Dawley rat	> 200	23.6 ± 1.9	> 8.5
NZW rabbit	13.4 ± 1.4	13.5 ± 3.8	1.0
Ferret	> 200	> 200	[1]
Gerbil	> 200	> 200	[1]
CF-1 mouse	0.74 ± 0.20	N.D.[a]	
Guinea pig	0.68 ± 0.15	N.D.	

[a]N.D., not determined.

In conscious dogs, SK&F 106760 and SK&F 107260 produced a dose-related inhibition of collagen-induced *ex vivo* whole blood platelet aggregation (Nichols *et al.*, 1990). Plasma concentrations of SK&F 106760 were measured by HPLC following the intravenous administration of 1 mg/kg SK&F 106760. The plasma concentration of SK&F 106760 at which *ex vivo* platelet aggregation was inhibited by 50% was estimated to be 0.46 μM, which is similar to the *in vitro* IC$_{50}$ value obtained in platelet-rich plasma and whole blood (Fig. 4). A comparison of the effect of 1 mg/kg SK&F 106760 given intravenously with 3 mg/kg given intraduodenally and intrajejunally on collagen-induced *ex vivo* whole blood platelet aggregation in conscious dogs demonstrated that SK&F 106760 was absorbed from the duodenum, but only to a limited extent (Fig. 5). An analysis of plasma levels from these experiments revealed a bioavailability of approximately 4–6%.

4. The Pharmacology and Therapeutic Applications of GPIIb/IIIa Antagonists

4.1. Unstable Angina

Unstable angina is characterized by transient episodes of myocardial ischemia at rest or at low work loads. Angiography and intravascular angioscopy have shown that thrombosis in the presence of a high-grade stenosis (> 70%) are the likely cause of the majority of cases of unstable angina (Ambrose *et al.*, 1985; Sherman *et al.*, 1986; Mizuno *et al.*, 1992). The precipitating factor for unstable angina appears to be atherosclerotic plaque fissuring or ulceration, leading to the

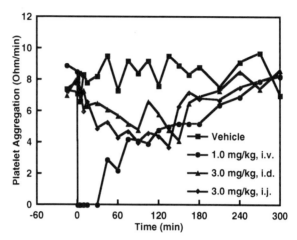

Figure 4. A comparison of the effects of SK&F 106760 given intravenously (1 mg/kg), intra-duodenally (3 mg/kg), and intrajejunally (3 mg/kg) on collagen-induced *ex vivo* whole blood platelet aggregation in conscious dogs.

exposure of a highly thrombogenic substrate. The intraluminal thrombus is prob-ably platelet-rich, in view of the following facts: (1) intravascular angioscopy reveals the thrombus to be predominantly white (Mizuno *et al.*, 1992), (2) fibrinolytic treatment has very little effect on unstable angina (Ambrose *et al.*, 1987), and (3) antiplatelet therapy, with either aspirin or ticlopidine, has been shown to be somewhat effective (Lewis *et al.*, 1983).

Folts has developed an animal model of unstable angina (Fig. 6), in which platelet-dependent coronary artery thrombosis is produced by endothelial dam-age of a coronary artery in the presence of a high-grade stenosis (Folts, 1991). Under these conditions, coronary artery blood flow is reduced as the platelet-rich thrombus forms in the region of the stenosis and gradually occludes the vessel. This thrombus can dislodge spontaneously or, under conditions of marked endothelial damage and with a very high degree of stenosis, can be dislodged mechanically to give rise to transient cyclic reductions in coronary artery blood flow (see Fig. 6). SK&F 106760 and SK&F 107260 abolish the cyclic blood flow reductions that are produced in endothelial damaged coronary arteries in the presence of a high-grade (> 90%) stenosis. This is shown in Fig. 7, which demonstrates that SK&F 106760 (1 mg/kg, i.v.) produces an immediate and complete inhibition of the platelet-dependent coronary artery cyclic flow reductions.

When administered at doses that produce at least 90% inhibition of *ex vivo* platelet aggregation, SK&F 106760 and SK&F 107260 always inhibit platelet-

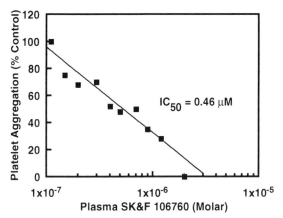

Figure 5. A plasma concentration–response curve for the SK&F 106760-mediated inhibition of collagen-induced *ex vivo* whole platelet aggregation in conscious dogs. Note the very close correlation between the *ex vivo* IC_{50} and the *in vitro* IC_{50} in Table I.

dependent coronary artery thrombosis. In contrast, aspirin at a dose sufficient to reduce *ex vivo* serum TxA_2 generation by more than 95% (5 mg/kg) was effective in only four out of nine animals (44%). Figure 8 shows such an animal, in which aspirin failed to inhibit the coronary artery cyclic flow reductions. In contrast, SK&F 106760 (0.3 mg/kg, i.v.) could immediately and completely abolish the cyclic flow reductions in animals that were unresponsive to aspirin.

Thrombosis can be reinduced in animals that are responsive to aspirin by increasing the degree of stenosis (Aiken *et al.*, 1981), such as that which occurs with acute rupture or hemorrhage into an atherosclerotic plaque or chronic progression of atherosclerosis, or by a mild elevation of plasma epinephrine levels by infusion to achieve plasma levels found during stress in humans (Folts and Rowe, 1988). In animals that were originally responsive to aspirin, but in which cyclic flow reductions could be reinduced by increasing the degree of stenosis or by an infusion of epinephrine, SK&F 106760 always produced an inhibition of the platelet-dependent thrombosis. It is our experience that whenever a GPIIb/IIIa antagonist is administered at doses sufficient to achieve plasma levels that are within the *in vitro* antiaggregatory range, coronary artery cyclic flow reductions are abolished. These data demonstrate two important phenomena: (1) aspirin is not universally effective at inhibiting platelet-dependent coronary artery thrombosis, since non-cyclooxygenase-dependent platelet aggregation commonly occurs *in vivo*, and (2) GPIIb/IIIa antagonists are universally effective as antiplatelet agents *in vivo*, since they inhibit the final common pathway of

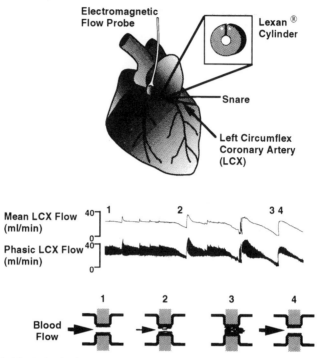

Figure 6. Model of platelet-dependent coronary artery thrombosis developed by Folts (1991). A coronary artery is exposed and the endothelium damaged by rubbing the vessel between a pair of forceps and a Lexan cylinder (1.0- to 2.0-mm i.d.) hard enough to produce a severe stenosis around the artery. Platelets adhere to the damaged subendothelium in the region of the stenosis and start to produce small oscillations in coronary arterial flow (a). Platelets then aggregate in the adherent platelets and start to occlude the vessel, causing coronary artery flow to decline (2). Eventually, the platelet thrombus fully occludes the artery and flow is reduced to zero (3). The thrombus then either spontaneously disrupts or is mechanically dislodged to allow flow to return (4). The stenosis is increased to the extent that the thrombus never spontaneously disrupts and must be mechanically dislodged, giving rise to a series of cyclic coronary artery flow reductions that repeats every 3–5 min.

platelet aggregation, and they work under conditions in which aspirin is ineffective. Thus, although aspirin is effective in treating unstable angina, a GPIIb/IIIa antagonist may be more effective, since activity would still be evident following acute rupture of the atherosclerotic plaque or chronic progression of the atherosclerotic lesion.

4.2. Prevention of Acute Myocardial Infarction

If left untreated, unstable angina often progresses to acute myocardial infarction, in which the coronary artery thrombus is large and stable enough to

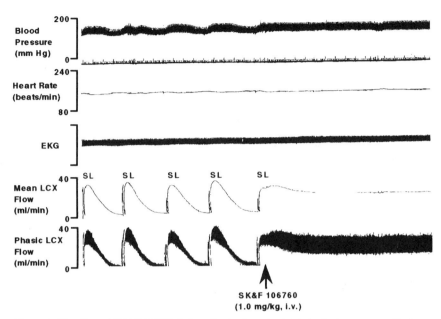

Figure 7. The effect of SK&F 106760 (1.0 mg/kg, i.v.) on platelet-dependent coronary cyclic flow reductions produced by a high degree of stenosis in an endothelium-damaged left circumflex coronary artery in an anesthetized dog. SL = mechanical disruption of the thrombus. Note that SK&F 106760 produced a complete and immediate inhibition of the cyclic flow reductions without producing any change in blood pressure or heart rate.

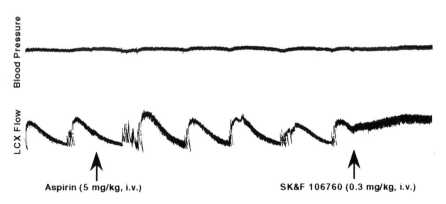

Figure 8. Contrasting effects of the cyclooxygenase inhibitor, aspirin (5 mg/kg, i.v.), with the GPIIb/ IIIa antagonist, SK&F 106760 (0.3 mg/kg, i.v.), on platelet-dependent coronary artery thrombosis in an anesthetized dog. Aspirin was ineffective at inhibiting the cyclic flow reductions, whereas SK&F 106760 produced an immediate and complete inhibition.

produce myocardial ischemia that is of sufficient duration and severity to cause irreversible myocyte death. The coronary artery thrombus in patients who die from such episodes consists of a platelet-rich head with a fibrin/erythrocyte-rich tail (Constantinides, 1966). Moreover, intravascular angioscopy has confirmed the presence of a mixed thrombus in the coronary arteries of patients while an acute myocardial infarction is in progress (Mizuno *et al.*, 1992). These data strongly suggest a role for platelets in the initiation, and perhaps the progression, of such thrombi. The role of platelets in the initiation of thrombosis that leads to acute myocardial infarction is further suggested by the correlation between the circadian variation in platelet aggregability and the incidence of the onset of acute myocardial infarction (Tofler *et al.*, 1987). Both parameters increase after waking and rising in the morning, with a peak approximately 4 hrs following waking.

Other factors probably also act with the increase in platelet aggregability. For example, blood pressure rises sharply on waking and rising (Miller-Craig *et al.*, 1978). This can cause an acute rupture of an atherosclerotic plaque into the vessel lumen, thereby reducing luminal diameter and increasing shear rate of blood flowing through the vessel and exposing highly thrombogenic sub-endothelial substrates (e.g., collagen). Such plaque ruptures are common in the coronary arteries of patients who die of acute myocardial infarction (Falk, 1983).

The morphology of the blood vessel and the thrombus both strongly suggest that platelets play an important role in acute myocardial infarction. Clinical studies with aspirin support this hypothesis. For example, in the Physician's Health Study, involving over 22,000 physicians in the United States, aspirin treatment over a period of 5 years reduced the annual incidence of acute myocardial infarction by 44%, from 0.4% to 0.2% (The Steering Committee of the Physician's Health Study Research Group, 1989). Interestingly, it has since been demonstrated that the beneficial effect of aspirin is confined to inhibiting the morning increase in the incidence of acute myocardial infarction that corresponds with the morning increase in platelet aggregability (Ridker *et al.*, 1990). In anesthetized dogs, GPIIb/IIIa antagonists have been shown to reduce the incidence and magnitude of the formation of fibrin/platelet-dependent thrombi produced by electrolytic damage to the endothelium of the left circumflex coronary artery in the presence of a critical, eccentric stenosis. Mickelson *et al.* (1989) demonstrated that the 7E3 monoclonal antibody prevented acute thrombotic occlusion of the coronary artery, and reduced the fibrin/platelet thrombus mass in this model. This was accompanied by a reduction in the deposition of platelets to the damaged vascular endothelium and within the formed thrombus that corresponded with an inhibition of *ex vivo* platelet aggregation. SK&F 106760 significantly reduced fibrin/platelet-dependent coronary artery thrombosis to the extent that thrombotic occlusion and subsequent acute myocardial infarction were abolished (Fig. 9). In contrast, aspirin delayed and reduced the magnitude of thrombosis, but only to the extent of producing a 20% reduction in the incidence of complete thrombotic occlusion, which reduced, but did not abolish, the severity of the acute myocardial infarction (Fig. 9). The precise mechanism by which GPIIb/IIIa antagonists prevent coronary artery thrombosis is unknown, but it probably involves the inhibition of platelet aggregation on the damaged vessel and the formation of the prothrombinase complex on the platelet surface by inhibiting platelet aggregation.

Despite the impressive effects of aspirin in the Physician's Health Study, aspirin did not reduce the incidence of acute myocardial infarction in a study of British physicians (Peto *et al.*, 1989). However, the cardiovascular event rate in the British study was 5 to 10 times higher than in the U.S. study, suggesting that the overall severity of coronary artery disease was greater in the British physicians. The apparent importance of the severity of coronary artery disease in determining the efficacy of aspirin in preventing acute myocardial infarction is supported by the Part 2 analysis of the Persantine–Aspirin Reinfarction Study (PARIS-II). This study showed that 1 year after an initial myocardial infarction, aspirin (in combination with dipyridamole) reduced the coronary event rate (reinfarction and death) by 48% in patients who had suffered an initial low-severity non-Q-wave myocardial infarction, but it only reduced the coronary event rate by 7% in patients with an initial high-severity Q-wave infarct (Klimt *et al.*, 1986). Thus, aspirin has not been shown to have any significant benefit in two patient populations in which coronary artery disease is more severe. This is analogous to the decrease in the efficacy of aspirin in inhibiting platelet-dependent coronary artery cyclic flow reductions in the Folts model as the degree of stenosis becomes more severe (see above). In view of the efficacy of GPIIb/IIIa antagonists in preventing acute myocardial infarction in experimental animal models and the maintained efficacy of GPIIb/IIIa antagonists under conditions in which aspirin becomes ineffective, it is likely that GPIIb/IIIa antagonists may be significantly more effective than aspirin in the primary and secondary prevention of acute myocardial infarction in patients with more advanced coronary artery disease.

4.3. Adjuvants to Thrombolysis

One of the major goals in thrombolytic research is to enhance the overall efficacy of coronary thrombolysis by increasing the incidence of reperfusion and eliminating reocclusion. One approach to this is to develop novel thrombolytics with pharmacological properties that are obviously distinct from those already available. An alternative approach is to increase the efficacy of those thrombolyt-

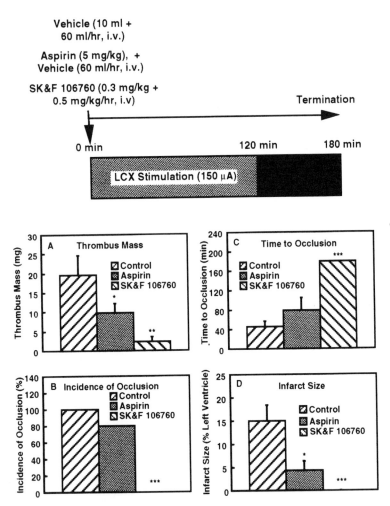

Figure 9. Contrasting effects of aspirin and SK&F 106760 on fibrin/platelet-dependent coronary thrombosis and subsequent acute myocardial infarction in anesthetized dogs. Note that while aspirin did reduce the extent of thrombosis and the severity of the acute myocardial infarction, the magnitude of the effect of SK&F 106760 was far greater, leading to a complete inhibition of the incidence of acute myocardial infarction. *$p < 0.05$, **$p < 0.01$, ***$p < 0.001$, relative to the vehicle-treated control animals.

ics that are currently available. GPIIb/IIIa antagonists may make it possible to achieve this without developing novel thrombolytics. The thrombolytic profile of t-PA or streptokinase can be enhanced by the concurrent administration of a GPIIb/IIIa antagonist in canine models of coronary thrombosis in which the thrombus consists primarily of erythrocytes enmeshed in a fibrin network with a significant platelet component. For example, the 7E3 monoclonal antibody (Gold *et al.*, 1988), the snake venom peptide, bitistatin (Shebuski *et al.*, 1990b), and the RGD peptide, SK&F 106760 (Nichols *et al.*, 1990), have been shown to reduce the time required to produce reperfusion, and to decrease the rate of reocclusion. Table IV demonstrates the dose-related enhancement of streptokinase-mediated coronary artery thrombolysis by SK&F 106760. Interestingly, the full benefit of a GPIIb/IIIa antagonist is only realized when it is given in combination with an anticoagulant, e.g., heparin. Figure 10 demonstrates that while either heparin or SK&F 106760 given alone will increase the incidence of reperfusion with t-PA, neither agent alone will significantly inhibit reocclusion. However, when heparin and SK&F 106760 are given in combination, the incidence of reocclusion is markedly reduced (Fig. 10). A similar interaction between heparin and a GPIIb/IIIa antagonist has been demonstrated with bitistatin (Shebuski *et al.*, 1990b).

It could be argued that GPIIb/IIIa antagonists improve thrombolytic efficacy because of an effect on the basic fibrinolytic activity (i.e., the ability to dissolve the fibrin component of the thrombus has been enhanced). However, even at very high concentrations (100 μM) SK&F 106760 does not potentiate streptokinase-mediated canine clot lysis *in vitro* (Nichols *et al.*, 1990), indicating

Table IV. The Effect of SK&F 106760 on the Thrombolytic Efficacy of Streptokinase[a] in the Anesthetized Dog

	Saline	SK&F 106760		
	(0.3 ml/kg)	0.3 mg/kg	1.0 mg/kg	3.0 mg/kg
Incidence of reperfusion[b]	12/16 (75%)	10/14 (71%)	11/12 (92%)	7/11 (64%)
Time to reperfusion (min)	34 ± 8	46 ± 9	17 ± 4	14 ± 3[c]
Incidence of reocclusion[d]	10/12 (83%)	7/10 (70%)	5/11 (45%)	1/7 (14%)[c]
Time to reocclusion (min)	27 ± 11	27 ± 7	63 ± 20	120[c]
Thrombus weight (mg)[e]	1.4 ± 0.4	2.2 ± 0.7	0.6 ± 0.2	0.3 ± 0.2[c]

[a]Streptokinase: 2000 IU/kg + 200 IU/kg/min, 60 min.
[b]Reperfusion is defined as a return of coronary blood flow to greater than 50% of control blood flow for longer than 5 min.
[c]$p < 0.05$.
[d]Reocclusion is defined as any return of coronary blood flow to zero following reperfusion.
[e]Only thrombus weights from animals that reperfused are included.

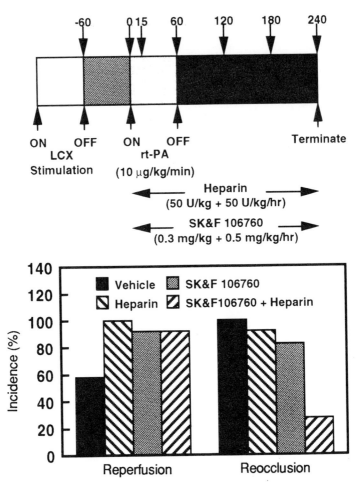

Figure 10. The ability of SK&F 106760 to enhance the thrombolytic profile of t-PA in the presence or absence of heparin in anesthetized dogs. Note that in order to maximize reperfusion and minimize reocclusion, both heparin and SK&F 106760 are required.

that the ability of SK&F 106760 to increase the thrombolytic efficacy of streptokinase is not caused by enhancement of the intrinsic fibrinolytic activity of streptokinase, but rather results from a mechanism independent of the intrinsic activity of streptokinase. Since GPIIb/IIIa antagonists enhance thrombolysis at doses that inhibit platelet aggregation, it must be assumed that this synergistic effect is related to the inhibition of an ongoing platelet-mediated dynamic remodeling of the thrombus.

The importance of preventing reocclusion following coronary thrombolysis has been demonstrated by the TAMI Study Group (Ohman and the TAMI Study Group, 1990) in an analysis of nearly 1000 patients receiving thrombolytic therapy for acute myocardial infarction. Thrombolysis was achieved with t-PA, urokinase, or a combination of both, and all patients received current optimal therapy, including aspirin. The in-hospital mortality rate increased 2.4-fold and the recovery of global and infarct zone cardiac function was significantly impaired in patients in whom coronary artery thrombolysis was followed by coronary artery reocclusion (Ohman and the TAMI Study Group, 1990). The TAMI Study Group concluded that reocclusion following successful reperfusion was associated with substantial morbidity and mortality rates, and that new strategies to prevent reocclusion, such as the use of GPIIb/IIIa antagonists, must be developed.

4.4. Transient Ischemic Attacks and Stroke

Cerebral transient ischemic attacks (TIAs) are associated with extracranial carotid artery disease (Rosenberg, 1989). The likelihood of stroke within the first 5 years of the initial TIA is approximately 40%, and there is a history of previous TIAs in approximately 75% of patients with a completed stroke (Millikan *et al.*, 1987). The pathological cause of TIA in the presence of carotid artery disease is believed to be transient thrombosis in the region of stenosis followed by cerebral embolization. Histologic studies suggest that platelets play an important role in carotid artery thrombosis (Millikan *et al.*, 1987). The critical role of platelets in TIA and the subsequent progression to stroke is demonstrated by clinical studies of aspirin and ticlopidine, in which both agents have been demonstrated to significantly reduce the incidence of stroke in high-risk patients (Gent *et al.*, 1989; Hass *et al.*, 1989). In view of the efficacy of SK&F 106760 in models of coronary artery thrombosis, and the evidence that SK&F 106760 would be superior to aspirin in the prevention and treatment of coronary artery thrombosis, we have studied SK&F 106760 in a model of extracranial carotid artery disease and compared its efficacy with the currently available optimal therapy, aspirin.

In a canine model of extracranial carotid artery disease, SK&F 106760 has been shown to inhibit thrombosis and subsequent stroke in 100% of animals with stenosed, endothelium-damaged carotid arteries (Willette *et al.*, 1992). In contrast, aspirin was effective in only 33% of animals. Moreover, in those animals in which aspirin prevented carotid arterial thrombosis, increasing the degree of stenosis rendered the aspirin ineffective (Willette *et al.*, 1992). Thus, SK&F 106760 has a superior antithrombotic profile relative to aspirin in this model of extracranial carotid artery disease, which suggests that GPIIb/IIIa antagonists may be significantly superior to aspirin in preventing TIA and stroke in high-risk patients.

5. Will Selective GPIIb/IIIa Antagonists Cause Bleeding?

The major concern about using agents, particularly antiplatelet drugs, to prevent thrombosis is their effect on normal hemostasis. The potential of drugs to cause bleeding is usually assessed both experimentally and clinically by determining their effect on template bleeding time. However, the utility of this test in determining the severity, location, or clinical consequences of drug-induced bleeding has been questioned (Rodgers and Levin, 1990). Aspirin prolongs template bleeding time by approximately twofold, and has been associated with an increase in intracerebral hemorrhage when used prophylactically to prevent acute myocardial infarction. In contrast, patients with Glanzmann's thrombasthenia, in which there is a total absence of functional GPIIb/IIIa, have prolonged bleeding times with no increased incidence of cerebral hemorrhage. These patients do, however, suffer from severe mucocutaneous bleeding, particularly from the skin and gums (George et al., 1990). Furthermore, very high doses of the 7E3 monoclonal antibody given to monkeys have been shown to produce gingival bleeding and ecchymoses, but no bleeding in the central nervous system (Cavagnaro et al., 1987). Thus, antiplatelet drugs can prolong bleeding time and induce spontaneous bleeding, but the clinical consequences of this bleeding vary with the drug.

We have assessed the effect of SK&F 107260 on template bleeding time in the anesthetized dog during infusion sufficient to abolish platelet-dependent coronary artery thrombosis. Table V demonstrates that at a dose of SK&F 107260 that resembles therapeutic administration of an antithrombotic dose, ex vivo platelet aggregation is abolished, but there is no significant increase in template bleeding time. In addition, there are no changes in the coagulation parameters, PT and APTT. A similar effect is also observed with SK&F 106760: at antithrombotic doses, ex vivo platelet aggregation is abolished, but template bleeding time is not prolonged. In contrast, aspirin, which is antithrombotic only in approximately 50% of animals, produces a significant twofold increase in template bleeding time, which is similar to the effect seen with clinical use. Thus, there is evidence that GPIIb/IIIa antagonists can be antithrombotic without increasing bleeding time. However, this is not the case with every GPIIb/IIIa antagonist. As described above, the 7E3 monoclonal antibody does prolong bleeding time, even at doses that do not completely abolish ex vivo platelet aggregation (Coller et al., 1989). In addition, the RGD-containing snake venom peptides, echistatin (Shebuski et al., 1990a) and bitistatin (Shebuski et al., 1989b), have also been shown to prolong bleeding time. The major difference between the GPIIb/IIIa antagonists that do or do not increase bleeding time may be their selectivity for GPIIb/IIIa relative to the vitronectin receptor. Thus, SK&F 106760 and SK&F 107260 both have a relatively high degree of selectivity for GPIIb/IIIa (see Table I), whereas the 7E3 monoclonal antibody (Charo

Table V. Effect on SK&F 107260 (0.1–0.2 mg/kg/hr, i.v.) on Platelet Aggregation, Coagulation Parameters, and Bleeding Time in Anesthetized Dogs

Parameter	Treatment	
	Control[a]	+ SK&F 107260[b]
In vivo platelet aggregation (time to thrombosis, min)	4.2 ± 0.3	Abolished
Ex vivo platelet aggregation (% control)	100 ± 0	2 ± 0
Prothrombin time (sec)	8.1 ± 0.2	8.2 ± 0.2
Activated partial thromboplastin time (sec)	10.6 ± 0.4	10.3 ± 0.4
Template bleeding time (sec)	100 ± 14	143 ± 31

[a]Measured just prior to initiating infusion of SK&F 107260.
[b]Measured when coronary cyclic flow reductions had just been abolished.

et al., 1986) and the snake venom peptides are also potent vitronectin receptor antagonists. We do not yet know the mechanism by which the combined GPIIb/IIIa and vitronectin receptor antagonism prolongs bleeding time, as compared with the selective GPIIb/IIIa blockade, which does not.

6. Conclusions

The acute treatment of cardiovascular and cerebrovascular diseases with selective GPIIb/IIIa antagonists appears to have potentially significant advantages over current treatment with aspirin, in terms of overall efficacy combined with the likelihood of reduced clinically significant bleeding. For this purpose, intravenous administration may be the most appropriate means of drug delivery, since it ensures the rapid onset of action. Monoclonal antibodies and peptides, large or small (such as the snake venoms or RGD analogues, respectively), would be suitable. Such compounds would not be appropriate for chronic therapy for primary or secondary prevention of thrombotic diseases, in view of their relatively short duration of action and low oral bioavailability. In the absence of the development of drug delivery systems to enhance the gastrointestinal absorption of proteins and peptides, or the discovery of orally active peptides, there is a need for new, nonpeptide GPIIb/IIIa antagonists.

References

Aiken, J. W., Shebuski, R. J., Miller, O. V., and Gorman, R. R., 1981, Endogenous prostacyclin contributes to the efficacy of a thromboxane synthetase inhibitor for preventing coronary artery thrombosis, *J. Pharmacol. Exp. Ther.* **219:**299–308.

Ali, F. E., Samanen, J. M., Calvo, R., Romoff, T., Yellin, T., Vasko, J., Powers, D., Stadel, J., Bennet, D., Berry, D., and Nichols, A., 1992, Potent fibrinogen receptor antagonists bearing conformational constraints, in: *Peptides: Chemistry and Biology* (J. A. Smith and J. E. Rivier, eds.), ESCOM, Leiden, pp. 761–762.

Ambrose, J. A., Winters, S. L., Stern, A., Eng, A., Teichholz, L. E., Gorlin, R., and Fuster, V., 1985, Angiographic morphology and the pathogenesis of unstable angina pectoris, *J. Am. Coll. Cardiol.* **5:**609–616.

Ambrose, J. A., Hjemdahl-Monsen, C., Borrico, S., Sherman, W., Cohen, M., Gorlin, R., and Fuster, V., 1987, Quantitative and qualitative effects of intracoronary streptokinase in unstable angina and non-Q wave infarction, *J. Am. Coll. Cardiol.* **9:**1156–1165.

Bennet, J. S., and Vilaire, G., 1979, Exposure of platelet fibrinogen receptors by ADP and epinephrine, *J. Clin. Invest.* **64:**1393–1401.

Bennet, J. S., Shattil, S. J., Power, J. W., and Gartner, T. K., 1988, Interaction of fibrinogen with its platelet receptor, *J. Biol. Chem.* **263:**12948–12953.

Cavagnaro, J., Serabian, M. A., Coller, B. S., and Iuliucci, J. D., 1987, Long term toxicologic evaluation of anti-platelet monoclonal antibody 7E3 F(ab')2 in monkeys, *Blood* **70**(Suppl. 1):337a.

Charo, I. F., Fitzgerald, L. A., Steiner, B., Rall, S. C., Jr., Bekeart, L. S., and Phillips, D. R., 1986, Platelet glycoproteins IIb and IIIa: Evidence for a family of immunologically and structurally related glycoproteins in mammalian cells, *Proc. Natl. Acad. Sci. USA* **83:**8351–8355.

Cheresh, D. A., Berlinger, S. A., Vicente, V., and Ruggeri, Z. M., 1989, Recognition of distinct adhesive sites on fibrinogen by related integrins on platelets and endothelial cells, *Cell* **58:**945–953.

Coller, B. S., 1980, Interaction of normal, thrombasthenic, and Bernard-Soulier platelets with immobilized fibrinogen: Defective platelet–fibrinogen interaction in thrombasthenia, *Blood* **55:**169–178.

Coller, B. S., 1985, A new murine monoclonal antibody reports an activation-dependent change in the conformation and/or microenvironment of the platelet glycoprotein IIb/IIIa complex, *J. Clin. Invest.* **76:**101–108.

Coller, B. S., Folts, J. D., Smith, S. R., Schudder, L. E., and Jordan, J., 1989, Abolition of *in vivo* platelet thrombus formation in primates with monoclonal antibodies to the platelet GPIIb/IIIa receptor, *Circulation* **80:**1766–1774.

Constantinides, P., 1966, Plaque fissures in human coronary thrombosis, *J. Atheroscler. Res.* **6:**1–17.

D'Souza, S. E., Ginsburg, M. H., Burke, T. A., Lam, S. C.-T., and Plow, E. F., 1988, Localization of an Arg-Gly-Asp recognition site within an integrin adhesion receptor, *Science* **242:**91–93.

D'Souza, S. E., Ginsburg, M. H., Burke, T. A., and Plow, E. F., 1990, The ligand binding site of the platelet integrin receptor GPIIb-IIIa is proximal to the second calcium binding domain of its α subunit, *J. Biol. Chem.* **265:**3440–3446.

Falk, E., 1983, Plaque rupture with severe pre-existing stenosis precipitating coronary thrombosis, *Br. Heart J.* **50:**127–134.

Folts, J., 1991, An *in vivo* model of experimental arterial stenosis, intimal damage, and periodic thrombosis, *Circulation* **83:**(Suppl. IV):IV-3–IV-14.

Folts, J. D., and Rowe, G. G., 1988, Epinephrine potentiation of *in vivo* stimuli reverses aspirin inhibition of platelet thrombus formation in stenosed canine coronary arteries, *Thromb. Res.* **50:**507–516.

Gent, M., Blakely, J. A., Easton, J. D., Ellis, D. J., Hachinski, V. C., Harbison, J. W., Panak, E., Roberts, R. S., Sicarella, J., Turpie, A. G. G., and the CATS group, 1989, The Canadian American ticlopidine study (CATS) in thromboembolic stroke, *Lancet* **1:**1215–1220.

George, J. N., Caen, J. P., and Nurden, A. T., 1990, Glanzmann's thrombasthenia: The spectrum of clinical disease, *Blood* **75**:1383–1395.

Gold, H. K., Coller, B. S., Yasuda, T., Saito, T., Fallon, J. T., Guerrero, J. L., Leinbach, R. C., Ziskind, A. A., and Collen, D., 1988, Rapid and sustained coronary artery recanalization with combined bolus injection of recombinant tissue-type plasminogen activator and monoclonal antiplatelet GPIIb/IIIa antibody in a canine preparation, *Circulation* **77**:670–677.

Gold, H. K., Gimple, L. W., Yasuda, T., Leinbach, R. C., Werner, W., Holt, R., Jordan, R., Berger, H., Collen, D., Coller, B. S., 1990, Pharmacodynamic study of F(ab′)$_2$ fragments of murine monoclonal antibody 7E3 directed against human platelet glycoprotein IIb/IIIa in patients with unstable angina pectoris, *J. Clin. Invest.* **86**:651–659.

Gunby, P., 1992, Cardiovascular diseases remain nation's leading cause of death, *Am. Med. Assoc.* **267**:335–336.

Hass, W. K., Easton, J. D., Adams, H. P., Pryse-Phillips, W., Molony, B. A., Anderson, S., Kamm, B., and the Ticlopidine Aspirin Stroke Study Group, 1989, A randomized trial comparing ticlopidine hydrochloride with aspirin for the prevention of stroke in high-risk patients, *N. Engl. J. Med.* **321**:501–507.

Hawiger, J., Kloczewiak, M., Bednarek, M. A., and Timmons, S., 1989, Platelet receptor recognition domains on the α chain of human fibrinogen: Structure–function analysis, *Biochemistry* **28**:2909–2914.

Klimt, C. R., Knatterud, G. I., Stamler, J., and Meier, P., 1986, Persantine–aspirin reinfarction study. Part II. Secondary coronary prevention with persantine and aspirin, *J. Am. Coll. Cardiol.* **7**:251–259.

Kloczewiak, M., Timmons, S., Lukas, T. J., and Hawiger, J., 1984, Platelet receptor recognition site on human fibrinogen. Synthesis and structure–function relationship of peptides corresponding to the carboxy-terminal segment of the γ chain, *Biochemistry* **23**:1767–1774.

Lewis, H. D., David, J. W., Archibald, D. G., Steinke, W. E., Smitherman, T. C., Doherty, J. E., Schnaper, H. W., LeWinter, M. M., Linares, E., Pouget, J. M., Sabharwal, S. C., Chesler, E., and DeMots, H., 1983, Protective effects of aspirin against acute myocardial infarction and death in men with unstable angina, *N. Engl. J. Med.* **309**:396–403.

Mickelson, J. K., Simpson, P. J., and Lucchesi, B. R., 1989, Antiplatelet monoclonal F(ab′)$_2$ antibody directed against the platelet GPIIb/IIIa receptor complex prevents coronary artery thrombosis in the canine heart, *J. Mol. Cell. Cardiol.* **21**:393–405.

Miller-Craig, M. W., Bishop, C. N., and Raftery, E. B., 1978, Circadian variation of blood-pressure, *Lancet* **1**:795–797.

Millikan, C. H., McDowell, F. H., and Easton, J. D., 1987, *Stroke,* Lea & Febiger, Philadelphia.

Mizuno, K., Satomura, K., Miyamoto, A., Arakawa, K., Shibuya, T., Arai, T., Kurita, A., Nakamura, H., and Ambrose, J. A., 1992, Angioscopic evaluation of coronary-artery thrombi in acute coronary syndromes, *N. Engl. J. Med.* **326**:287–291.

Mustard, J. F., Kinlough-Rathbone, R. L., Packham, M. A., Perry, D. W., and Pai, K. R. M., 1979, Comparison of fibrinogen association with normal and thrombasthenic platelets on exposure to ADP or chymotrypsin, *Blood* **54**:987–993.

Nichols, A. J., Vasko, J., Koster, P., Smith, J., Barone, F., Nelson, A., Stadel, J., Strohsacker, M., Rhodes, G., Bennett, D., Berry, D., Romoff, T., Calvo, R., Ali, F., Sorenson, E., and Samanen, J., 1990, SK&F 106760, a novel GPIIB/IIIA antagonist: Antithrombotic activity and potentiation of streptokinase-mediated thrombolysis, *Eur. J. Pharmacol.* **183**:2019.

Ohman, E. M., and the TAMI Study Group, 1990, Consequences of reocclusion after successful therapy in acute myocardial infarction, *Circulation* **82**:781–791.

Parise, L. V., Helgersons, S. L., Steiner, B., Nannizzi, L., and Phillips, D. R., 1987, Synthetic peptides derived from fibrinogen and fibronectin change the conformation of purified platelet glycoprotein IIb-IIIa, *J. Biol. Chem.* **262**:12597–12602.

Peto, R., Gray, R., Collins, R., Wheatley, K., Hennekens, C., Jamrozik, K., Warlow, C., Hafner, B., Thompson, E., Norton, S., Gilliland, J., and Doll, R., 1989, Randomised trial of prophylactic daily aspirin in British male doctors, *Br. Med. J.* **299**:1247–1250.

Phillips, D. R., and Agin, P. P., 1977, Platelet membrane defects in Glanzmann's thrombasthenia, *J. Clin. Invest.* **60**:535–545.

Pierschbacher, M. D., and Ruoslahti, E., 1984, Cell attachment activity of fibronectin can be duplicated by small synthetic fragments of the molecule, *Nature* **309**:30–33.

Plow, E. F., Pierschbacher, M. D., Ruoslahti, E., Marguerie, G., and Ginsburg, M. H., 1985, The effect of Arg-Gly-Asp-containing peptides on fibrinogen and von Willebrand factor binding to platelets, *Proc. Natl. Acad. Sci. USA* **82**:8057–8061.

Ridker, P. M., Manson, J. E., Buring, J. E., Muller, J. E., and Hennekens, C. H., 1990, Circadian variation of acute myocardial infarction and the effect of low-dose aspirin in a randomized trial of physicians, *Circulation* **82**:897–902.

Rodgers, R. P. C., and Levin, J., 1990, A critical reappraisal of the template bleeding time, *Thromb. Hemost.* **16**:1–20.

Rosenberg, N., 1989, *CRC Handbook of Carotid Artery Surgery: Facts and Figures,* CRC Press, Boca Raton, Fla.

Ruoslahti, E., and Pierschbacher, M. D., 1984, New perspectives in cell adhesion: RGD and integrins, *Science* **238**:491–497.

Samanen, J., Ali, F., Romoff, T., Calvo, R., Sorenson, E., Vasko, J., Storer, B., Berry, D., Bennett, D., Strohsacker, M., Powers, D., Stadel, J., and Nichols, A., 1991, Development of a small RGD-peptide fibrinogen receptor antagonist with potent antiaggregatory activity *in vitro, J. Med. Chem.* **34**:3114–3125.

Shebuski, R. J., Berry, D. E., Bennett, D. B., Romoff, T., Storer, B. L., Ali, F., and Samanen, J., 1989a, Demonstration of Ac-Arg-Gly-Asp-Ser-NH$_2$ as an antiaggregatory agent in the dog by intracoronary administration, *Thromb. Haemost.* **61**:183–188.

Shebuski, R. J., Ramjit, D. R., Bencen, G. H., and Polokoff, M. A., 1989b, Characterization of platelet inhibitory activity of bitistatin, a potent arginine-glycine-aspartic acid-containing peptide from the venom of the viper Bitis arientans, *J. Biol. Chem.* **264**:21550–21556.

Shebuski, R. J., Ramjit, D. R., Sitko, G. R., Lumma, P. K., and Garski, V. M., 1990a, Prevention of canine coronary artery thrombosis with echistatin, a potent inhibitor of platelet aggregation from the venom of the viper, Echis carinatus, *Thromb. Haemost.* **64**:576–581.

Shebuski, R. J., Stabilito, I. J., Sitko, G. R., and Polokoff, M. H., 1990b, Acceleration of recombinant tissue-type plasminogen activator-induced thrombolysis and prevention of reocclusion by the combination of heparin and Arg-Gly-Asp-containing peptide bitistatin in the canine model of coronary thrombosis, *Circulation* **82**:169–177.

Sherman, C. T., Litvack, F., Grunfest, W., Lee, M., Hickey, A., Chauz, A., Kass, R., Blanche, C., Matloff, J., Morgenstern, L., Ganz, W., Swan, H. J. C., and Forrester, J., 1986, Coronary angioscopy in patients with unstable angina pectoris, *N. Engl. J. Med.* **315**:913–919.

The Steering Committee of the Physician's Health Study Research Group, 1989, Final report on the aspirin component of the ongoing physician's health study, *N. Engl. J. Med.* **321**:129–135.

Timmons, S., Bednarek, M., Kloczewaik, M., and Hawiger, J., 1989, Antiplatelet "hybrid" peptide analogues to receptor recognition domains on γ and α chains of human fibrinogen, *Biochemistry* **28**:2919–2923.

Tofler, G. H., Brezinski, D. A., Schafer, A., Czeisler, C. A., Rutherford, J. D., Willich, S. N., Gleason, R. E., Williams, G. H., and Muller, J. E., 1987, Concurrent morning increase in platelet aggregability and the risk of myocardial infarction and sudden cardiac death, *N. Engl. J. Med.* **316**:1514–1518.

Weiss, H. J., Hawiger, J., Ruggeri, Z. M., Turitto, V. T., Thiagarajan, P., and Hoffmann, T., 1989, Fibrinogen-independent platelet adhesion and thrombus formation on subendothelium mediated by glycoprotein IIb-IIIa complex at higher shear rate, *J. Clin. Invest.* **83:**288–297.

Willette, R. N., Sauermelch, C. F., Rycyna, R., Sarkar, S., Feuerstein, G. Z., Nichols, A. J., and Ohlstein, E. H., 1992, Antithrombotic effects of a platelet fibrinogen receptor antagonist in a canine model of carotid artery thrombosis, *Stroke* **23:**703–711.

III

TARGETING ADHESION MOLECULES—THERAPEUTIC POTENTIAL

14

The Pathological Consequences of Bacterial Adhesion to Medical Devices
A Practical Solution to the Problem of Device-Related Infections

J. W. COSTERTON

1. Introduction

Bacterial adhesion is something of an anomaly in this symposium, which addresses the elaborate processes whereby human tissues and cells interact by means of very specific molecular ligands. Bacteria do have very specific adhesion systems, usually involving pili (Deneke *et al.*, 1981), which react with cell surface proteins or with fibronectin to mediate very specific pathogenic interactions—as in the adhesion of cells of enterotoxigenic *E. coli* (ETEC) to endothelial cells of certain regions of the intestine of newborn calves and lambs. However, bacteria live in a wide variety of environments, and rigorous specificity of the adhesion process that is pivotal to their ecological success would be counterproductive because, in order to be successful, they must adhere to a wide variety of inert and living surfaces (Costerton *et al.*, 1978). Bacterial pathogens that adhere to specific human tissues, such as the adhesion of *Salmonella typhi* to crypt cells in the intestine, have been largely controlled by modern antibiotics and by vaccines that are often designed to counteract their specific adhesion and toxigenic mechanisms. Modern bacterial pathogens, such as *Pseudomonas*

J. W. COSTERTON • Department of Biological Sciences, University of Calgary, Calgary, Alberta, Canada T2N 1N4. *Present address:* Center for Biofilm Engineering, Montana State University, Bozeman, Montana 59717.

Cellular Adhesion: Molecular Definition to Therapeutic Potential, edited by Brian W. Metcalf, Barbara J. Dalton, and George Poste. Plenum Press, New York, 1994.

aeruginosa and *Staphylococcus aureus,* are ubiquitous in their distribution: the former species is actually the most numerous organism on earth because it predominates in all salt and freshwater ecosystems. It would be very difficult to attribute specificity to the adhesion process carried out so efficiently by cells of *P. aeruginosa* because this organism adheres equally well to plastics, metals, wood, rocks, plant tissues, and most tissues of the human body. Cells of *P. aeruginosa* produce only one class of exopolysaccharide molecule (Sutherland, 1977); this alginate is a polymer of guluronic and mannuronic acid that is acetylated and/or pyruvylated to various degrees. Planktonic (floating) cells of *P. aeruginosa* are surrounded by relatively small amounts of alginate, which mediate the extremely avid initial adhesion of these organisms to surfaces (Marshall *et al.,* 1971). Alginate synthesis is rapidly derepressed in response to the adhesion event, so that adherent cells quickly become buried in this polymer, which cements them firmly, but nonspecifically, to the substratum (Fig. 1). These adherent cells continue to produce alginate that mediates their adhesion to the substratum and to each other until a slime-enclosed biofilm is formed (Fig. 2), yielding a confluent population on the colonized surface (Fig. 3). Confocal scanning laser (CSL) microscopy now allows us to visualize this process by examining living, fully hydrated biofilms (Lawrence *et al.,* 1991). We have determined that the alginate matrix occupies as much as 85% of the volume of a

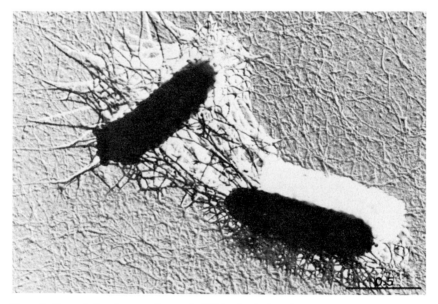

Figure 1. Transmission electron micrograph (TEM) of a shadowed preparation of cells of *Pseudomonas aeruginosa* adhering to a surface by means of their alginate glycocalyx fibers. Bar = 0.5 μm.

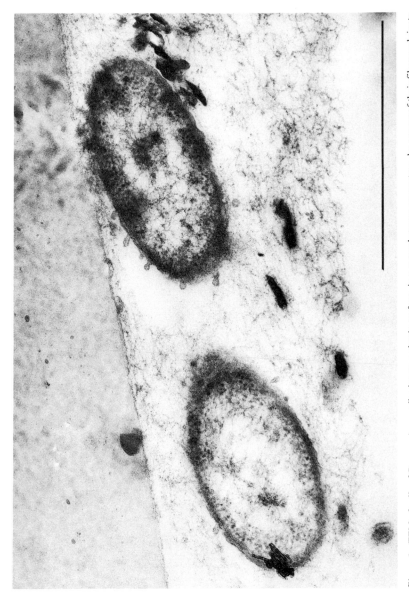

Figure 2. TEM of cells of *P. aeruginosa* adhering to a plastic surface, in a natural stream ecosystem, by means of their fibrous alginate glycocalyx. Bar = 1 μm.

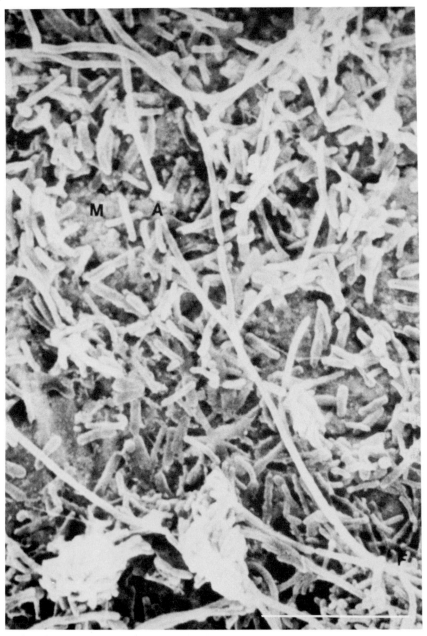

Figure 3. Scanning electron micrograph (SEM) of a metal surface colonized by cells of *P. aeruginosa*, on which these organisms have formed a confluent biofilm within which these bacteria cluster (M) and elongate (A). Bar = 5 μm.

mature biofilm, and that the cells themselves grow in microcolonies within this matrix (Fig. 4). The use of diffusion probes of various sizes has shown that the biofilm is traversed by relatively open water channels (Fig. 4), and that the microcolonies of bacterial cells are surrounded by a much denser polyanionic alginate matrix, which limits the diffusion of charged ions and molecules. This exciting new CSL microscope has therefore yielded an accurate image of the structure of the living biofilm, which can serve as the rational basis for our continuing examination of these complex surface populations and their enigmatic, inherent resistance to antibacterial agents, including antibiotics.

2. The Predominance of Bacterial Biofilms in Chronic Bacterial Infections

We have used the same methods of direct observation employed to detect biofilms in natural and industrial ecosystems to examine the plastic and metal surfaces of medical devices that were removed because they were obvious foci of device-related bacterial infections. All of these devices, from simple sutures to complex cardiac pacemakers (Fig. 5), were found to be colonized by bacterial biofilms in which the adherent bacteria were embedded in very large amounts of extracellular matrix material that cemented them firmly to the colonized surfaces. After demonstrating the biofilm mode of growth of the bacteria that cause device-related infections, we systematically examined medical devices that had been used for a finite length of time without any overt signs of associated infection. Any devices that traversed the skin (e.g., vascular and peritoneal catheters) or passed from a nonsterile organ to a sterile organ (e.g., urinary catheters and IUDs) were found to be largely covered by bacterial biofilms (Figs. 6 and 7) that were not readily distinguishable from those associated with infections. All peritoneal catheters in place for more than 6 weeks were found to be heavily colonized by bacterial biofilms (Dasgupta *et al.*, 1987), as were all Hickmann subclavian catheters used for chemotherapy (Tenney *et al.*, 1986), and all IUD devices withdrawn from the uterus (Marrie and Costerton, 1983). It soon became obvious that the biofilm mode of growth protects bacteria from humoral and cellular host defense mechanisms, and enables these organisms to persist and to cause chronic device-related infections (Costerton *et al.*, 1987). This hypothesis was tested by colonizing the surfaces of plastic devices (Ward *et al.*, 1992) with a newly formed biofilm of cells of *P. aeruginosa*, and implanting these colonized devices in the peritoneum of normal rabbits and rabbits that had been immunized with the same bacterial strain prior to device implantation. These plastic devices were colonized with very young biofilms, which had covered less than 1% of their surface areas, and were only composed of $\pm$ 1.0 $\times$ 10^3 cells/cm^2, but these nascent biofilms resisted the activated host defenses of the

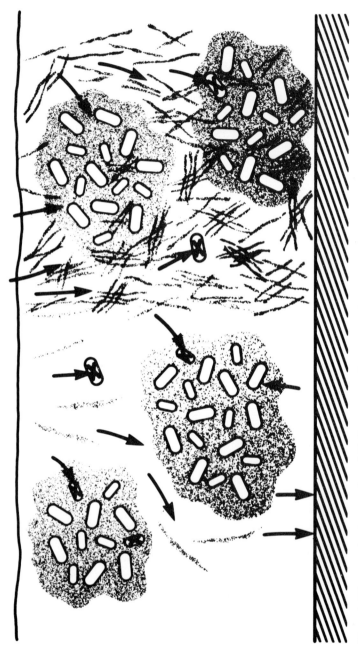

Figure 4. Diagrammatic representation of biofilm structure, based on confocal scanning laser microscopy, that shows bacteria growing in microcolonies surrounded by dense matrix material, while the intervening water channels that admit solutes (arrows) are composed of much less dense matrix material.

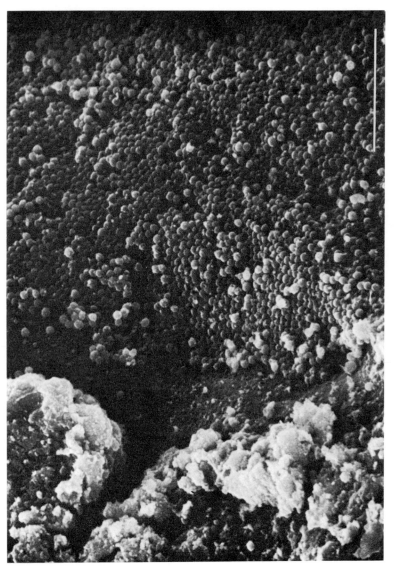

Figure 5. SEM of the surface of a cardiac pacemaker lead that served as a nidus of recurrent *Staphylococcus aureus* infection. Note the growth of these spherical bacterial cells in a very thick slime-enclosed biofilm that resisted clearance by very high sustained concentrations of antibiotics. Bar = 5 μm.

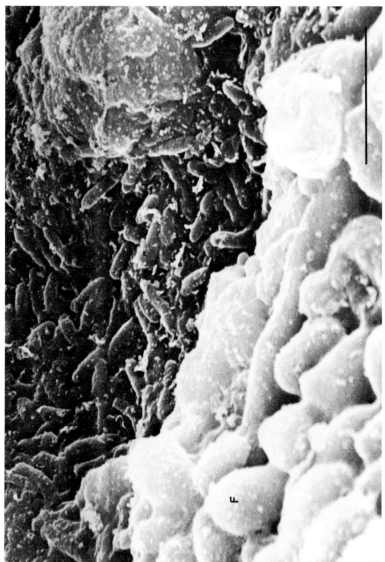

Figure 6. SEM of the surface of a Tenckhoff peritoneal catheter whose entire intraperitoneal surface was colonized by a biofilm composed of many different microbial species. Here we see cells of *Candida albicans* (F) and a gram-negative bacterium. This patient had not experienced peritonitis and had obviously accommodated this multispecies biofilm without pathogenic consequences. Bar = 5 μm.

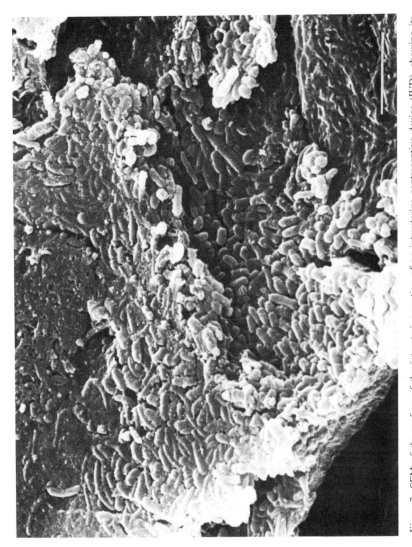

Figure 7. SEM of the surface of the intrauterine portion of an intrauterine contraceptive device (IUD), showing its occlusion by a thick multispecies biofilm. The formation of this biofilm has no identifiable pathological effects in this organ, and it may be partially responsible for the efficacy of these devices. Bar = 5 μm.

peritoneum and burgeoned to produce mature biofilms (1×10^7 to 1×10^8 cells/cm^2) in every case. We conclude that the formation of biofilms by bacteria confers on these sessile cells a very large measure of resistance to host defense mechanisms which, in turn, enables them to persist and cause chronic infections. This exopolysaccharide-enclosed mode of growth confers protection on the pathogen, and often causes these organisms to persist without causing overt disease, perhaps because biofilm cells release fewer toxins to damage host tissues and only occasionally mobilize an acute planktonic cell attack from their defended slime-enclosed positions on colonized surfaces. These acute exacerbations may be short-lived, because planktonic bacteria are susceptible to host defense mechanisms and to antibiotic chemotherapy, but our clinical success in treating these acute phases of device-related infections did not extend to any measure of control over the basic foci of biofilm bacteria that constitute their continuing protected niduses (Gristina et al., 1988; Khoury and Costerton, 1988). Colonized medical devices must usually be removed, with their accretions of biofilm bacteria, in order to allow the resolution of device-related bacterial infections by antibiotic chemotherapy.

The direct morphological examination of non-device-related chronic bacterial infections has revealed that their causative agents have adopted the protected biofilm mode of growth. Bacteria in chronic osteomyelitis (Gristina and Costerton, 1984) grow in slime-enclosed biofilms on the surfaces of dead bone (Fig. 8); the cells of P. aeruginosa that cause 20- to 30-year pneumonias in cystic fibrosis patients (Lam et al., 1980) grow in slime-protected microcolonies (Fig. 9); and the causative agents of chronic bacterial prostatitis in the human (Nickel and Costerton, 1992) also adopt this highly protected mode of growth. These non-device-related chronic bacterial infections all exhibit acute exacerbations that can be controlled by conventional antibiotic chemotherapy, but even prolonged high-dose chemotherapy almost always fails to kill the causative biofilm bacteria and fully resolve the infection. Subacute bacterial endocarditis is an exception to this dismal pattern in which the causative bacteria live in well-developed biofilms called "vegetations"; these infections are often cured by sustained, very-high-dose antibiotic chemotherapy.

3. The Inherent Resistance of Biofilm Bacteria to Antimicrobial Agents

The inherent resistance of bacteria within biofilms to biocides and antibiotics has been demonstrated in a wide variety of industrial (Ruseska et al., 1982) and medical (Nickel et al., 1985) systems. This resistance does not involve the traditional types of antibiotic resistance, which depend on adaptive or genetic changes in the cell to alter the metabolic targets of these antibacterial agents; this is a simple matter of biofilm cells being protected from antibacterial agents as

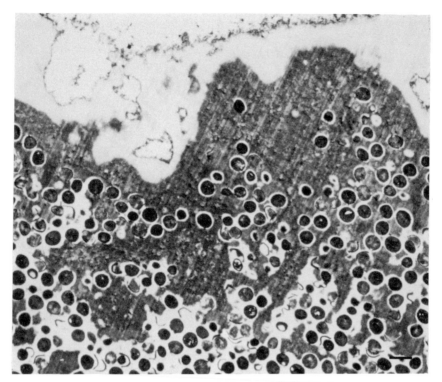

Figure 8. TEM of a biofilm composed of cells of *S. aureus,* embedded in very large amounts of their exopolysaccharide slime, growing on the surface of dead bone. These infections are very refractory to antibiotic chemotherapy. Bar = 1 μm.

long as they are embedded in the biofilm matrix. Cells released from biofilms, in which they are protected from concentrations of biocides and antibiotics up to 5000 times greater than those that will kill planktonic cells, are just as susceptible as ordinary free-floating cells when they are released from their protected biofilm mode of growth (Nickel *et al.*, 1985). Young (2 day) biofilms afford considerably less protection from antibiotics, when compared with old (7 day) biofilms of cells of the same species (Anwar *et al.*, 1992), and cells that produce antibiotic-degrading enzymes (e.g., β lactamases) are almost completely protected from antibiotics (Giwercman *et al.*, 1991) when they grow in slime-enclosed micro-colonies within mature biofilms.

Two basic mechanisms have been suggested to account for this inherent resistance of biofilm bacteria to biocides and antibiotics, and our analysis of the data in this field leads us to invoke both of these mechanisms to explain this important phenomenon. We have suggested (Costerton *et al.*, 1987) that the hydrated polyanionic matrix that surrounds the bacterial cells within biofilms

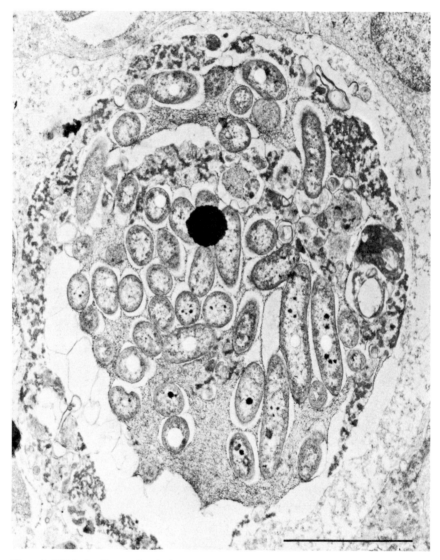

Figure 9. TEM of a slime-enclosed microcolony of *P. aeruginosa* growing in the lung in an animal model of cystic fibrosis pneumonia. These protected microcolonies are highly resistant to host defense factors and to antibiotic chemotherapy. Bar = 1 μm.

(Fig. 4) may limit the diffusion of molecules, especially charged molecules, from the bulk fluid through the open water channels and to the bacterial cells within microcolonies. As our diffusion probes in CSL microscopy have become increasingly sophisticated, we have shown that even very large (more than 50,000 Da) molecules can penetrate the water channels of the biofilm. We have also used Fourier transform infrared (FTIR) spectroscopy to show that antibiotics added to the bulk fluid can reach the colonized substratum in less than 1 min. Thus, the water channels of the biofilm are shown to be highly permeable (Fig. 4) to antibacterial agents, but kill data clearly show that these molecules do not penetrate the dense glycocalyx surrounding the bacterial microcolonies, nor kill their component bacteria, until much higher concentrations are used. Studies of the matrix materials (e.g., alginate) that surround bacteria in biofilms indicate that these exopolysaccharides behave like ion exchange resins, in that incoming charged molecules must "satisfy" large numbers of potential binding sites before they can saturate the system and reach their targets inside the matrix (Costerton *et al.*, 1987).

The other proposed mechanism that may contribute to the inherent resistance of biofilm bacteria to antimicrobial agents involves the phenotypic plasticity that Brown and Williams (1985) have described so brilliantly during the past decade. It is now clear that bacteria can adapt their phenotype to precisely suit them for growth in specific microenvironments, and that these adaptations often involve cell envelope components that play pivotal roles in antibiotic resistance. Bacteria living in microcolonies in biofilms may exist in many different physiological and structural states, and may even be growing at several very different rates within the same biofilm. Because growth rate, the structure of the bacterial cell envelope, and the metabolic activity of the bacterial cell all exert profound influences on the sensitivity of bacteria to antibiotics, it is probable that at least some of the cells growing in a given biofilm at a given time are resistant to a specific antibiotic by virtue of their altered physiological state.

While both of these mechanisms may contribute to the inherent resistance of biofilm bacteria to biocides and antibiotics, and while other mechanisms may also contribute to this inherent resistance, the central fact remains that bacterial biofilms are routinely resistant to concentrations of these agents 5000 times greater than those that will readily kill planktonic cells of the same species. In this context, it is sobering to recall that all of the biocides and antibiotics currently available for use in controlling bacteria were selected for their ability to kill planktonic bacteria growing in monospecies batch cultures in media that do not necessarily resemble body fluids. It is highly probable that if a program to design and select antimicrobial agents was targeted to killing biofilm bacteria, it would yield new agents that would be especially effective in treating device-related infections and other chronic bacterial infections.

Because biofilm bacteria are resistant to humoral and cellular host defense mechanisms, and because they tend to grow progressively along the surfaces that they colonize (Nickel et al., 1992), they eventually colonize most transcutaneous devices and most devices that pass or extend from a colonized organ (e.g., vagina) to a normally sterile organ (e.g., uterus). Once biofilms have formed on the surface of a medical device, this colonized device may or may not become the focus of an overt, occasionally acute, bacterial infection, or it may continue for an extended period of time as a protected nidus from which acute infections are mobilized at intervals, but which is generally quiescent. Similarly, the slime-protected bacterial microcolonies in non-device-related infections constitute protected niduses from which acute exacerbations can be mobilized at intervals. Generally, unless a direct infusion of antibiotics to the biofilm-colonized surface is possible, systemic treatment with conventional doses of currently available antibiotics fails to kill the biofilm and thus to clear these protected reservoirs of bacteria. We have recently seen some exceptions, in which very high doses of new fluoroquinolone antibiotics have been able to kill biofilm bacteria, but these instances are still very rare.

4. The Bioelectric Initiative

Our attempts to kill bacteria within mature biofilms with antimicrobial agents, which have not been notably successful, actually constitute a crude frontal attack on a biological problem using ever higher concentrations of chemicals designed to kill unprotected planktonic organisms. This fixed mind-set has led us to somewhat of a nadir, in that we persist in an unsuccessful strategy and use doses that we know to be ineffectual because we should be seen to be doing something for the increasing number of elderly and/or debilitated patients affected by these chronic bacterial infections. We reasoned that it might be possible to invoke physical forces to solve what is essentially a problem of limited diffusion. We thus decided to examine the effects of electric fields (Fig. 10) on the ability of antimicrobial agents to penetrate biofilms and kill their component bacteria. At first we used moderate DC electric fields (field strength 10 V/cm, current density 15 mA/cm^2), and found that these fields had a very dramatic effect on the efficacy of both biocides and antibiotics. While very high concentrations [ca. 500 times the minimal inhibitory concentration (MIC)] of biocides such as glutaraldehyde, quaternary ammonium compounds, and isothiazalone are usually required to kill bacteria within well-established biofilms, we found that concentrations as low as 0.5 MIC could kill all of the bacterial cells within mature biofilms if these agents were used within these electric fields. The electric fields alone had no adverse affect on bacteria as they colonized surfaces and formed biofilms within the test chambers, and biocides alone were similarly ineffective, but the combination of low concentrations (0.5–4.0 MIC) of these

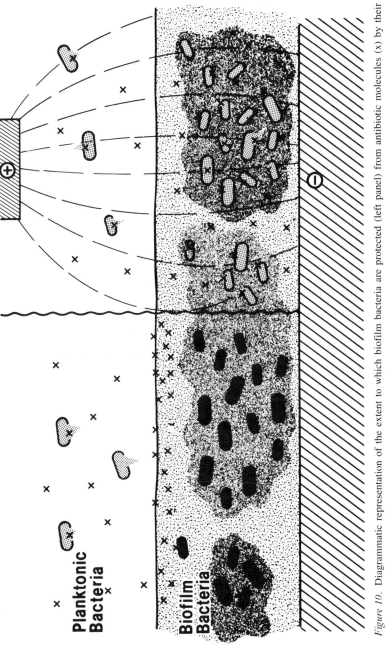

Figure 10. Diagrammatic representation of the extent to which biofilm bacteria are protected (left panel) from antibiotic molecules (x) by their growth in densely slime-enclosed microcolonies. When these same molecules (x) are present within a DC electric field (right panel), they readily penetrate the biofilm matrix and kill the biofilm bacteria.

Planktonic Bacteria

Biofilm Bacteria

agents with the electric field killed all of the biofilm bacteria (Blenkinsopp *et al.,* 1992). When we treated mature biofilms of several species of gram-positive bacteria, gram-negative bacteria, and a fungus (*Candida albicans*) with low concentrations of antibiotics (0.5–4.0 MIC) in the presence of a much lower strength electric field (field strength 1.5 V/cm, current density 15 μA/cm^2), we were able to kill all of the bacteria and fungi within these biofilms in 4–8 hr (Costerton *et al.,* 1993). Because we were confident that these very low strength electric fields would not cause tissue damage, we colonized a prototype device with a mature biofilm, placed it in the peritoneum of a rabbit, and treated the colonized device with antibiotic perfused through the device and an electric field generated on the device itself. The bacteria in these biofilms were completely killed by the bioelectric effect, *in vivo,* and the experimental animals experienced no distress or perceptible tissue damage.

The mechanism of the bioelectric effect is not established, but most evidence points to an electrophoretic mechanism. We note that DC fields are necessary, and that the polarity of the field must be maintained in one direction for more than 30 sec. This suggests that the electric field may move charged antibacterial agent molecules through the matrix of the biofilm by an electrophoretic mechanism (Fig. 10), but cell leakage data and recent observations of the depletion of divalent cations in biofilms subjected to electric fields suggest that electroporation and cation depletion may both contribute to the effect.

We have already begun to use the new bioelectric technology to accelerate the sterilization of medical devices in antiseptic solutions (e.g., glutaraldehyde), and we find that this electric enhancement of penetration enables us to kill protected biofilm bacteria and even protected spores in a very short time. As we develop more effective ways of establishing low-level electric fields at the surfaces of medical devices, we expect to be able to sterilize recently implanted medical devices, perioperatively *in situ,* in order to prevent the future development of chronic biofilm-mediated infections, and to be able to enhance the efficacy of antibiotic chemotherapy in the treatment of well-developed chronic infections focused on medical devices. The addition of a physical force to our antibacterial armamentarium may enable us to use chemical agents more effectively to solve a serious biological problem—the inherent resistance of biofilm bacteria to antibacterial agents.

References

Anwar, H., Strap, J. L., and Costerton, J. W., 1992, Establishment of aging biofilms: A possible mechanism of bacterial resistance to antimicrobial therapy, *Antimicrob. Agents Chemother.* **36:**1347–1351.

Blenkinsopp, S. A., Khoury, A. E., and Costerton, J. W., 1992, Electrical enhancement of biocide efficacy against *Pseudomonas aeruginosa* biofilms, *Appl. Environ. Microbiol.* **58:**3770–3773.

Brown, M. R. W., and Williams, P., 1985, The influence of environment on envelope properties affecting survival of bacteria in infections, *Annu. Rev. Microbiol.* **39**:527–556.

Costerton, J. W., Geesey, G. G., and Cheng, K.-J., 1978, How bacteria stick, *Sci. Am.* **238**:86–95.

Costerton, J. W., Cheng, K.-J., Geesey, G. G., Ladd, T. I., Nickel, J. C., Dasgupta, M., and Marrie, T. J., 1987, Bacterial biofilms in nature and disease, *Annu. Rev. Microbiol.* **41**:435–464.

Costerton, J. W., Ellis, B., Lam, K., Johnson, F., and Khoury, A. E., 1993, Electrical enhancement of the efficacy of antibiotics in killing biofilm bacteria, *Antimicrob. Agents Chemother.* in press.

Dasgupta, M. K., Bettcher, K. B., Ulan, R. A., Burns, V., Lam, K., Dossetor, J. B., and Costerton, J. W., 1987, Relationship of adherent bacterial biofilms to peritonitis in chronic ambulatory peritoneal dialysis, *Peritoneal Dialysis Bull.* **7**:168–173.

Deneke, C. F., Thorne, G. M., and Gorbach, S. L., 1981, Serotypes of attachment pili from enterotoxigenic *Escherichia coli* isolated from humans, *Infect. Immun.* **32**:1254–1260.

Giwercman, B., Jensen, E. T., Hoiby, N., Kharazmi, A., and Costerton, J. W., 1991, Induction of β-lactamase production in *P. aeruginosa* biofilm, *Antimicrob. Agents Microbiol.* **35**:1008–1010.

Gristina, A. G., and Costerton, J. W., 1984, Bacterial adherence and the glycocalyx and their role in musculoskeletal infection, *Orthop. Clin. North Am.* **15**:517–535.

Gristina, A. G., Dobbins, J. J., Giammara, B., Lewis, J. C., and DeVries, W. C., 1988, Biomaterial-centered sepsis and the total artificial heart: Microbial adhesion vs. tissue integration, *J. Am. Med. Assoc.* **259**:870–877.

Khoury, A. E., and Costerton, J. W., 1988, Biofilms in nature and medicine, in: *Proceedings of the Third ECCM Symposium on Foreign-Body Related Infections*, Elsevier, Amsterdam, pp. 2–15.

Lam, J. S., Chan, R., Lam, K., and Costerton, J. W., 1980, The production of mucoid microcolonies by *Pseudomonas aeruginosa* within infected lungs in cystic fibrosis, *Infect. Immun.* **38**:546–556.

Lawrence, J. R., Korber, D. R., Hoyle, B. D., Costerton, J. W., and Caldwell, D. E., 1991, Optimal sectioning of bacterial biofilms, *J. Bacteriol.* **173**:6558–6567.

Marrie, T. J., and Costerton, J. W., 1983, A scanning and transmission electron microscopic study of the surfaces of intrauterine contraceptive devices, *Am. J. Obstet. Gynecol.* **146**:384–394.

Marshall, K. C., Stout, R., and Mitchell, R., 1971, Mechanisms of the initial events in the sorption of marine bacteria to surfaces, *J. Gen. Microbiol.* **68**:337–348.

Nickel, J. C., and Costerton, J. W., 1992, Coagulase-negative staphylococcus in chronic prostatitis, *J. Urol.* **147**:398–401.

Nickel, J. C., Ruseska, I., and Costerton, J. W., 1985, Tobramycin resistance of cells of *Pseudomonas aeruginosa* growing as a biofilm on urinary catheter material, *Antimicrob. Agents Chemother.* **27**:619–624.

Nickel, J. C., Downey, J. A., and Costerton, J. W., 1992, Movement of *Pseudomonas aeruginosa* along catheter surfaces, *Urology* **39**:93–98.

Ruseska, I., Robbins, J., Lashen, E. S., and Costerton, J. W., 1982, Biocide testing against corrosion-causing oilfield bacteria helps control plugging, *Oil Gas J.* **1982**:253–264.

Sutherland, I. W., 1977, Bacterial exopolysaccharides—Their nature and production, in: *Surface Carbohydrates of the Prokaryotic Cell* (I. W. Sutherland, ed.), Academic Press, New York, pp. 27–96.

Tenney, J. H., Moody, M. R., Newman, K. A., Schimpff, S. C., Wade, J. C., Costerton, J. W., and Reed, W. P., 1986, Adherent microorganisms on lumenal surfaces on long-term intravenous catheters: Importance of *Staphylococcus epidermidis* in patients with cancer, *Arch. Intern. Med.* **146**:1949–1954.

Ward, K. H., Olson, M. E., Lam, K., and Costerton, J. W., 1992, Mechanism of persistent infection associated with peritoneal implants, *J. Med. Microbiol.* **36**:406–413.

15

Peptide Mimetics as Adhesion Molecule Antagonists

JAMES SAMANEN, FADIA ALI, JOHN BEAN, JAMES
CALLAHAN, WILLIAM HUFFMAN, KENNETH KOPPLE,
CATHERINE PEISHOFF, and ANDREW NICHOLS

1. Introduction: Peptide and Protein Platelet Fibrinogen Receptor Antagonists

1.1. The Search for Adhesion Molecule Antagonists with Oral Activity in the Area of Platelet Fibrinogen Receptor Antagonists

As a class, drugs that modulate cell adhesion as a mechanism of action are in their infancy. Although several drugs are in clinical development as intravenous agents, none have reached the market. Even in this early period, however, the pursuit of adhesion molecule antagonists that may be administered orally has begun within the arena of platelet fibrinogen receptor antagonists (also called GPIIb/IIIa antagonists). As one of the more mature areas of antiadhesion, platelet fibrinogen receptor antagonists have advanced from the native protein fibrinogen to small potent inhibitory peptides, and more recently to potent semipeptide and nonpeptide antagonists. Although nonpeptide ligands to GPIIb/IIIa have been discovered through compound data-base screening, the path to non-

JAMES SAMANEN, FADIA ALI, JOHN BEAN, JAMES CALLAHAN, and WILLIAM HUFF-MAN • Department of Medicinal Chemistry, SmithKline Beecham Pharmaceuticals, King of Prussia, Pennsylvania 19406.   *KENNETH KOPPLE and CATHERINE PEISHOFF* • Department of Physical and Structural Chemistry, SmithKline Beecham Pharmaceuticals, King of Prussia, Pennsylvania 19406.   *ANDREW NICHOLS* • Department of Cardiovascular Pharmacology, SmithKline Beecham Pharmaceuticals, King of Prussia, Pennsylvania 19406.
Cellular Adhesion: Molecular Definition to Therapeutic Potential, edited by Brian W. Metcalf, Barbara J. Dalton, and George Poste. Plenum Press, New York, 1994.

peptides through peptide mimetics is also being avidly pursued. This chapter will focus on the development of an RGD peptide pharmacophore model for the design of novel nonpeptide mimetic ligands to GPIIb/IIIa. It will be instructive, however, to examine first the path that led to the discovery of small potent inhibitory peptide GPIIb/IIIa antagonists, focusing on the work at SmithKline Beecham.

1.2. Peptide and Protein Platelet Fibrinogen Receptor Antagonists

1.2.1. Fibrinogen Fragments

In the mid-1980s, researchers realized that it might be possible to inhibit platelet aggregation with peptide fragments from fibrinogen that contained the elements for GPIIb/IIIa binding. Groups at Harvard University, Scripps Clinic, and the University of Pennsylvania identified three minimal peptide fragments that bound to GPIIb/IIIa (Fig. 1):

- The C-terminal dodecapeptide from the gamma chain, Fg-γ(400–411), containing the sequence HHLGGAKQAGDV (Kloczewiak *et al.*, 1984, 1989),
- Two similar sequences from the alpha chain, Fg-α(572–575) (Gartner and Bennett, 1985; Ginsberg *et al.*, 1985 and Fg-α(95–98) (Andrieux *et al.*, 1989; Plow *et al.*, 1987), containing the sequence -Arg-Gly-Asp or RGD in single-letter code.

In our lab, RGDS, Fg-α(572–575), displayed higher affinity than the dodecapeptide, FG-γ(400–411). Its smaller size attracted us as a starting point for structural studies.

Ac-RGDS-NH$_2$ inhibited platelet aggregation, whereas Ac-RGES-NH$_2$, which contains glutamic instead of aspartic acid, had no effect (Shebuski *et al.*, 1989a). Thus, Ac-RGDS-NH$_2$ is an antiaggregatory agent, albeit with low potency. On the basis of this observation we felt it worthwhile to demonstrate the efficacy of Ac-RGDS-NH$_2$ in the Folts model (Folts *et al.*, 1976), an animal model of thrombosis, before launching an extensive effort to develop novel antithrombotic agents from Ac-RGDS-NH$_2$. The model (described in detail by Nichols *et al.*, this volume) mimics the process of unstable angina and makes it possible to measure the ability of an antithrombotic agent to inhibit thrombus formation *in vivo*. We found that Ac-RGDS-NH$_2$ could inhibit thrombus formation in the Folts model only by direct infusion into the coronary artery at a molar rate equivalent to 19 mg/min (Shebuski *et al.*, 1989a). Ac-RGES-NH$_2$ was also infused at the same concentration, and did not cause inhibition. The inhibition by Ac-RGDS-NH$_2$, however, was dose-dependent and reversible. The effect by Ac-RGDS-NH$_2$ was thus specific for the peptide. These experiments suggested that

Fg-γ (400-411): HHLGGAKQAGDV

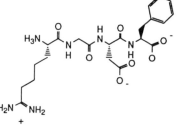

Fg-α (572-575): RGDS

Fg-α (95-98): RGDF

Figure 1. Fibrinogen fragments that inhibit platelet–fibrinogen binding.

Ac-RGDS-NH$_2$ could be considered a lead for the further development of novel antithrombotic agents. We obviously had a long way to go from Ac-RGDS-NH$_2$, since this peptide was at best a weak inhibitor both *in vitro* and *in vivo* (Samanen *et al.*, 1991).

1.2.2. Monoclonal Antibodies and Disintegrins

Two other approaches to platelet fibrinogen receptor antagonists have the immediate advantages of potency and duration: the monoclonal antibodies developed by Coller and others (Table I; Coller, 1985; Coller *et al.*, 1991; Hanson *et al.*, 1988; Shattil *et al.*, 1985); and the family of snake venom inhibitors, which contain RGD or KGD sequences (Table II; Dennis *et al.*, 1989; Gan *et al.*, 1988;

Table I. Antibodies and Antibody Fragments That Inhibit Platelet–Fibrinogen Binding and Platelet Aggregation

Name	Type	Reference	Inhibits Fg and RGDS binding to GPIIb/IIIa	Inhibits platelet aggregation
10E5	IgG	Coller et al. (1983)	Yes	Yes
7E3	IgG	Coller (1985)	Yes	Yes
PAC-1	IgG	Shattil et al. (1985)	Yes	Yes
AP-2	IgG	Hanson et al. (1988)	Yes	Yes
7E3	F(ab')$_2$, Fab	Coller et al. (1991)	Inhibit thrombus formation in the dog	
			Inhibit thrombus formation in baboons receiving an arterial graft	
			Inhibit platelet aggregation *ex vivo* in humans	
PAC-1	H-CDR3 peptide: Arg-Ser-Pro-Ser-Tyr-Tyr-**Arg-Tyr-Asp**-Gly-Ala-Gly-Pro-Tyr-Tyr-Ala-Met-Asp-Tyr IC$_{50}$ platelet aggregation 40 μM	Taub et al. (1989)		
	Arg-Ser-Pro-Ser-Tyr-Tyr-**Arg-***Gly*-**Asp**-Gly-Ala-Gly-Pro-Tyr-Tyr-Ala-Met-Asp-Tyr IC$_{50}$ platelet aggregation ~4 μM			

Table II. Disintegrins: Snake Venom Inhibitors of Platelet Aggregation

Name	Partial disintegrin structure	Platelet aggregation		Binding Fg/GPIIb/IIIa	
		hPRP IC$_{50}$ (nM)	dprp IC$_{50}$ (nM)	ELISA IC$_{50}$ (nM)	Hgfp K_i (nM)
Echistatin-α1,2 Merck COR Ther.		30 (hGFP) 160			8.0
Echistatin-α2 (Genentech)	-Cys-Lys-Arg-Ala-**Arg-Gly-Asp**-Asp-Met-Asp-Asp-Tyr-Cys-	555		2.7	
Trigramin -α (Temple U.)	-Cys-Arg-*Ile*-Ala-**Arg-Gly-Asp**-Asp-Leu-Asp-Asp-Tyr-Cys-		130		21
-β1 (Genentech)		300		3.0	
-β2 (Genentech)		170		2.3	
-γ (Genentech)		240		2.2	
Bitan-α (Genentech)	-Cys-Arg-Arg-Ala-**Arg-Gly-Asp**-Asp-Leu-Asp-Asp-Tyr-Cys-	108		1.8	
Kistrin (Genentech)	-Cys-Arg-*Ile*-Ala-**Arg-Gly-Asp**-Asp-*Trp-Asn*-Asp-Asp-Ar<u>g</u>-Cys-	128		2.7	
Bitistatin 1 (Merck)	-Cys-Arg-*Ile*-Ala-**Arg-Gly-Asp**-Asp-*Trp-Asn*-Asp-Asp-Tyr-Cys-	237			
Barbourin (COR Ther.)	-Cys-Arg-*Val*-Ala-**Lys-Gly-Asp**-Asp-*Trp-Asn*-Asp-Asp-*Thr*-Cys-	309		15	
Tergeminin (COR Ther.)	-Cys-Arg-*Val*-Ala-**Arg-Gly-Asp**-Asp-*Trp-Asn*-Asp-Asp-*Thr*-Cys-	192		25	
compared to:	H-Gly-**Arg-Gly-Asp**-Ser-OH	225,000		205	
	H-**Arg-Gly-Asp**-Phe-OH	6,000			

Garsky *et al.*, 1989; Huang *et al.*, 1987, 1989; Scarborough *et al.*, 1991; Shebuski *et al.*, 1989b). Much of the early justification for the exploration of fibrinogen receptor antagonists was established by Coller and his collaborators
through *in vitro* and *in vivo* experiments with his 7E3 and 10E5 antibodies,
F(ab′)$_2$ and Fab fragments. A fragment from the PAC-1 antibody was found to
display moderate potency as an antiaggregatory agent. On the hunch that an RYD
motif in that fragment might be responsible for receptor binding, Taub *et al.*
(1989) prepared an analogue in which the RYD was replaced with RGD. As seen
in Table I, the RGD analogue displayed a tenfold enhancement in potency. The
Fab fragment of Coller's 7E3 antibody has been humanized and is currently
undergoing clinical trials at Centocor (Anderson *et al.*, 1992; Kleiman *et al.*,
1991). Researchers at Temple University, Merck, Genentech, COR Therapeutics, and Biogen have explored a family of RGD-containing snake venom inhibitors, collectively called "disintegrins" (see Table II). All of these peptides display
high affinity and low selectivity, with the exception of Barbourin, which displays
enhanced selectivity for GPIIb/IIIa over other integrins.

The small peptide and peptide mimetic fibrinogen receptor antagonists have
potential advantages over the monoclonal antibody or disintegrin derived agents
with regard to the issues of cost and immunogenicity. The small peptides are also
more suitable as potential leads for the ultimate goal of designing orally active
fibrinogen receptor antagonists.

1.2.3. RGD Peptide Analogues

Ac-RGDS-NH$_2$ has served as an initial starting point for drug development
in several laboratories. Its affinity (K_i 4200 nM) is lower than fibrinogen (K_i 43
nM), probably because this small 474-Da peptide lacks the conformational constraints imposed on the RGD sequence in the protein by secondary structure, and
Ac-RGDS-NH$_2$ undergoes fewer receptor interactions than the 420-kDa protein
fibrinogen. One would have to improve receptor affinity dramatically in order to
circumvent the competition of 6 μM plasma fibrinogen (Plow *et al.*, 1986). One
option would be to "go larger": to reach out to find other binding sites to increase
affinity (Fig. 2). In contrast to this "regional" approach, we chose the "local"
approach, seeking enhancement of affinity by refining the interaction between
Ac-RGDS-NH$_2$ with its specific receptor site. We sought lead peptides that
would be small, since such leads would be more amenable to the future development of orally active antithrombotic agents. Using synthetic organic chemistry, it
would be possible to incorporate structural elements into Ac-RGDS-NH$_2$ that are
not available from the native amino acids. The nonnative structural elements
could allow for enhanced or even novel receptor interactions that would not be
available to the native peptide.

In our initial structure–activity studies (Ali *et al.*, 1990; Samanen *et al.*,

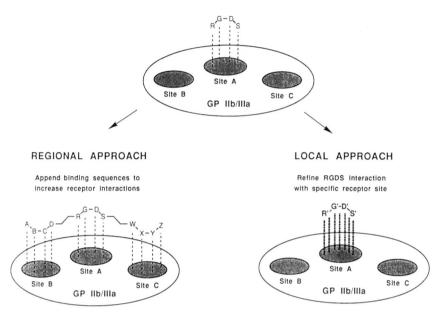

Figure 2. The regional approach versus the local approach to fibrinogen receptor antagonists.

1991), we found that modifications to the linear Ac-RGDS-NH$_2$ peptide sequence gave rise to analogues with only modest enhancements in affinity. Although modifications to Arg, Gly, or Asp tended to give analogues with reduced activity (especially to the guanidine and carboxylate groups on Arg and Asp, respectively), Arg could be replaced with D-Arg or homoarginine without a loss of activity. We began to see significant enhancements in affinity, however, when we enclosed the tripeptide sequence into a five-membered ring containing a disulfide-bridged pair of cysteines (see Table III) (Samanen *et al.*, 1991). Substitution of the β-branched penicillamine for the C-terminal cysteine and methylation of the alpha amine of arginine gave rise to SK&F 106760 **9**, which displayed submicromolar antiaggregatory activity. With SK&F 106760, we obtained an analogue with affinity that rivaled that of native fibrinogen. Although collectively these modifications increased affinity, the possibility for further enhancements in affinity remained. We were attracted to modifications that reduced flexibility in the relatively mobile Cys-Pen tether with the more constrained diaryldisulfide group we refer to as "Mba-Man" (so named for its two components: mercaptobenzoic acid and mercaptoaniline) (Fig. 3). With the diaryldisulfide, at least two torsion angles in the disulfide bridge have been fixed. The resultant semipeptide SK&F 107260 **14** displays a 2 nM K_i in the binding assay and a 90 nM IC$_{50}$ in our antiaggregatory assay (Table IV) (Ali *et al.*, 1992).

Table III. The Development of SK&F 106760[a]

	Structure	Mol. wt.	Antiagg. dog PRP IC_{50} (μM)	Human GPIIb/IIIa K_i (μM)	Human Platelets K_i (μM)
				Inhibition of [^{125}I]-Fg binding	
1	Fibrinogen	450,000		0.043	0.707
2	Fg-γ(400–411) Ac-HHLGGAKQAGDV-OH	1,231	>200	14.5	
3	Fg-α(572–575) Ac-Arg-Gly-Asp-Ser-NH$_2$	474	91.3	4.2	62
4	Ac-Arg-Gly-Asp-**Val**-NH$_2$	487	55.5	<10	
5	Ac-Cys-Arg-Gly-Asp-Ser-Cys-NH$_2$ (S——S)	679	32.7	5.3	
6	Ac-Cys-Arg-Gly-Asp-Cys-NH$_2$ (S——S)	592	16.2	0.78	
7	Ac-Cys-Arg-Gly-Asp-**Pen**-Cys-NH$_2$ (S——S)	619	4.1		
8	Ac-Cys-D-Arg-Gly-Asp-Pen-NH$_2$ (S——S)	619	4.1	0.72	
9	Ac-Cys-(**NMe**)Arg-Gly-Asp-Pen-NH$_2$ (S——S) SK&F 10670	634	0.36	0.058	0.062

[a] Antiagg., inhibition of platelet aggregation; dog PRP, dog platelet-rich plasma. Inhibition of [^{125}I]-Fg, binding either to purified human GPIIb/IIIa or to human gel-filtered platelets.

The combination of Mba and Man was not predictable. Single substitutions in **11** and **12** each enhanced activity, but together the substitutions gave a synergistic enhancement in affinity with **13**. The affinity for SK&F 107260 **14** was an order of magnitude higher than might be expected from its antiaggregatory potency. The affinity displayed in Table IV was measured against a tritiated version of itself, but its K_i against [^{125}I] fibrinogen was identical. This abnormally high affinity complicates the comparison of analogues to SK&F 107260.

Other groups have pursued the development of cyclic peptides related to RGD; the types of cyclic peptide they have discovered are shown in Table V.

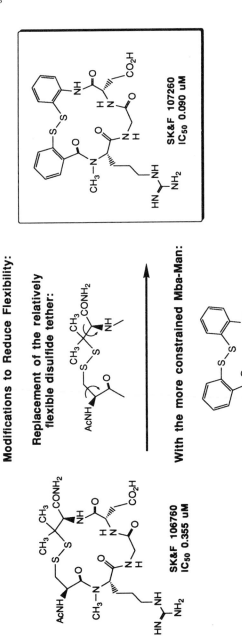

Figure 3. Modifications of SK&F 106760 leading to SK&F 107260.

Table IV. The Development of SK&F 107260[a]

	Structure	Antiagg. dog PRP IC$_{50}$ (μM)	Human GPIIb/IIIa K_i (μM)

9 SK&F 106760	Ac-Cys-(NMe)Arg-Gly-Asp-Pen-NH$_2$
	    | |
	    S —————— S
		0.36	0.058

10	Ac-Cys-(NH)-Arg-Gly-Asp-**Cys**-NH$_2$
	    | |
	    S —————— S
		16.2	0.780

11	CO-(NH)-Arg-Gly-Asp-Cys-NH$_2$ | S —————— S	8.08	3.30
12	Ac-Cys-(NH)-Arg-Gly-Asp-NH | S —————— S	9.71	0.280
13	CO-(NH)-Arg-Gly-Asp-NH S —————— S	0.290	0.047
14 SK&F 107260	CO- (NMe)Arg-Gly-Asp-NH S —————— S	0.090	0.002

[a]Antiagg., inhibition of platelet aggregation; dog PRP, dog platelet-rich plasma. Inhibition of [^{3}H]-SK–F 107260 binding either to purified human GPIIb/IIIa (Stadel et al., 1992; Wong et al.. 1992).

Both COR Therapeutics and Merck have described potent heptapeptide analogues **15–17** (Nutt et al., 1992). Merck has also developed hexapeptides, e.g., **19** (Nutt et al., 1992), and both Genentech (Barker et al., 1992) and SmithKline Beecham (Samanen et al., 1991; Ali, et al., 1992) have described potent pentapeptides **20, 9,** and **14**. The potency and simplicity of structure of SK&F 107260 make it as effective as any of the other peptides in Table V.

In contrast to the previously cited transient inhibition of cyclic flow reductions following direct coronary infusions of Ac-RGDS-NH$_2$, a simple bolus injection of SK&F 106760 produces a prolonged inhibition of cyclical flow reductions in the Folts model (Nichols et al., 1990, 1992). This inhibition is dose-dependent. In contrast to other antithrombotic agents, the peptide has no effect on blood pressure, heart rate, and platelet count (Nichols et al., 1990).

The fibrinogen receptor antagonist SK&F 106760 has also been compared with aspirin in the Folts model (Nichols et al., this volume). It was shown that SK&F 106760 completely blocks thrombus formation in dogs, whereas aspirin, which blocks the thromboxane pathway to GPIIb/IIIa activation, is only partially

Table V. Peptide Fibrinogen Receptor Antagonists

			X		IC$_{50}$ (nM), inhibition of platelet aggregation[a,b]	
					hWP	hPRP
Heptapeptides	Merck	15	Arg	Ac-Cys-Asn-Dtc-X-Gly-Asp-Cys-OH (S—S)	40	
		16	Amf = *p*-aminomethyl-Phe	L367073, MK852	26	
	COR Ther.	17	Ph(C=NH)$^{\epsilon}$Lys	Mpr-X-Gly-Asp-Trp-Pro-Pen-NH$_2$ (S—S)		500
		18	hArg			100
Hexapeptides	Merck	19		Arg-Gly-Asp-Trp-Pro (Aha)	31	
Pentapeptides	Genentech	20		CO-Gly-Arg-Gly-Asp-NH-CH-OH (CH$_2$—S—CH$_2$)		80
	SmithKline Beecham	9	SK&F 106760	Ac-Cys-(NMe)Arg-Gly-Asp-Pen-NH$_2$ (S—S)	230	356 (dPRP)
		14	SK&F 107260	CO-(NMe)Arg-Gly-Asp-NH (S—S)	60	90 (dPRP)

[a] hWP, human washed platelets; hPRP, human platelet-rich plasma.
[b] The IC$_{50}$ measured in hWP are lower than in hPRP since micromolar concentrations of competing fibrinogen and von Willebrand factor have been removed in hWP.

effective. Furthermore, SK&F 106760 continues to work in the presence of aspirin. In a variant of the Folts model as a model of thrombotic stroke (Willette *et al.*, 1992), SK&F 106760 was completely effective, whereas aspirin was less effective than SK&F 106760. The latter has also been demonstrated to be an effective adjunct to thrombolytic therapy. These experiments in animals suggest that fibrinogen receptor antagonists should be investigated in humans as an acute treatment of a number of thrombotic conditions. Indeed, several agents are currently undergoing clinical investigation for such acute indications. Consequently, research has shifted, focusing on the discovery of oral fibrinogen receptor antagonists. These could be used in the primary prevention of acute myocardial infarction in individuals at high risk, i.e., for patients with unstable angina and in the secondary (or chronic) prevention of acute myocardial infarction and stroke.

2. The Development of an RGD Peptide Pharmacophore Model for the Design of Peptide Mimetic Ligands

In general, peptides and semipeptides (which contain some nonnative structural elements) do not display adequate oral bioavailability. An important exception is the recent report that there is adequate oral bioavailability in certain semipeptide renin inhibitors (Boyd *et al.*, 1992). Until more experience is gained in the development of peptides and semipeptides with adequate oral bioavailability, the pursuit of small organic nonpeptide ligands will continue to be the most promising approach to oral drugs that act at peptide receptors. The screening of compound data bases and biological extracts has often yielded nonpeptide leads that have been developed into orally active drugs (Freidinger, 1989; see also references in Collins *et al.*, 1992). These compounds may be allosteric inhibitors. As we will describe in Section 3, at least two classes of nonpeptide GPIIb/IIIa antagonists have been developed from screening leads.

Peptide receptors are capable of binding any type of ligand that contains an appropriate set of binding groups in a specific topological relationship. Alternatively, one could approach the design of a nonpeptide ligand by considering the receptor-bound conformation of the peptide. This conformation could serve as a model in which the peptide structure is replaced with nonpeptide structures that retain the appropriate ensemble of binding groups. Ideally, the receptor-bound conformation could be obtained from an X-ray crystal structure or an NMR solution structure of an RGD peptide bound to GPIIb/IIIa. In reality, such structures are not available, primarily because GPIIb/IIIa is a membrane-bound receptor. In the absence of such a structure, one must develop an understanding of the *peptide pharmacophore,* a term that refers to a topological conceptualization of the space that the peptide ligand may occupy at a receptor (Fig. 4). A peptide

CONFORMATIONALLY CONSTRAINED LIGANDS   PHARMACOPHORE

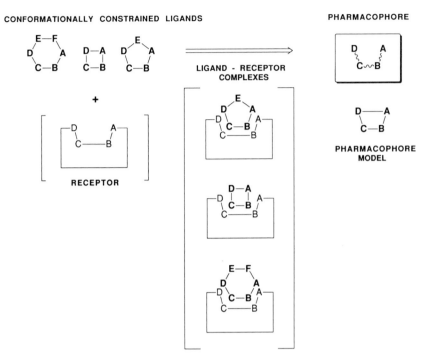

Figure 4. Conceptualization of the peptide pharmacophore and a peptide pharmacophore model via conformationally constrained peptides. Unknown structures, the receptor site, and structures of receptor-bound ligands are in parentheses.

pharmacophore model would be a discrete chemical structure, such as a cyclic peptide, in a conformation that fits the pharmacophore. These models can be derived from studies of peptide SAR and peptide conformation. It would be optimal to obtain knowledge of the biologically active or receptor-bound conformation(s), but this is not available in the case of RGD. Unfortunately, the tripeptide RGD is inherently flexible (see references in Bogusky *et al.*, 1992). Furthermore, the crystal structures of RGD-containing peptides and proteins display a wide variety of conformations (Table VI) (Eggleston and Feldman, 1990). We chose, therefore, to investigate a number of conformationally constrained peptide and semipeptide analogues of Ac-RGDS-NH$_2$. We sought peptides that displayed high affinity for GPIIb/IIIa and were constrained enough to yield relatively discrete models of their solution conformation through a combination of NMR spectroscopy and molecular modeling. Ultimately, such models could be employed in the design of nonpeptide ligands to GPIIb/IIIa.

Nonpeptides developed in this manner can be considered peptide mimetics, since they were designed to bind to the receptor in the manner of the peptide

Table VI. Backbone Torsion Angles for -Arg-Bly-Asp Sequences in X-Ray Crystal Structures of Peptides and Proteins[a]

	Ψ^1	Φ^2	Ψ^2	Φ^3
H-Arg-Gly-Asp-OH	160.3	−85.5	−129.7	−102.6
Thermolysin	111.4	53.4	−131.4	−87.8
γ-II crystallin	153.9	142.5	170.7	−100.4
Alpha-lytic protease	133.4	105.7	−22.3	−77.7

[a] Eggleston and Feldman (1990).

(although this may be unprovable). The term *peptide mimetic* has also been used to describe structural elements that are not derivable from native amino acids. Peptide mimetics range from *semipeptides* [containing some mimetic elements of peptide structure, typically structural elements that mimic certain types of peptide secondary structure (see Section 2.2) to *nonpeptides* (in which the peptide structure is no longer recognizable; i.e., the entire structure is a peptide mimetic that retains the geometric ensemble of binding groups]. The design of an RGD semipeptide bearing a secondary structure mimetic will be described in Section 2.3. This semipeptide is an important first step toward the rational design of a nonpeptide from a peptide.

2.1. The Development of an RGD Peptide Pharmacophore Model through the Evaluation of Conformationally Constrained Peptides

SK&F 106760 and SK&F 107260 are two examples of high-affinity ligands for GPIIb/IIIa developed from Ac-RGDS-NH$_2$ through the addition of conformational constraints. SK&F 107260 displays 1000-fold enhancement in affinity and a corresponding enhancement in antiaggregatory potency over RGDS. In SK&F 106760 and SK&F 107260, the critical Arg and Asp sidechains have not been constrained. Nonetheless, one might ask how much the backbone conformation is constrained in these peptides. Is SK&F 107260 more rigid than SK&F

106760? If these compounds are conformationally constrained, what are the backbone conformations for these peptides in solution? Answers to these questions were obtained through a study of the ¹H NMR spectra of both compounds (Kopple *et al.*, 1992). Interproton distances (nOe's) from the ¹H NMR spectra were incorporated into an nOe-constrained distance geometry study (Kopple *et al.*, 1992). For SK&F 107260, two slowly exchanging conformations were determined from the ¹H NMR spectrum at 203 K in methanol; they differed chiefly in the orientation of the amide between aspartic acid and the C-terminal anilide group (Fig. 5). Both conformations are characterized by an extended glycine, flanked by an Arg residue in a conformation roughly consistent with the *i*+2 position of a β turn and an Asp residue in two different turn conformations, depending on the disposition of the Asp-anilide amide. In the major component, the aspartic acid occurs in a γ-turn conformation. In the minor component, the aspartic acid approximates a right-handed α-helical conformation. These conformations will be referred to as "turn–extended-turn" conformations. The nitrate salt of SK&F 107260 can be crystallized from aqueous ethanol to determine its X-ray crystal structure (Kopple *et al.*, 1992). Interestingly, the structure is nearly identical to the minor component of the solution conformation determined in methanol.

Figure 6 displays the X-ray crystal structure of SK&F 107260 as an ORTEP model, clearly indicating the turn–extended-turn conformation about Arg-Gly-Asp. The side view is informative: it shows that the aromatic groups in the disulfide tether are parallel to each other and perpendicular to the plane of the RGD peptide backbone. The ¹H-NMR spectrum of SK&F 106760 appears to correspond to a mixture of conformations, as evidenced by line broadening at lower temperatures, although individual conformers could not be observed at −70°C. Nonetheless, the NMR data are consistent with the major conformation determined for SK&F 107260. Figure 7 displays molecular models of the two predominant backbone conformations determined for SK&F 107260 and the dominant component of SK&F 106760. In this figure the Arg and Asp side chains in these molecules are placed in purposefully different orientations, since there were no NMR data to serve as input data in the conformational analysis to help position these side chains in each molecule. Superpositioning both molecules shows how similar the backbone conformations of the two compounds appear in solution (Fig. 7). This is surprising, since the two different disulfide tethers might have been expected to torque the molecules into two different conformations.

Bogusky *et al.*, (1992) have examined the solution conformations of Ac-S,S-cyclo-Cys-Arg-Gly-Asp-Cys-OH, which lacks the constraining elements of the *N*-methyl group on the Cys-Arg amide and the β-branched Pen. This peptide seems to be less well defined conformationally than SK&F 106760 or SK&F 107260.

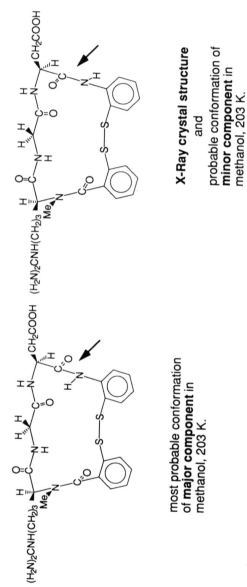

most probable conformation of major component in methanol, 203 K.

X-Ray crystal structure and probable conformation of minor component in methanol, 203 K.

Figure 5. The conformations of SK&F 107260 (Kopple *et al.*, 1992). This figure highlights the chief similarities and differences between the most probable conformations determined in an nOe-constrained distance geometry conformational study of the major and minor components of the NMR spectrum for SK&F 107260 determined in methanol at 203 K. The most probable conformation of the major component is on the left. The most probable conformation of the minor component is on the right. The X-ray crystal structure is comparable to the minor component.

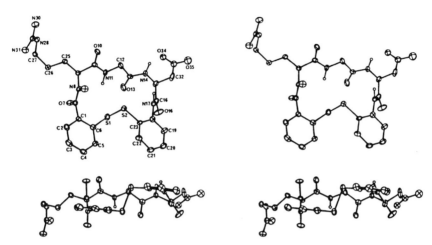

Figure 6. X-ray crystal structure of SK&F 107260. Two views of the X-ray crystal structure of SK&F 107260, displayed in an ORTEP model (Kopple *et al.*, 1992): Shown from above the plane of the macrocycle and in front of the RGD backbone.

The fact that both SK&F 106760 and SK&F 107260 display such similar solution conformations and interact with GPIIb/IIIa with a high degree of affinity suggests that these conformations might define a peptide pharmacophore model for GPIIb/IIIa. To probe GPIIb/IIIa for other potential high-affinity conformations, we turned from *heterodetic peptides,* cyclic peptides in which the macrocycle includes the bridging side chains, to *homodetic peptides,* cyclic peptides in which the macrocycle involves only the peptide backbone. We used homodetic peptides that have been extensively studied by NMR spectroscopy and crystal structure analysis to design cyclic peptides that contained RGD in different types of secondary structure, e.g., β sheet (fully extended) and β turn. Gramicidin S

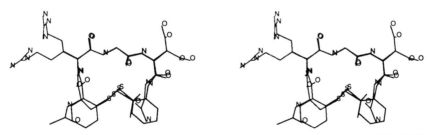

Figure 7. Comparison of the most probable conformations of SK&F 107260 and SK&F 106760 (Kopple *et al.*, 1992). Models for the most probable conformations of each molecule are superimposed and displayed in stereo to highlight the high degree of similarity between the peptide backbone conformations.

20, for instance, is a cyclic octapeptide, in which the Val-Orn-Leu sequences reside in fully extended β-sheet conformations and the D-Phe-Pro groups occupy the corners of type II' β turns (Fig. 8A). An analogue of gramicidin **21** was prepared (Peishoff *et al.,* 1992) in which Val-Orn-Leu was replaced with Arg-Gly-Asp (Fig. 8B). It was confirmed that **21** had the same backbone conformation as the native gramicidin **20.** The affinity and *in vitro* potency of **21** will be discussed below.

It is well known that cyclic hexapeptides tend to adopt boxlike conformations with two opposing β turns, and that prolines tend to occupy a turn position (see references in Bean *et al.,* 1992). It was reasoned that two adjacent prolines could be arranged to form one turn and force a β turn among the residues opposite them (Fig. 9). To reduce the prevalence of a *cis*-amide bond between the two prolines, the preceding proline was incorporated with the D configuration. On that basis, we hoped to obtain two different hexapeptides bearing Arg-Gly-Asp and D-Pro-Pro sequences: one with Arg-Gly in a β turn **22** and another with Gly-Asp in a β turn **23** (Bean *et al.,* 1992). Both peptides were synthesized (Ali and Samanen, 1992); so was a third hexapeptide in which a glycine was positioned between D- Pro and Pro to give analogue **24.**

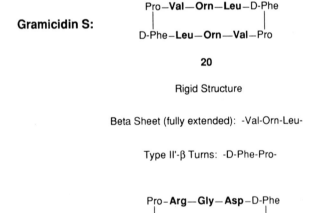

Gramicidin S:

Pro—Val—Orn—Leu—D-Phe
| |
D-Phe—Leu—Orn—Val—Pro

20

Rigid Structure

Beta Sheet (fully extended): -Val-Orn-Leu-

Type II'-β Turns: -D-Phe-Pro-

Pro-**Arg**—**Gly**—**Asp**–D-Phe
| |
D-Phe-**Asp**—**Gly**—**Arg**–Pro

21

Figure 8. Design of an RGD analogue based on gramicidin S. Limited [1]H-nmr study of this compound confirms similar conformation, i.e., Beta sheet (**fully extended**): -Arg-Gly-Asp-

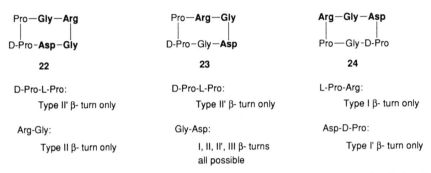

Figure 9. Three cyclic hexapeptide RGD analogues. The resulting conformations are also displayed (Peishoff *et al.*, 1992).

Employing the nOe-constrained distance geometry search procedures used for the cyclic disulfides, the solution conformations of all three compounds were examined (Fig. 9) (Bean *et al.*, 1992). As had been anticipated, the D-Pro-Pro sequence was found exclusively in a type II' β turn in both peptides. In **23**, the data for the Gly-Asp of RGD were consistent with any of four types of β turns. In **22** the Arg-Gly of RGD occurred in a type II β turn. The D-Pro-Gly-Pro peptide **24** adopted a conformation in which Pro-Arg occupied a type I β turn and Asp-D-Pro occupied a type I' β turn. In other words, **24** also adopted a turn–extended-turn conformation.

Cyclic pentapeptides tend to display conformations containing a β turn and an opposing γ turn (Fig. 10). All residues in the cyclic peptide may adopt any of the five positions in this conformation. Thus, five conformations could be present in solution. Kessler and colleagues have shown that incorporating a D-amino acid residue fixes that residue into the $i+1$ position of the β turn (Aumailley *et al.*, 1991). They recently showed that the analogue of cyclo-RGDV bearing a D-Asp presents a solution conformation in which the D-Asp and Val occupy the $i+1$ and $i+2$ positions of the type II' β turn and Gly is in a γ turn (Aumailley *et al.*, 1991). The analogue bearing D-Val presents a "frame-shifted" conformation in which D-Val-Arg now occupies the $i+1$ and $i+2$ positions of the type II' β turn.

Table VII compiles the biological data for the D-Phe-Val analogue **26** and various cyclic peptides that have been discussed. The table is constructed to allow easy comparison of the types of RGD conformations displayed by these peptides, with their resultant activities. The D-Pro-Gly-Pro peptide **24,** with the turn–extended-turn RGD conformation that differs from SK&F 107260, displayed reduced affinity and potency relative to SK&F 106760, **9.** The gramicidin analogue **21** with the fully extended RGD conformation displayed even lower

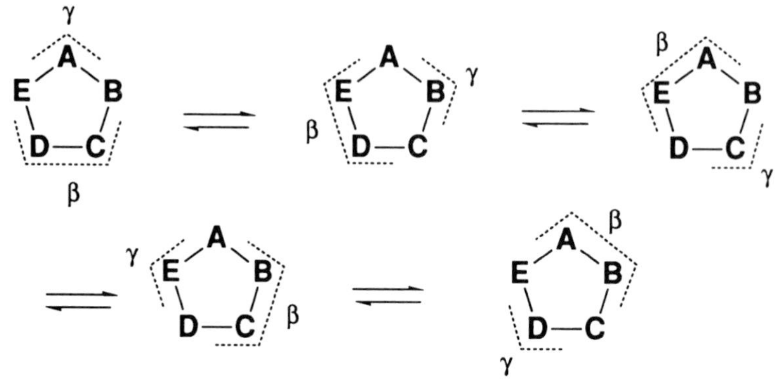

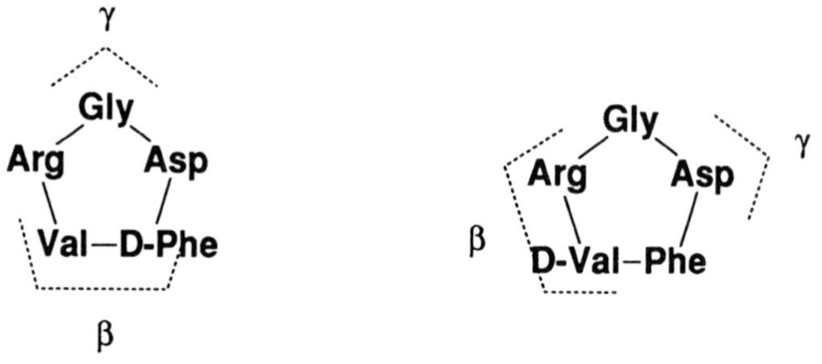

26 **27**

Figure 10. Two cyclic pentapeptide RGD analogues (Aumailley *et al.*, 1991).

activity, as did the D-Pro-Pro analogue **22** with the Arg-Gly β turn. The D-Pro-Pro analogue **23** with the Gly-Asp β turn, on the other hand, showed moderate affinity and antiaggregatory activity. The activity of this latter analogue is comparable to a cyclic decapeptide **25**, which is described as also containing a Gly-Asp β turn (Siahaan *et al.*, 1990). The cyclic pentapeptide **26**, which displays a γ turn about Gly, displays low affinity for GPIIb/IIIa.

At face value, the data in the table may suggest that GPIIb/IIIa appears to recognize only peptides with specific RGD conformations. The receptor, however, must recognize more than just the RGD portion of these peptides, since it can readily distinguish between three different peptides **9**, **14**, and **24**, all of which display turn–extended-turn conformations. Thus, a number of structural and conformational differences between the peptides in Table VII could contribute to the differences in activities (Peishoff *et al.*, 1992). For instance, the low activity of peptide **24**, bearing the alternate turn–extended-turn RGD conforma-

Table VII. Activities of Conformationally constrained RGD Analogues[a]

No.	Structure	Probable conformation	Antiagg. dog PRP IC_{50} (μM)	[3H]-107260 human GPIIb/IIIa K_i (μM)
9	Ac-Cys-(NMe)**Arg-Gly-Asp**-Pen-NH$_2$ ⌊_____⌋	— R | G | — D	0.36	0.058
24	**Arg-Gly-Asp** | | Pro-Gly-D-Pro	— R | G | — D	76.6	>100
21	Pro-**Arg-Gly-Asp**-D-Phe | | D-Phe-**Asp-Gly-Arg**-Pro	| R | G | D |	>200	
22	Pro-Gly-**Arg** | | D-Pro-**Asp-Gly**	— R | β — D — G	>200	
23	Pro-**Arg-Gly** | | D-Pro-Gly-**Asp**	— R — G | β — G — D	5.26	1.9
25	H-Gly-Pen-Gly-**Arg-Gly** | | HO-Ala-Cys-Pro-Ser-**Asp**	— R — G | β — S — D	10.1	
26	**Arg-Gly-Asp** | | Val——D-Phe	╱ R ╲ G γ ╲ D ╱		41.8

[a] Antiagg, inhibition of platelet aggregation; dog PRP, dog platelet-rich plasma. Inhibition of [3H]-SK&F 107260 binding to purified human GPIIb/IIIa (Stadel *et al.*, 1992; Wong *et al.*, 1992). Probable conformations about RGD as discussed in the text.

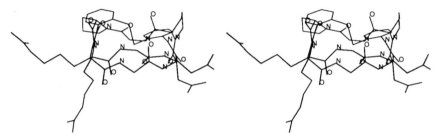

Figure 11. Comparison of SK&F 107260 with the alternate turn-extended turn peptide **24**. Models of the probable conformations of both compounds are superimposed and displayed in stereo.

tion, may be explained in part from a comparison of the conformation of SK&F 107260, **14** with that of peptide **24** (Fig. 11). It is apparent that in **14** the Arg and Asp side chains adopt pseudo-*equatorial* orientations relative to the peptide macrocycle. The Arg and Asp side chains in the homodetic peptide **24**, however, adopt pseudo-*axial* orientations relative to the peptide macrocycle, which shortens the distance between the guanidine and carboxylate groups and potentially orients these important groups into different regions than on **14**. These and other structural features could be contributing to the differences in activities between peptide **24** and SK&F 107260, **14** (Peishoff *et al.*, 1992).

A comparison of the turn–extended-turn conformation, found in SK&F 106760, **9,** with the Gly-Asp β-turn conformation, exemplified by analogue **23** (Fig. 12), suggests that the turn around the central glycine in **23** may force Gly to extend out in a manner that interacts negatively with the receptor. The guanidine-to-carboxylate distance in **23** is also shorter than in SK&F 106760. Either of

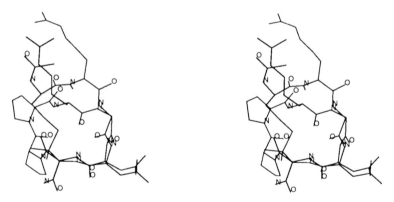

Figure 12. Comparison of SK&F 106760 with the Gly-Asp β-turn peptide **22**. Models of the probable conformations of both compounds are superimposed and displayed in stereo.

these factors or other structural features may contribute to the reduced activity of **22** relative to SK&F 106760 (Peishoff *et al.*, 1992). Our data suggest collectively that two rather different types of RGD conformations may give rise to high-affinity interactions with GPIIb/IIIa in certain structural contexts. With the data in hand, it would appear that RGD analogues with the turn–extended conformation display higher affinity, at least within the context of cyclic disulfide penta-peptides.

2.2. Support for an RGD Peptide Pharmacophore Model from a Peptide Mimetic for GPIIb/IIIa

With SK&F 106760 and SK&F 107260, we have two peptides with high affinity that both display nearly identical, highly defined conformations. These are compelling reasons to explore these peptides as potentially defining a pharmacophore model for GPIIb/IIIa. We sought an alternate approach, however, which would avoid the problem of receptor interaction with the portion of the structure that brings the ends of the RGD sequence into a cyclic structure.

In theory, the conformation of the RGD sequence can be constrained through either *external constraints* or *internal constraints* (Fig. 13). The afore-mentioned cyclic peptides succeeded in applying external constraints upon the RGD sequence into a variety of different conformations. As noted above, how-ever, an external constraint may not only influence receptor interaction with RGD by affecting the RGD conformation, but may also undergo receptor interac-tion in either a positive or a negative manner. An alternate approach to retaining the RGD conformation of SK&F 107260 would be to use internal constraints, or functional groups within the RGD sequence itself that serve to constrain confor-

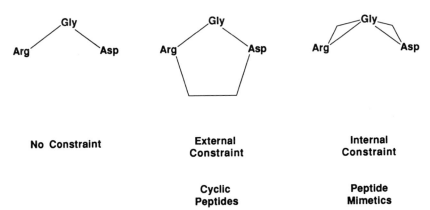

No Constraint **External Constraint** **Internal Constraint**

Cyclic Peptides **Peptide Mimetics**

Figure 13. Hypothetical representations of RGD peptides bearing no constraints, external con-straints, and internal constraints.

mation. As noted earlier, such modified portions of peptide sequence that mimic a particular type of secondary structure are also called peptide mimetics. With this alternate approach, the ambiguity of receptor interactions with the external constraint can be avoided.

In order to investigate aspects of the turn–extended-turn conformation for RGD in greater detail, we have also investigated the incorporation of peptide mimetics into RGD peptides. We have evaluated these mimetic-containing peptides for affinity and potency to discern whether that aspect of conformation incorporated into the mimetic is important for high-affinity receptor interaction. With these semipeptide analogues, we hope to further refine the peptide pharmacophore to a point where nonpeptide ligands for GPIIb/IIIa can be designed.

As noted earlier, a C_7 or γ turn was observed about aspartic acid in the predominant conformation for SK&F 107260. This suggested the use of a C_7-turn mimetic that had been investigated in other peptide series (Huffman et al., 1988, 1989). In this conformational mimetic the elements of the hydrogen bond were replaced with an ethylene bridge and the Gly-Asp amide was replaced with a trans olefin to give the seven-membered lactam structure shown in Fig. 14. The mimetic was incorporated into linear analogues of SK&F 107260, ones in which the conformation was not constrained by a disulfide bridge, to determine whether the turn mimetic was a sufficient conformational constraint to retain the affinity and potency of SK&F 107260 (Callahan et al., 1992). Figure 15 displays an overlay of the basic seven-membered lactam mimetic derived from previous structural studies (Callahan et al., 1992) onto the γ turn around aspartic acid in SK&F 107260. It is evident that the seven-membered lactam ring follows the backbone conformation of SK&F 107260 remarkably well. Two linear peptides bearing the mimetic were prepared (Callahan et al., 1992) with either an N-ter-

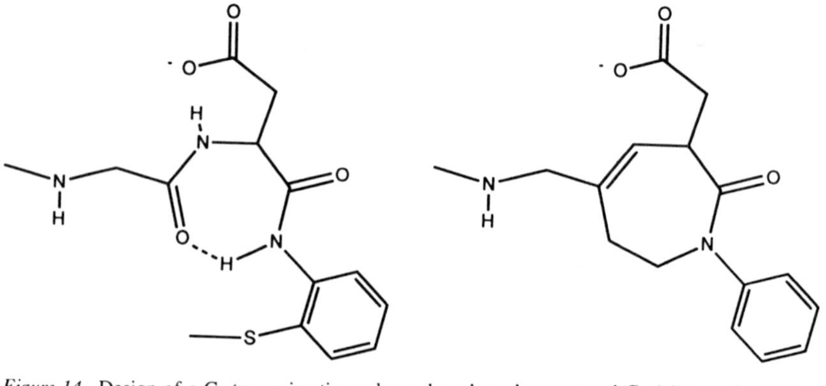

Figure 14. Design of a C_7-turn mimetic analogue based on the proposed C_7 (γ) turn about Asp in SK&F 107260 (Callahan et al., 1992).

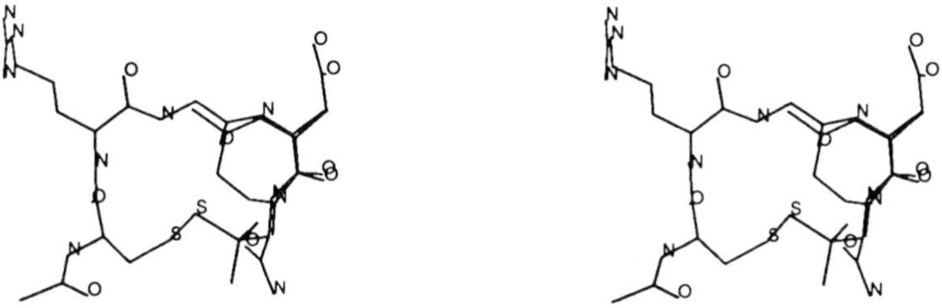

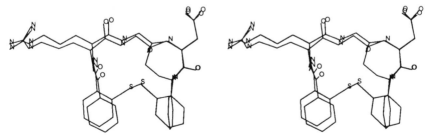

Figure 15. Comparison of the lactam C_7-turn mimetic structure with the γ turn about Asp in the probable conformation for SK&F 106760 (Callahan *et al.*, 1992).

minal benzoyl-(Nme)arginine **27** or acetyl-(Nme)arginine **28**. Figure 16, which displays the *N*-benzoyl-turn mimetic analogue **27** aligned with SK&F 107260, reveals a high degree of similarity between the two compounds.

Peptides **27** and **28** were evaluated for GPIIb/IIIa affinity and antiaggregatory activity (Table VIII) (Callahan *et al.*, 1992). Both analogues retained a level of affinity and potency that is quite comparable to the fibrinogen receptor antagonist SK&F 106760. It is important to note that these activities are high for linear RGD peptides. In the synthesis of these mimetics, the chirality of the acetic acid side chain was not controlled; both isomers were thus obtained. The diastereomers of **28** displayed nearly identical activity with the racemate **28**. SK&F 107260, however, displayed greater potency or affinity than these linear semipeptides. At least two possible explanations may be considered for these differences in activity. It is possible that the mimetic contains certain structural elements that are somewhat less favorable to the receptor. In addition, the added constraint of the disulfide bridge may be necessary for high affinity. Nonetheless, from the activities of the peptides bearing the C_7-turn mimetic, it appears that GPIIb/IIIa is capable of recognizing RGD analogues in which aspartic acid is constrained in a C_7 turn. This work, then, supports at least one portion of the

Figure 16. Comparison of SK&F 107260 with the liinear benzoyl-(NMe)Arg-C_7-turn mimetic analogue **27**.

Table VIII. Activities of C_7-Turn Mimetic Analogues[a]

No.	Structure	Antiagg. dog PRP IC_{50} (μM)	[³H]-107260 human GPIIb/IIIa K_i (μM)
9	Ac-Cys-(NMe)Arg-Gly-Asp-Pen·NH₂	0.36	0.175 (T) 0.058
14	[structure: CO-(NMe)Arg-Gly-Asp-NH with two S-linked rings]	0.090	0.004 (T) 0.0024 (I)
27	[structure with CO₂H, CO-(NMe)Arg-N]	1.11	0.106 (T)
28	[structure: CH₃-CO-(NMe)Arg-N with CO₂H]	0.72	0.271 (T)

[a] Antiagg, inhibition of platelet aggregation; dog PRP, dog platelet-rich plasma. (I) = inhibition of I-Fg binding to purified human GPIIb/IIIa. (T) = inhibition of [³H]-SK&F 107260 binding to purified human GPIIb/IIIa (Stadel et al., 1992; Wong et al., 1992).

pharmacophore model. We are presently exploring other semipeptides containing secondary structure mimetics in the Arg-Gly-Asp sequence in order to further define our peptide pharmacophore model and validate it as a tool for the design of novel nonpeptide ligands for GPIIb/IIIa.

3. Semipeptide and Nonpeptide Mimetics of RGD as Candidates for Oral Antithrombotic Agents

SK&F 106760 is a potent antithrombotic agent that displays measurable antiaggregatory activity *ex vivo* on intraduodenal administration. The intraduodenal bioavailability of SK&F 106760 has been determined to be approximately 4% in the dog (Nichols *et al.*, this volume). This level of bioavailability is remarkable for a peptide, but it was judged insufficient for the development of SK&F 106760 as an oral antithrombotic agent. Pharmacokinetic analysis of plasma levels after intravenous and intraduodenal administration suggests that the low level of oral activity is the result of a lack of gastrointestinal absorption.

If one assumes that these peptides cross the gut by passive diffusion, the requirements for both hydrophilic cationic and anionic groups in high-affinity GPIIb/IIIa antagonists suggest that compounds must be sought in which the relatively hydrophilic peptide template is replaced with a more hydrophobic template. It is highly probable, therefore, that fibrinogen receptor antagonists with adequate oral bioavailability may be compounds that bear little or no apparent structural resemblance to a peptide.

A number of groups are investigating semipeptide and nonpeptide fibrinogen receptor antagonists as well (Table IX). Searle has described a guanidino-octanoyl derivative of aspartame **29** (Tjoeng *et al.*, 1992), and more recently, an amidinophenyl analogue, **30** (Garland *et al.*, 1992) reported to display enhanced activity. Merck has recently described tyrosine derivatives, e.g., **31** and **32,** with fibrinogen receptor antagonist activity, discovered initially through data-base screening (Hartman *et al.*, 1992). Roche has also described a series of nonpeptides, e.g., **33** and **34,** which were discovered by data-base screening (Alig *et al.*, 1989, 1990). These potent compounds contain an amidinophenyl group that was first employed as a mimetic of arginine in thrombin inhibitors (Voigt and Wagner, 1986). None of these compounds has been reported to display oral activity.

In relation to SK&F 107260, the compounds developed by screening in an Fg-GPIIb/IIIa binding assay could potentially bind to GPIIb/IIIa in some other manner. Hartman has proposed that the butylsulfonamide group in the Merck tyrosine analogue binds to a novel receptor site or exo-site (Hartman *et al.*, 1992). The fact that these compounds retain amine and carboxylic acid groups separated by the same distance as the guanidine and carboxylate groups in SK&F 107260 supports a mimetic binding mode. As evidence that the solution conformation of SK&F 107260 may contain the elements of a suitable pharmacophore model, these compounds can be aligned with a model of SK&F 107260 to provide a good overlay between the corresponding cationic and anionic groups.

4. Conclusion

In less than a decade since the discovery of the minimally inhibitory RGD-containing peptide fragments to the integrins (Pierschbacher and Ruoslahti, 1984a,b), potent antithrombotic agents have been discovered that are demonstrably superior to aspirin. The research effort is rapidly closing in on the discovery of fibrinogen receptor antagonists with oral antithrombotic activity. It is highly probable that fibrinogen receptor antagonists with adequate oral bioavailability will advance toward the clinic before a decade of research has passed.

Finally, it should be noted that fibrinogen receptor antagonists may be the first class of antiadhesion drugs to enter the clinic, but exciting advances in other

Table IX. Activities of Semipeptide and Nonpeptide Fibrinogen Receptor Antagonists

| | | | IC$_5$0 (nM), inhibition of platelet aggregation | |
			hWP[a]	d,hPRP[b]
Searle	29	SC 4992		2000 (h)
	30			53 (d)
Merck	31		2600	
	32	L 700462	11	~230 (d)[c]
SmithKline Beecham	28	CH$_3$·CO-(NMe)Arg-N		720 (d)[c]
Roche	33			70 (h)[c] 1.05 (d)[c]
	34			30 (h)[c] 70 (d)[c]

[a] hWP, human washed platelets.
[b] d,hPRP, dog or human platelet-rich plasma.
[c] Data obtained at SmithKline Beecham.

areas described in this book suggest that RGD-related compounds could be developed as antimetastatic agents (Cheresh *et al.*, this volume), as modulators of osteoporosis (Bertolini *et al.*, this volume), and in many other areas.

ACKNOWLEDGMENTS. It is important to emphasize that this chapter reviews the collective work of many talented colleagues at SmithKline Beecham, who have contributed to the advance of fibrinogen receptor antagonists. As noted in the text, this work has begun to appear in print. The relationships between biological activity and structure that are described in this chapter could not have been delineated without the expert structural verifications by amino acid analysis performed by Robert Sanchez and by mass spectrometry performed by Mark Bean, Steven Carr, and Walter Johnson. The authors also thank Sidney Hecht, Brian Metcalf, and Robert Ruffolo, Jr. for their support in these endeavors.

References

Ali, F. E., and Samanen, J. M., 1992, Synthesis of protected cyclic homodetic peptides by solid phase peptide synthesis, in: *Innovations and Perspectives in Solid Phase Peptide Synthesis and Related Technologies* (R. Epton, ed.), SPPC Ltd., Birmingham, pp. 333–335.

Ali, F. E., Calvo, R., Romoff, T., Samanen, J., Nichols, A., and Storer, B., 1990, Structure–activity studies toward the improvement of antiaggregatory activity of Arg-Gly-Asp-Ser (RGDS), in: *Peptides, Chemistry and Biology, Proc. 11th Amer. Peptide Symp.* (J. E. Rivier and G. R. Marshall, eds.), ESCOM, Leiden, pp. 94–96.

Ali, F. E., Samanen, J. M., Calvo, R., Romoff, T., Yellin, T., Vasko, J., Powers, D., Stadel, J., Bennett, D., Berry, D., and Nichols, A., 1992, Potent fibrinogen receptor antagonists bearing conformational constraints, in: *Peptides, Chemistry and Biology: Proceedings of the Twelfth American Peptide Symposium* (J. A. Smith and J. E. Rivier, eds.), ESCOM, Leiden, pp. 761–762.

Alig, L., Edenhofer, A., Muller, M., Trzeciak, A., and Weller, T., 1989, European Patent Application, EP 0,372,468.

Alig, L., Edenhofer, A., Muller, M., Trzeciak, A., and Weller, T., 1990, European Patent Application, EP 0,381,033.

Anderson, H. V., Revana, M., Rosales, O., Brannigan, L., Stuart, Y., Weisman, H., and Willerson, J., 1992, Intravenous administration of monoclonal antibody to the platelet GPIIb/IIIa receptor to treat abrupt closure during coronary angioplasty, *Am. J. Cardiol.* **69:**1373–1376.

Andrieux, A., Hudry-Clergeon, G., Ryckewaert, J.-J., Chapel, A., Ginsberg, M. H., Plow, E. F., and Marguerie, G., 1989, Amino acid sequences in fibrinogen mediating its interaction with its platelet receptor, GPIIbIIIa, *J. Biol. Chem.* **264:**9258–9265.

Aumailley, M., Gurrath, M., Muller, G., Calvete, J., Timpl, R., and Kessler, H., 1991, Arg-Gly-Asp constrained within cyclic pentapeptides, strong and selective inhibitors of cell adhesion to vitronectin and laminin fragment P1, *FEBS Lett.* **291:**50–54.

Barker, P. L., Bullens, S., Bunting, S., Burdick, D. J., Chan, K. S., Deisher, T., Eigenbrot, C., Gadek, T. R., Gantzos, R., Lipari, M. T., Muir, C. D., Napier, M. A., Pitti, R. M., Padua, A., Quan, C., Stanley, M., Struble, M., Tom, J. Y. K., and Burnier, J. P., 1992, Cyclic RGD peptide analogues as antiplatelet antithrombotics, *J. Med. Chem.* **35:**2040–2048.

Bean, J. W., Kopple, K. D., and Peishoff, C. E., 1992, Conformational analysis of cyclic hexapeptides containing the D-Pro-L-Pro sequence to fix β-turn positions, *J. Am. Chem. Soc.* **114:**5328–5334.

Bogusky, M. J., Naylor, A. M., Pitzenberger, S. M., Nutt, R. F., Brady, S. F., Colton, C. D., Sisko, J. T., Anderson, P. S., and Veber, D. F., 1992, NMR and molecular modeling characterization of RGD containing peptides, *Int. J. Pept. Protein Res.* **39:**63–76.

Boyd, S. A., Fung, A. K. L., Baker, W. R., Mantei, R. A., Armiger, Y. L., Stein, H. H., Cohen, J., Egan, D. A., Barlow, J. L., Klinghofer, V., Verburg, K. M., Martin, D. L., Young, G. A., Polakowski, J. S., Hoffman, D. J., Garren, K. W., Perun, T. J., and Kleinert, H. D., 1992, C-Terminal modifications of nonpeptide renin inhibitors: Improved oral bioavailability via modification of physicochemical properties, *J. Med. Chem.* **35:**1735–1746.

Callahan, J. F., Bean, J. W., Burgess, J. L., Eggleston, D. S., Hwang, S. M., Kopple, K., Koster, P. F., Nichols, A., Peishoff, C., Samanen, J., Vasko, J. A., Wong, A., and Huffman, W., 1992, Design and synthesis of a C7 mimetic for the predicted γ-turn conformation found in several constrained RGD antagonists. *J. Med. Chem.* **35:**3970–3972.

Coller, B. S., 1985, A new murine monoclonal antibody reports an activation-dependent change in the conformation and/or microenvironment of the platelet GPIIb/IIIa complex, *J. Clin. Invest.* **76:**101–108.

Coller, B. S., Scudder, L. E., Beer, J., Gold, H. K., Folts, J. D., Cavagnaro, J., Jordan, R., Wagner, C., Iuliucci, J., Knight, D., Ghrayeb, J., Smith, C., Weisman, H. F., and Berger, H., 1991, Monoclonal antibodies to platelet glycoprotein IIb/IIIa as antithrombotic agents, *Ann. N.Y. Acad. Sci.* **614:**204–213.

Collins, J. L., Dambek, P. J., Goldstein, S. W., and Faraci, W. S., 1992, CP-99,711: A nonpeptide glucagon receptor antagonist, *Bioorg. Med. Chem. Lett.* **2:**915–918.

Dennis, M. S., Henzel, W. J., Pitti, R. M., Lipari, M. T., Napier, M. A., Deisher, T. A., Bunting, S., and Lazarus, R., 1989, Platelet glycoprotein IIb-IIIa protein antagonists from snake venoms: Evidence for a family of platelet-aggregation inhibitors, *Proc. Natl. Acad. Sci. USA* **87:**2471–2475.

Eggleston, D. S., and Feldman, S. H., 1990, Structure of the fibrinogen binding sequence: Arginylglycylaspartic acid (RGD), *Int. J. Pept. Protein Res.* **36:**161–166.

Folts, J. D., Crowell, E. D., and Rowe, G. G., 1976, Platelet aggregation in partially obstructed vessels and its elimination with aspirin, *Circulation* **54:**365–370.

Freidinger, 1989, Non-peptide ligands for peptide receptors, *Trends Pharm. Sci.* **10:**270–274.

Gan, Z.-R., Gould, R. J., Jacobs, J. W., Friedman, P. A., and Polokoff, M. A., 1988, Echistatin: A potent platelet aggregation inhibitor from the venom of the viper, Echis carinatus, *J. Biol. Chem.* **63:**19827–19832.

Garland, R. B., Miyano, M., and Zablocki, J. A., 1992, European Patent Application, EP 0,502,536.

Garsky, V. M., Lumma, P. K., Freidinger, R. M., Pitzenberger, S. M., Randall, W. C., Veber, D. F., Gould, R. J., and Friedman, P. A., 1989, Chemical synthesis of echistatin, a potent inhibitor of platelet aggregation from Echis carinatus: Synthesis and biological activity of selected analogs, *Proc. Natl. Acad. Sci. USA* **86:**4022–4026.

Gartner, T. K., and Bennett, J. S., 1985, the tetrapeptide analogue of the cell attachment site of fibronectin inhibits platelet aggregation and fibrinogen binding to activated platelets, *J. Biol. Chem.* **260:**11891–11894.

Hanson, S. R., Pareti, F. I., Ruggeri, Z. M., Marzec, U. M., Kunicki, T. J., Montgomery, R. R., Zimmerman, T. S., and Harker, L. A., 1988, Effects of monoclonal antibodies against the platelet glycoprotein IIb/IIIa complex on thrombosis and hemostasis in the baboon, *J. Clin. Invest.* **81:**149–158.

Hartman, G. D., Egbertson, M. S., Halszenko, W., Laswell, W. L., Duggan, M. E., Smith, R. L., Naylor, A. M., Manno, P. D., Lynch, R. J., Zhang, G., Chang, C. T. C., and Gould, R. J.,

1992, Non-peptide fibrinogen receptor antagonists. 1. Discovery and design of exosite inhibitors, *J. Med. Chem.* **35**:4640–4642.

Huang, T.-F., Holt, J. C., Lukasiewicz, H., and Niewiarowski, S., 1987, Trigramin: A low molecular weight peptide inhibiting fibrinogen interaction with platelet receptors expressed on glycoprotein IIb/IIIa complex, *J. Biol. Chem.* **262**:16157–16163.

Huffman, W. F., Callahan, J. F., Eggleston, D. S., Newlander, K. A., Takata, D. T., Codd, E. E., Walker, R. F., Schiller, P. W., Lemieux, C., Wire, W. S., and Burks, T. F., 1988, Reverse turn mimics, in: *Peptides, Chemistry and Biology, Proc. 10th Amer. Peptide Symp.* (G. R. Marshall, ed.), ESCOM, Leiden, pp. 105–108.

Huffman, W. F., Callahan, J. F., Codd, E. E., Eggleston, D. S., Lemieux, C., Newlander, K. A., Schiller, P. W., Takata, D. T., and Walker R. F., 1989, Mimics of secondary structural elements of peptides and proteins, in: *Synthetic Peptides: Approaches to Biological Problems, UCLA Colloquium on Molecular and Cellular Biology* (J. P. Tam and E. T. Kaiser, eds.), Liss, New York, pp. 257–266.

Kleiman, N. S., Ohman, E. M., Keriakas, D. J., Ellis S. G., Weisman, H. F., and Topol, E. J., 1991, Profound platelet inactivation with 7E3 shortly after thrombolytic therapy for acute myocardial infarction: Preliminary results of the TAMI 8 trial, *Circulation Suppl. II* **84**:11–522.

Kloczewiak, M., Timmons, S., Lukas, T. J., and Hawiger, J., 1984, Recognition site for the platelet receptor is present on the 15-residue fragment of the γ chain of human fibrinogen and is not involved in the fibrin polymerization reaction, *Biochemistry* **23**:1767–1774.

Kloczewiak, M., Timmons, S., Bednarek, M. A., Sakon, M., and Hawiger, J., 1989, Platelet receptor recognition domain on the γ chain of human fibrinogen and its synthetic peptide analogues, *Biochemistry* **28**:2915–2919.

Kopple, K. D., Baures, P. W., Bean, J. W., D'Ambrosio, C. A., Hughes, J. L., Peishoff, C. E., and Eggleston, D. S., 1992, Conformations of Arg-Gly-Asp containing heterodetic cyclic peptides: Solution and crystal structures, *J. Am. Chem. Soc.* **114**:9615–9623.

Nichols, A. J., Vasko, J. A., Koster, P., Smith, J., Barone, F., Nelson, A. J. S. Strohsacker, M., Rhodes, G., Bennett, D., Berry, D., Romoff, T., Calvo, R., Ali, F., Sorenson, E., and Samanen, J., 1990, SK&F 106760, a novel GPIIb/IIIa antagonist: Antithrombotic activity and potentiation of streptokinase-mediated thrombolysis, *Eur. J. Pharmacol.* **183**:2019.

Nichols, A. J., Ruffolo, R. R., Jr., Huffman, W. F., Poste, G., and Samanen, J., 1992, Development of GPIIb/IIIa antagonists as antithrombotic drugs, *Trends Pharm. Sci.* **13**:413–417.

Nutt, R. F., Brady, S. F., Colton, C. D., Sisko, J. T., Ciccarone, T. M., Levy, M. R., Duggan, M. E., Imagire, I. S., Gould, R. J., Anderson, P. S., and Veber, D. F., 1992, Development of novel, highly selective fibrinogen receptor antagonists as potentially useful antithrombotic agents, in: *Peptides, Chemistry and Biology, Proc. 12th Amer. Peptide Symp.* (J. A. Smith and J. E. Rivier, eds.), ESCOM, Leiden, pp. 914–916.

Peishoff, C. E., Ali F. E., Bean, J. W., Calvo, R., D'Ambrosio, C. A., Eggleston, D. S., Kline, T. P., Koster, P. F., Nichols, A., Powers, D., Romoff, T., Samanen, J. M., Stadel, J., Vasko, J. A., Wong, A., and Kopple, K. D., 1992, Investigation of conformational specificity at GPIIb/IIIa: Evaluation of conformationally constrained RGD peptides, *J. Med. Chem.* **35**:3962–3969.

Pierschbacher, M. D., and Ruoslahti, E., 1984a, Cell attachment activities of fibronectin can be duplicated by small synthetic fragments of the molecule, *Nature* **309**:30–33.

Pierschbacher, M. D., and Ruoslahti, E., 1984b, Variants of the cell recognition site of fibronectin that retain attachment-promoting activity, *Proc. Natl. Acad. Sci. USA* **81**:5985–5988.

Plow, E. F., Ginsberg, M. H., and Marguerie, G. A., 1986, Expression and function of adhesive proteins on the platelet surface, in: *Biochemistry of Platelets*, (D. R. Phillips and M. R. Shuman, eds.), Academic Press, New York, pp. 225–256.

Plow, E. F., Pierschbacher, M. D., Ruoslahti, E., Marguerie, G., and Ginsberg, M. H., 1987, Arginyl-glycyl-aspartic acid sequences and fibrinogen binding to platelets, *Blood* **70:**110–115.

Samanen, J., Ali, F. E., Romoff, T., Calvo, R. Sorenson, E., Vasko, J., Berry, D., Bennett, D., Strohsacker, M., Powers, D., Stadel, J. S., and Nichols, A., 1991, Development of a small RGD-peptide fibrinogen receptor antagonist with potent antiaggregatory activity in vitro, *J. Med. Chem.* **34:**3114–3125.

Scarborough, R. M., Rose, J. W., Hsu, M. A., Phillips, D. R., Fried, V. A., Campbell, A. M., Nunnizzi, L., and Charo, I. F., 1991, Barbourin, a GPIIb-IIIa-specific integrin antagonist from the venom of Sistrurus M. Barbouri, *J. Biol. Chem.* **266:**9359–9362.

Shattil, S. J., Hoxie, J. A., Cunningham, M., and Brass, L. F., 1985, Changes in the platelet membrane glycoprotein IIb-IIIa complex during platelet activation, *J. Biol. Chem.* **260:**11107–11114.

Shebuski, R. J., Berry, D. E., Bennett, D. B., Romoff, T., Storer, B. L., Ali, F. E., and Samanen, J., 1989a, Demonstration of Ac-Arg-Gly-Asp-Ser-NH$_2$ as an antiaggregatory agent in the dog by intracoronary administration, *Thromb. Haemost.* **61:**183–188.

Shebuski, R. J., Ramjit, D. R., Bencen, G. H., and Polokoff, M. A., 1989b, Characterization and platelet inhibitory activity of bitistatin, a potent arginine-glycine-aspartic acid- containing peptide from the venom of the viper Bitis arietans, *J. Biol. Chem.* **264:**21550–21556.

Siahaan, T., Lark, L. R., Pierschbacher, M., Ruoslahti, E., and Gierasch, L. M., 1990, A conformationally constrained 'RGD' analog specific for the vitronectin receptor: A model for receptor binding, in: *Peptides: Chemistry, Structure and Biology, Proceedings of the 11th American Peptide Symposium* (J. E. Rivier and G. R. Marshall, eds.), ESCOM, Leiden, pp. 699–701.

Stadel, J. M., Powers, D. A., Bennett, D., Nichols, A., Heys, R., Ali, F., and Samanen, J., 1992, [^{3}H]SK&F 107260, a novel radioligand for characterizing the fibrinogen receptor, $\alpha_{IIb}\beta_3$, of human platelets, *J. Cell Biochem.* **16F:**153.

Taub, R., Gould, R. J., Garsky, V. M., Ciccarone, T. M., Hoxie, J., Friedman, P. A., and Shattil, S. J., 1989, A monoclonal antibody against the platelet fibrinogen receptor contains a sequence that mimics a receptor recognition domain in fibrinogen, *J. Biol. Chem.* **264:**259–265.

Tjoeng, F. S., Fok, K. F., Zupec, M. E., Garland, R. B., Miyano, M., Panzer-Knodle, S., King, L. W., Taite, B. B., Nicholson, N. S., Feigen, L. P., and Adams, S. P., 1992, Peptide mimetics of the RGD sequence, in: *Peptides, Chemistry and Biology, Proceedings of the 12th American Peptide Symposium* (J. A. Smith and J. E. Rivier, eds.), ESCOM, Leiden, pp. 752–754.

Voigt, B., and Wagner, G., 1986, Synthesis of Nα-(arylsulfonyl-L-prolyl)- and Nα-(benzyloxycarbonyl-L-prolyl)-D,L-4-amidinophenylalanine amides as inhibitors of thrombin, *Pharmazie* **41:**233–235.

Willette, R. N., Sauermelch, C. F., Rycyna, R., Sarkar, S., Feuerstein, G. Z., Nichols, A. J., and Ohlstein, E. H., 1992, Antithrombotic effects of a platelet fibrinogen receptor antagonist in a canine model of carotid artery thrombosis, *Stroke* **23:**703–711.

Wong, A., Hwang, F. M., Johanson, K., Stadel, J., Powers, D. A., Bennett, D., Heys, R., Ali, F., Bondinell, W., Ku, T., and Samanen, J., 1992, Cation-dependent binding of [^{3}H]-SK&F 107260, a cyclic Arg-Gly-Asp (RGD) peptide to glycoprotein IIIb/IIIa: Competitive inhibition by fibrinogen, Fg γ-dodecapeptide and cyclic RGD peptides, *J. Cell Biochem.* **16F:**181.

16

Derivation of Therapeutically Active Humanized and Veneered Anti-CD18 Antibodies

GEORGE E. MARK III, MELVIN SILBERKLANG, D. EUAN MACINTYRE, IRWIN I. SINGER, and EDUARDO A. PADLAN

1. Introduction

The identification and production of murine monoclonal antibodies (Mabs) has led to numerous therapeutic applications of these exquisitely specific molecules to human disease. The technologies of molecular biology have further expanded the utility of many antibodies by allowing the creation of class-switched molecules, whose functionality has been improved by the acquisition or loss of complement fixation. The size of the bioactive molecule may also be reduced so as to increase the tissue target availability of the antibody, either by changing the class from an IgM to an IgG, removing most of the heavy chain constant region in the creation of a F(ab')$_2$, or severely truncating both heavy and light chain constant regions with the formation of an Fv antibody. Common to all of these potentially therapeutic forms of antibody are the requisite CDRs

GEORGE E. MARK III, and MELVIN SILBERKLANG • Department of Cellular and Molecular Biology, Merck Research Laboratories, Rahway, New Jersey 07065. *D. EUAN MACINTYRE* • Department of Cellular and Molecular Immunology, Merck Research Laboratories, Rahway, New Jersey 07065. *IRWIN I. SINGER* • Department of Biochemical and Molecular Pathology, Merck Research Laboratories, Rahway, New Jersey 07065. *EDUARDO A. PADLAN* • National Institute of Diabetes and Digestive and Kidney Diseases, National Institutes of Health, Bethesda, Maryland 20892.

Cellular Adhesion: Molecular Definition to Therapeutic Potential, edited by Brian W. Metcalf, Barbara J. Dalton, and George Poste. Plenum Press, New York, 1994.

(complementarity-determining regions), which guide the molecule to its ligand, and the framework residues (FRs), which support the latter structures and dictate the disposition of the CDRs relative to one another.

Crystallographic analyses of numerous antibody structures reveal that the combining site is composed almost entirely of the CDR residues arranged in a limited number of loop motifs (Amit et al., 1986; Bruccoleri et al., 1988; Chothia et al., 1989). The need for the CDRs to form these structures, combined with the appreciated hypervariability of their primary sequence, leads to a great diversity in the antigen combining site, but it is a site with a finite number of possibilities. Thus, hypermutability and a limited primary sequence repertoire for each CDR would suggest that the CDRs derived for a given antigen from one species of animal would be the same when derived from another species. Hence, the CDRs should be poorly immunogenic, if at all, when presented to a recipient organism in a nonforeign context.

MAb-producing hybridomas have been most readily obtained from immunized rodents. Attempts to develop similar reagents from human sources have been impaired by the current inability to maintain long-term cultures of cells that produce sufficient quantities of antibody. There are additional problems, from the regulatory standpoint, when cells of human origin are employed to produce agents for human use. These considerations have led to the widespread use of rodent MAbs for the imaging and treatment of malignancy (Hale et al., 1988; Seemann et al., 1990; Blend, 1991; Eary, 1991; Larson, 1991; Waldmann, 1991), as a prophylactic to guard against septic shock (Gorelick et al., 1990; Ziegler et al., 1991), for modification of graft rejection episodes, and to temper acute inflammatory reactions (Gorman et al., 1991; Gomoll et al., 1991). When completely or partially rodent (i.e., rodent/human chimeras) antibodies have been used for therapy, the recipients have often elicited an immune response directed toward the antibody (LoBuglio et al., 1989; Tjandra et al., 1990). These reactions have limited the duration and effectiveness of the therapy.

Various attempts have been made to minimize or eliminate the immunogenicity of nonhuman antibodies while preserving their antigen-binding properties. The first chimeric antibodies constructed contained the rodent variable regions, and their associated CDRs, fused to human constant domains (Morrison et al., 1984; Boulianne et al., 1984; Neuberger et al., 1985). These proved to be less immunogenic, but approximately half of the recipients mounted an immune response to the rodent variable region FRs. Further reduction of the "foreign" nature of the chimeric antibodies has been achieved by grafting the CDRs from the rodent monoclonal into a human supporting framework prior to fusing with an appropriate human constant domain (Jones et al., 1986; Verhoeyen et al., 1988; Reichmann et al., 1988; Queen et al., 1989; Tempest et al., 1991; Co et al., 1991; Gorman et al., 1991; Kettleborough et al., 1991; Güssow and Seemann, 1991). The procedures employed to accomplish CDR grafting often result in imperfectly humanized antibodies: either the resultant antibody has lost avidity or, in an attempt to retain its original avidity, a significant number of the murine FRs have replaced the corresponding residues of the human framework (Queen *et al.*, 1989; Tempest *et al.*, 1991; Co *et al.*, 1991; Gorman *et al.*, 1991; Kettleborough *et al.*, 1991). In the latter case, the immunogenicity of the modified humanized antibody is difficult to anticipate *a priori*.

The ligand binding characteristics of an antibody combining site are determined primarily by the structure and relative disposition of the CDRs, although some neighboring residues are also involved in antigen binding (reviewed by Davies *et al.*, 1990). Fine specificity can be preserved in a humanized antibody only if the CDR structures, their interaction with each other, and their interaction with supporting FRs are strictly maintained. Many of the key FRs may represent interior and interdomain contact residues.

Some procedures to achieve humanization have been biased by the following premise: either (1) all frameworks are equivalent (Reichmann *et al.*, 1988), or (2) human frameworks must be chosen from the sequences of naturally occurring heavy and light chain pairs, into which selected human framework residues must be replaced by their murine counterpart (Queen *et al.*, 1989).

The method described below is employed to create numerous humanized versions of a murine anti-CD18 MAb (IB4), demonstrating that the optimally humanized antibody need not contain reverted murine residues in order to maintain its specificity and avidity.

2. Recombinant Methodology of CDR Transfer

2.1. PCR-Mediated Construction of Humanized V-Regions

Several approaches to constructing CDR-grafted variable regions have been employed. The most common approach involves the preparation of a single-stranded DNA template from an M13 vector containing the human variable region. Three oligodeoxynucleotides are synthesized so that centrally located murine CDR residues are flanked by approximately 12 bases of DNA complementary to the adjacent sequences of the chosen human FR. These are combined and used in an oligodeoxynucleotide-based *in vitro* mutagenesis protocol with the single-stranded variable region DNA template and DNA polymerase (Taylor *et al.*, 1985; Nakamaye and Eckstein, 1986; Reichmann *et al.*, 1988; Verhoeyen *et al.*, 1988; Tempest *et al.*, 1991; Kettleborough *et al.*, 1991). Recombinant clones are selected and submitted to DNA sequencing to identify those clones in which all three CDRs have been mutagenized to the desired sequence. (This may represent as little as 5–20% of the recombinant clones.) We have used an alternative PCR-based method that provides for flexibility, ease, and rapidity, and results in high-frequency generation of the desired CDR-grafted variable region fragment (Daugherty *et al.,* 1991; DeMartino *et al.,* 1991).

We have used two general strategies involving PCR recombination to graft the murine CDRs into their appropriate human frameworks. When the human framework is available as a cDNA sequence, we synthesize short oligodeoxynucleotide primers containing terminal complementarity and all or part of the murine CDR sequence. Following the annealing of each primer pair to the human template V-region cDNA and subsequent PCR amplification, the resultant fragments are themselves combined by PCR to generate the humanized CDR-grafted variable region (Fig. 1). When the human framework template is not available, we synthesize long oligodeoxynucleotides (ca. 80–100 bases in length) of alternating polarity to contain both the human FR and the CDR sequences, as well as terminal regions of complementarity (Fig. 2). PCR amplification of the combined oligodeoxynucleotides with two short terminal amplifying primers results in a DNA fragment encoding the murine CDR-grafted human V-region framework. In either case, the CDR-grafted FRs are subsequently combined with additional PCR-generated fragments representing, in part, the immunoglobulin signal peptide and a portion of the intron 3′ of the human heavy or light chain J-regions. When the total number of PCR amplification cycles is kept below 45, approximately 90% of the V-region clones are found to be error-free. Finally, these molecules are cloned into expression vectors containing an insert encoding either the human light chain constant region or the human heavy chain constant region.

Selection of Human Frameworks

In order to identify human framework sequences that would be compatible with the CDRs of murine IB4 MAb (mIB4), we identified human frameworks with a high degree of sequence similarity to the mIB4. Sequence similarity was measured using identical residues as well as evolutionarily conservative amino acid substitutions (Schwartz and Dayhoff, 1979; Risler *et al.,* 1988). Similarity searches (Devereux *et al.,* 1984) were performed using the mIB4 framework sequence from which the CDR sequences had been removed. This sequence was used to query a data base of human immunoglobulin sequences culled from multiple nucleic acid and protein sources (George *et al.,* 1986; Kabat *et al.,* 1987). Human variable region sequences with a high degree of identity to the murine sequences were then evaluated individually for their potential as humanizing framework sequences. We focused on the amino acid residues that were most likely to interact with other such residues. These potentially interactive residues ("packing" residues) contain side chains whose influences are completely or partially buried (Padlan, 1991). Two heavy and light chain frameworks were selected for CDR-grafting: "Gal" and "Jon" represented the murine heavy

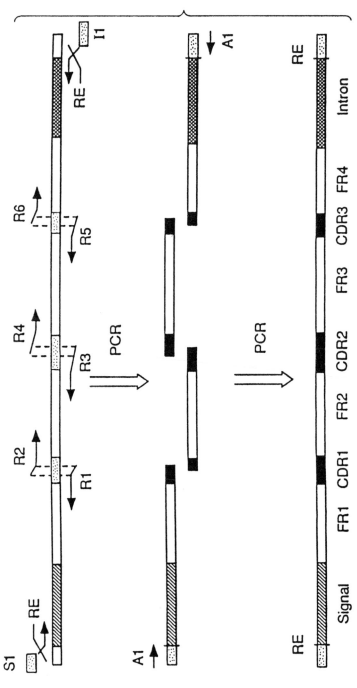

Figure 1. PCR amplification and recombination for the construction of CDR-grafted V-regions: Rei light chain grafting using short oligodeoxynucleotides.

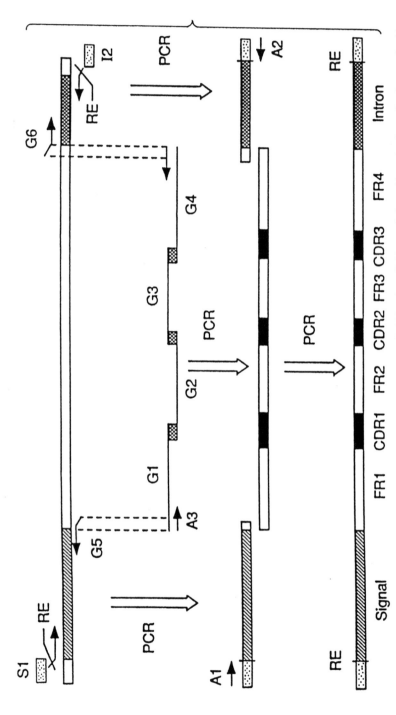

Figure 2. PCR amplification and recombination for the construction of CDR-grafted V-regions: Gal heavy chain grafting using long oligodeoxynucleotides.

Heavy Chain

```
NEW: QVQLQESGPGLVRPSQTLSLTCTVSGFTFS  [NDYYT]  WVRQPP
IB4: DVKLVESGGDLVKLGGSLKLSCAASGETES  [DYYMS]  WVRQTP
Jon: DVQLVESGGGLVKPGGSLRLSCAASGFTFS  [TAWMK]  WVRQAP
Gal: EVQLVESGGDLVQPGRSLRLSCAASGFTFS  [BLGMT]  WVRQAP
mGal:                             G
```

```
GRGLEWIG  [YVFYHGTSDDTTPLRS-]      RFTMLVDTSKNQFSLRL
EKRLELVA  [AIDNDGGSISYPDTVKG]      RFTISRDNAKNTLYLQM
GKGLEWVV  [WRVEQVVEKAFANSVNG]      RFTISRNDSKNTLYLQM
GKGLEWVA  [NIKZBGSZZBYVDSVKG]      RFTISRDNAKNSLYLQM
   L
```

```
                                                        %Id
SSVTAADTAVYYCAR  [---NLIAGCIDV]WGQGSLVTVSS...             55
SSLRSEDTALYYCAR  [-QGRLRRDYFDY]WGQGTTLTVSS...
ISVTPEDTAVYYCAR  [VPLYGBYRAFNY]WGQGTPVTVSS...             78
NSLRVEDTALYYCAR  [-----GWGGGD-]WGQGTLVTVST...             82
                                         L               85
```

Light Chain

```
REI: DIQLTQSPSSLSASVGDRVTITC  [RASGNIHNYLA------]WY
IB4: DIVLTQSPASLAVSLGQRATISC  [RASESVDSYGNSFMH--]WY
Len: DIVMTQSPNSLAVSLGERATINC  [KSSQSVLYSSNSKNYLA]WY
```

```
QQKPGKAPKLLIY  [YTTTLAD]  GVPSRFSGSGSGTDFTFTISSL
QQKPGQPPKLLIY  [RASNLES]  GIPARFSGSGSRTDETLTINPV
QQKPGQPPKLLIY  [WASTRES]  GVPDRFSGSGSGTDFTLTISSL
```

```
                                                %Id
QPEDIATYYC  [QHFWSTPRT]  FGQGTKVVIKR...          69
EADDVATYYC  [QQSNEDPLT]  FGAGTKLELKR...
QAEDVAVYYC  [QQYYSTPYS]  FGQGTKLEIKR...          81
```

%Id: percent identity to IB4 FRs
Packing residues are underlined

Figure 3. Sequence comparison of heavy and light chain frameworks.

chain framework, and "Rei" and "Len" represented the murine light chain framework (Fig. 3). This approach goes beyond that of Reichmann *et al.*, (1988), since any and all V-regions within current databases are potential recipients for a set of murine CDRs. While this approach is similar to that of Queen *et al.*, (1989), it differs in two important ways. First, any combination of human heavy and light chain frameworks may be chosen to receive the murine CDRs, even if they have never existed as a heterodimer. Second, it is not necessary to undertake the molecular model construction of the antibodies involved, since choices are based on primary sequence information and identification of the relevant interactive residues. We thus selected the human homologues that provided the murine CDRs with the structural support most similar to their native murine framework for the subsequent construction of the humanized variable regions.

2.2. Humanization of mIB4

This conversion was separated into three steps in order to systematically evaluate the process by which the murine IB4 MAb was humanized. First, a CDR-grafted version of the human light chain was expressed with a chimeric version of the mIB4 heavy chain, in order to determine the relative importance of the light chain variable region. Second, this grafted light chain was coexpressed with CDR-grafted heavy chain variable regions derived from several human frameworks. This allowed us to select the best scaffolding for the murine CDRs, and assisted us in understanding the variable region elements that contributed to the successful transfer of CDRs. Third, mutagenesis of the FRs was undertaken to clarify the role of these residues in the positioning of the transposed CDRs. It should be noted that although we used molecular modeling to assess the consequences of several mutational possibilities, the humanization procedures described here do not rely on structural modeling to choose suitable human frameworks for CDR-grafting.

The CDR-grafted human light chain V-region was constructed from a cDNA encoding the Rei light chain framework (Reichmann *et al.*, 1988) by substituting the mIB4 CDR sequences for its resident sequences. We used PCR amplification of this cDNA template with primers incorporating the desired murine CDR sequences. The DNA sequence of the CDR-grafted V-region was confirmed before it was inserted into an adenovirus major late promoter driven expression vector that contained the human kappa constant region. A CDR-grafted version using the human Len light chain FR was constructed from long oligodeoxynucleotides and placed into an identical expression vector.

We obtained the murine heavy chain V-region required for constructing a chimeric murine/human gamma-4 heavy chain through PCR amplification of the murine FR1 to FR4 sequences, followed by the PCR-mediated attachment to this product of DNA fragments containing the signal peptide (including its intron) and 3' J–C intronic sequences (Fig. 4). This V-region was inserted into the expression vector containing the coding region of the human gamma-4 constant domain. In order to obtain enough recombinantly expressed antibody to measure its avidity, monkey kidney cells (CV1P) were cotransfected with heavy and light chain vectors, and the protein secreted into serum-free medium was collected. The hemichimeric grafted antibody (chimeric heavy chain and grafted light chain) was assayed for its ability to compete with [^{125}I]mIB4 for the CD18

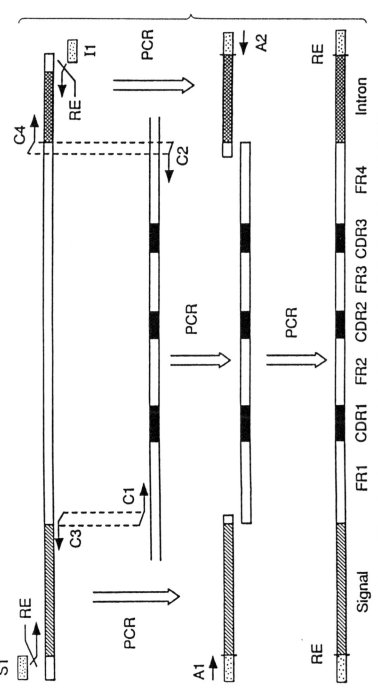

Figure 4. PCR amplification and recombination for the construction of a hemichimeric heavy chain V-region.

Table I. Summary of Competitive Binding Activity of Murine IB4 and
Recombinant Human IB4 Antibodies

Heavy chain	Light chain	IC_{50} (nM)
Murine IB4	Murine IB4	0.52 ± 0.20
Murine IB4	Grafted IB4/Rei	0.46 ± 0.08
Veneered mIB4	Veneered mIB4	0.39 ± 0.07
Grafted IB4/mGal	Grafted IB4/Rei	0.67 ± 0.08
Grafted IB4/Gal	Grafted IB4/Rei	1.68 ± 0.26
Grafted IB4/Gal	Grafted IB4/Len	2.80 ± 1.04
Grafted IB4/Jon	Grafted IB4/Rei	5.88 ± 0.13
Grafted IB4/New	Grafted IB4/Rei	7.99 ± 0.73

receptors on activated polymorphonuclear (PMN) leukocytes (Table I). There was no measurable loss in avidity when we grafted the murine CDRs onto the human framework Rei.

The human heavy chain frameworks were chosen in accordance with the hypothesis stated above, and compared with the "New" framework preferred by several investigators (Jones et al., 1986; Verhoeyen et al., 1988; Reichmann et al., 1988; Tempest et al., 1991; Gorman et al., 1991; Güssow and Seemann, 1991). The CDR-grafted versions of the Gal and Jon heavy chain domains were constructed using long oligodeoxynucleotides and PCR amplification (Daugherty et al., 1991; DeMartino et al., 1991). The murine CDRs of IB4 were inserted into the human New framework by using a cDNA encoding this V-region (Reichmann et al., 1988) and the same PCR procedures described for the grafted light chain construction. The sequence-verified heavy chain variable regions were inserted into the heavy chain expression vector in place of the chimeric V-region. These plasmids were cotransfected into CV1P cells with the CDR-grafted Rei kappa light chain vector, and the secreted antibody was purified from the serum-free conditioned medium. Competitive binding curves were again used to evaluate the avidities of the various antibodies for the CD18 ligand on the activated PMNs (Table I). Although each heterodimeric antibody contains the same six CDRs, they do not exhibit the same avidity for the CD18 ligand. Thus, some of the biological properties of an antibody molecule (i.e., its avidity) rely significantly on the variable region framework structure supporting the CDR loops. In the case of this murine MAb, the choice of light chain framework appeared to be less critical than the choice of the heavy chain structure. The human Gal framework was the best of those chosen for humanization, and resulted in the synthesis of a fully grafted antibody whose avidity was not markedly diminished from that of the murine IB4 MAb. The Jon and New versions of these CDRs created antibodies that might be, respectively, marginally useful and completely useless for human therapy.

Although the Len light chain V-region framework sequences show more identical and similar residues when aligned to the murine IB4 framework than do the Rei light chain frameworks, this has little, if any, impact on the measured antibody/antigen interactions. A comparison of the presumed three-dimensional structure of these two light chain V-regions shows that the alpha carbon trace of the IB4 CDRs residing within these FRs are superimposable, again suggesting that both FRs identically support these CDRs.

3. Antibody Reshaping

Can individual FR residues be altered to increase the avidity of the humanized antibody for its ligand? We noted mismatches between the Gal and murine FRs by comparing the murine and human heavy chain FR packing residues. We made a mutated form of the CDR-grafted Gal variable region by placing the altered codons in the oligodeoxynucleotide primers used to amplify and recombine DNA fragments generated using the grafted Gal DNA as template. When the avidity of the recombinant antibody secreted by CV1P cells transformed by mutant Gal (mGal) and Rei expression vectors was measured by competitive binding, it was found to be equivalent to that of the parent murine antibody (Table I). This improvement in binding revealed that subtle changes in FR packing residues may substantially affect the way that CDRs are displayed within the antigen binding site.

Mammalian cell clones may be grown in numerous types of bioreactors. Very large (more than 500-liter fermentors) manufacturing can most reliably be accomplished using a suspension cell process. This requires the development of mammalian cell expressor clones capable of anchorage-independent growth, narrowing the choice to either NS/0 or CHO cells as the ultimate source of recombinant protein. Cell clones expressing humanized IB4 have been identified from both cell types, and have been compared for suitability of scaleup.

4. Manufacturing Recombinant Antibodies

4.1. Selection of Cell Line for Manufacturing

Whether they are of human derivation or humanized by CDR-grafting or veneering, recombinant MAbs require very efficient manufacturing processes to offset the combination of their anticipated high efficacious dose (1–10 mg/kg) and the relatively expensive mammalian cell fermentation process. Numerous suspension-adaptable host cells have been investigated in order to find one that is acceptable to regulatory agencies and that will produce the largest amount of antibody in serum-free medium (Page and Sydenham, 1991). There are two approaches to expressing and manufacturing antibodies for human use. Murine cell lines (hybridomas, myelomas, or heterohybridomas) have been the overwhelming choice for many of the antibodies currently in clinical trials. This, however, may reflect the availability of the producer cell line (i.e., it may be a university-derived murine monoclonal or human heterohybridoma), or the lack of commitment (or skill) of the manufacturer in creating a recombinant antibody.

It is only recently that the pharmaceutical industry has undertaken the manufacturing of recombinant proteins. The precedent for the production and licensure of mammalian cell-derived biologicals would suggest the use of Chinese hamster ovary (CHO) cells for the manufacturing of recombinant antibodies. We compared the utility of CHO cells and NS/0 (murine myeloma) cells during the development of the IB4 humanized recombinant antibody. Both cell lines were employed to obtain numerous expression clones, which were subsequently evaluated in laboratory-scale bioreactors (20 liters) for process and product characteristics. The NS/0 myeloma and CHO cell lines were compared in conjunction with the glutamine synthetase (GS) amplification system (Cockett *et al.*, 1990; Bebbington *et al.*, 1992), obtained from Celltech, Ltd. Previous experiments with CHO cells and DHFR-mediated amplification all resulted in productivity levels that were economically unacceptable for large-scale manufacturing. The CDR-grafted versions of the Gal heavy chain gamma-4 constant region and the Rei light chain kappa constant region were both inserted, behind hCMVIE promoters, into a plasmid vector expressing GS. Although it is considered preferable to place the light chain encoding gene before that of the heavy chain, we chose to construct this expression vector with the opposite orientation.

Multiple electroporations of the two cell types were performed, and several hundred clones were isolated and characterized (Table II). Only 3.2–1.3% of the CHO clones that grew out in selective medium were sufficiently productive to warrant subsequent amplification. Two versions of the GS cistron were evaluated; one contained intervening sequences (minigene), and the other was composed of a completely spliced gene (cDNA). Cell lines selected with either GS cDNA or mDHFR (mutant DHFR) demonstrated equivalent frequency and pro-

Table II. Summary of Stable Cell Lines

Cell	Promoter	Marker	No. of clones examined	Expresssion (pg/cell/day)	
				Initial	Amplified
CHO	HIV-LTR	GS cDNA	4/135	0.5–3.0	5–12
CHO	CMVIE	GS cDNA	4/125	0.5–1.5	3–6
		GS minigene	4/298	0.1–0.5	None
		mDHFR	3/205	0.5–1.5	5–9
NS/0	CMVIE	GS cDNA	7/15	4–12	20–70

ductivity characteristics, while none of the GS minigene-derived cell lines amplified sufficiently to justify further manipulations. In contrast, 7 of the initial 15 selected NS/0 cell lines produced as much or more recombinant antibody as the amplified CHO cell lines. Amplification of the best of these clones resulted in cell lines whose productivity measured between 20 and 70 pg/cell per day. The best CHO and NS/0 amplified cell lines were chosen for adaptation to serum-free suspension growth and fed-batch process development. We compared CHO and NS/0 processes and characterized the secreted antibodies in order to choose the eventual manufacturing process (summarized in Table III).

NS/0 and CHO cell lines demonstrated essentially equal growth rates in serum-containing medium prior to their adaptation to a serum-free suspension process. NS/0 and CHO cells differed in their ease of initial scalability and fermentation. The former cells were readily adaptable to serum-free growth (usually requiring 2 weeks) and growth in suspension (4 weeks from shake flask to Bellco spinner). CHO cells required several weeks (3–5) to adapt their growth from anchorage-dependent to suspension, and an equivalent time to adapt to serum-free suspension growth. Once both cell types were ready for fermentation, however, no differences were observed in their ability to be scaled from laboratory scale to pilot plant scale (200 liters). Following adaptation, NS/0 cell lines were found to be capable of reaching a peak cell density twice that of the CHO cell line, and simple batch fermentation processes typically yielded two to three times more antibody from the NS/0 culture.

The glutamine synthetase selection system was used with both cell lines. One would expect the CHO cell process to be more genetically unstable, since removing the selective drug removes all selective pressure from the cell/plasmid

Table III. Comparison of CHO and NS/0 as Hosts for the Production of Humanized Antibodies

	NS/0	CHO
Scalability	No difficulties	No difficulties
Also secretes	Few other proteins	Many other proteins
Expected adaptation to protein-free medium	Yes	Probably not
Productivity	300 mg/liter	100 mg/liter
Factor of improvement expected	4×	3×
Expression amplification system	Glutamine synthetase (GS)	Glutamine synthetase (GS) mDHFR
Vector stability	GS—very stable	GS—unstable mDHRF—unstable
Glycoforms	Complex, biantennary	Complex, biantennary
Retroviruses	Yes; A- & C-Type; easily identified & quantitated	Yes; mostly A-type; identification & quantitation difficult

interaction, and genetic instability usually leads to a significant loss in expression levels of the culture. As a result, the selective drug methionine sulfoximine (MSX) must be included in all of the cell expansion steps preceding the actual fermentation run. Not unexpectedly, the CHO cell clones were found to require MSX throughout their expansion and fermentation. A similar requirement was found when mDHFR amplification was used in place of the GS system. In contrast, there are some NS/0 cell clones whose productivity is not affected by the removal of the selective drug MSX.

One would expect small differences in the glycoforms attached to the recombinant antibody secreted by the two cell types. Humanized IB4 was purified from NS/0 and CHO suspension cell cultures, and N-glycan mapping data were obtained by using high-pH anion-exchange chromatography coupled with pulsed amperometric detection after digestion with N-glycanase. Similar patterns of complex, biantennary galactosylated and fucosylated glycans were found, with more terminally sialylated and fewer high-mannose-type structures present on the CHO-derived MAb. It is difficult to say whether the observed glycoforms of one are more "natural" than the other, and the consequences that may arise from these small differences are unpredictable.

Murine retroviruses, in isolation, do not appear to pose a risk to human health. There may be some small risk associated with the potential for recombination between endogenous retroviruses in human/mouse heterohybridomas. It is still necessary to characterize, inactivate, and remove any and all retroviruses from the bulk MAb product. Viral clearance studies are most reliably performed using the retrovirus produced by the manufacturing cell line. CHO and NS/0 cell lines each express endogenous retrovirus(es) (Anderson *et al.*, 1991; Leuders, 1991). Electron micrographs of CHO cells expressing the recombinant humanized MAb have demonstrated the presence of intracellular A-type retrovirus(es). To date, this variety of retrovirus has not been found to be infectious. The nonreplicative aspect of this contaminant complicates the eventual process validation, since the contaminating entity is not available (cannot be grown) and thus cannot be used to spike the various purification steps in order to demonstrate its clearance or inactivation. The same electron micrograph assay, as well as host range and sero-neutralization assays, have been applied to NS/0 cell lines producing the identical recombinant MAb. As expected, both A- and C-type retrovirus(es) were observed, and eventual assays indicated the C-type titers to be very low. (Amplification of the viruses by serial passage was required prior to their quantitation by "plaque" assays.)

4.2. NS/0 Process Development

As a result of the above considerations, NS/0 cell lines have clearly been favored for the manufacturing of recombinant antibodies. Following the adaptation of eight clones expressing humanized IB4 to serum-free suspension growth in shake flasks, the most productive clone (43 pg/cell per day with a doubling time of 28–43 hr) was subjected to fed-batch process development. Initially, the relatively low peak cell densities of $1.1–1.3 \times 10^6$ cells/ml obtained within 4–5 days postseeding were followed by rapid cell death, and yielded only 150–180 mg/liter of recombinant antibody. An analysis of spent medium revealed that only low amounts of glucose were consumed, and no ammonia had accumulated (undoubtedly the consequence of the ammonia scavenging nature of the expressed glutamine synthetase gene). Quantitation of the amino acid content revealed the almost complete depletion of glutamine, asparagine, and aspartic acid, and significant utilization of half of the remaining amino acid constituents. The basal medium was enriched for the metabolized amino acids, and subsequent fermentations were supplemented with cocktails of the depleted amino acids once a cell density of 1×10^6 cells/ml was reached. The final cell density increased twofold, and the yields rose proportionally. An iterative process involving amino acid supplementations and medium depletion analyses at various times during the fermentation led to an optimized fed-batch process: viability could be maintained for 16–18 days and yields typically reached 600–700 mg/liter. Further refinement of this protocol, including the addition of lipoproteins, resulted in recombinant antibody yields of approximately 1 g/liter before the cell viability dropped below 30% (Fig. 5). These exceptionally high fed-batch productivities are not unique to this antibody clone: similar accumulations have been reached early in the process development of another therapeutic antibody.

5. Veneering

The veneering approach to the humanization procedure has the potential to simultaneously reduce the immunogenicity of the rodent MAb while preserving its ligand binding properties in their entirety. Since the immunogenicity of a protein is primarily dependent on its displayed surface, the reaction to a murine antibody could be reduced by replacing the exposed residues, which differ from those usually found in human antibodies. This judicious replacement of exterior residues should have little or no effect on the interior domains, or on the interdomain contacts. Thus, ligand binding properties should be unaffected by alterations limited to the surface-exposed variable region framework residues. We refer to this procedure of humanization as "veneering," since only the skin of the antibody is altered, while the supporting residues remain undisturbed.

There are two steps in the process of veneering. First, the framework of the mouse variable domains is compared with the human variable region database. The most homologous human variable regions are identified and subsequently compared residue for residue with the corresponding murine regions. Second, the residues in the mouse framework that differ from its human homologue are replaced by the residues present in the human homologue. This switching occurs

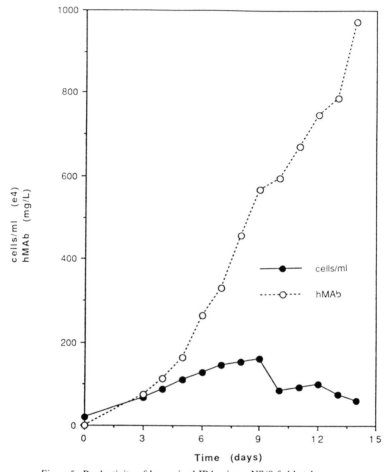

Figure 5. Productivity of humanized IB4 using a NS/0 fed-batch process.

only with those residues that are at least partially exposed (Padlan, 1991). The following are retained in the veneered mouse antibody: CDRs, residues adjoining the CDRs, residues defined as completely or mostly buried, and residues believed to be involved with interdomain contacts (Padlan, 1991). Attention is also paid to the N-termini of the heavy and light chains, since they are often contiguous with the CDR surface and are in a position to be involved in ligand binding. Care should likewise be exercised in the placement of proline, glycine, and charged amino acids, since they may have significant effects on tertiary structure and electrostatic interactions of the variable region domains.

We determined an appropriate human framework by using the criteria discussed above. In practice, we found that the human light chain variable region framework with significant homology to the mIB4 framework was the human Len framework (with a similarity of 90% and an identity of 81%). In order to demonstrate the veneering process, we used an IB4 CDR-grafted version of the Len light chain variable region as the template into which mutations were placed to easily create the veneered framework sequence. Specific amino acid residues within the human Len framework were replaced with residues found in the murine IB4 framework, so that the final light chain V-region appeared as it would if the murine V-region were the starting material for the veneering process (Fig. 6). The veneered heavy chain portion of the recombinant antibody was

Heavy Chain

```
                         √                              √
vIB4: DVKLVESGGDLVKPGGSLKLSCAASGFTFS [DYYMS] WVRQAP
mIB4: DVKLVESGGDLVKLGGSLKLSCAASGFTFS [DYYMS] WVRQTP
 Gal: EVQLVESGGDLVQPGRSLRLSCAASGFTFS [BLGMT] WVRQAP


√ √                                        √
GKGLELVA [AIDNDGGSISYPDTVKG] RFTISRDNSKNTLYLQM
EKRLELVA [AIDNDGGSISYPDTVKG] RFTISRDNAKNTLYLQM
GKGLEWVA [NIKZBGSZZBYVDSVKG] RFTISRDNAKNSLYLQM


√    √                              √
NSLRAEDTALYYCAR [-QGRLRRDYFDY] WGQGTLLTVSS...
SSLRSEDTALYYCAR [-QGRLRRDYFDY] WGQGTTLTVSS...
NSLRVEDTALYYCAR [-----GWGGGD-] WGQGTLVTVST...
```

Light Chain

```
              √    √    √
vIB4: DIVMTQSSNSLAVSLGERATISC [RASESVDSYGNSFMH--] WY
mIB4: DIVLTQSPASLAVSLGQRATISC [RASESVDSYGNSFMH--] WY
 Len: DIVMTQSSNSLAVSLGERATINC [KSSQSVLYSSNSKNYLA] WY


                    √          √          √√
QQKPGQPPKLLIY [RASNLES] GIPDRFSGSGSGTDFTLTISSV
QQKPGQPPKLLIY [RASNLES] GIPARFSGSGSRTDFTLTINPV
QQKPGQPPKLLIY [WASTRES] GVPDRFSGSGSGTDFTLTISSL


√                      √    √
EAEDVATYYC [QQSNEDPLT] FGQGTKLEIKR...
EADDVATYYC [QQSNEDPLT] FGAGTKLELKR...
QAEDVAVYYC [QQYYSTPYS] FGQGTKLEIKR...
```

Figure 6. Heavy and light chain V-region sequence alterations performed to create veneered surfaces.

derived by mutating the murine IB4 heavy chain variable region so that it contained only human surface-exposed residues. In this case, the human Gal framework was used as the template for surfacer residue comparisons (see Fig. 6). The figure shows the sequences of the veneered heavy and light chain variable regions (annotated vIB4). The locations of the point mutations placed to convert an exposed murine residue to a human-appearing residue are indicated with a check mark ($\sqrt{}$). In most instances, the corresponding residues in the human templates (Gal or Len) are used to substitute for undesired murine residues at analogous positions. The two exceptions are residues 74 and 84 in the heavy chain, where a more preferred amino acid is used. The plasmids encoding the IB4 veneered IgG4 heavy chain and the IB4 veneered kappa light chain were cotransfected, as before, into the monkey kidney cell line CV1P or the human embryonic kidney cell line 293. The veneered recombinant antibody secreted into the culture supernatants was purified by protein A chromatography. Avidities of the anti-CD18 antibodies were determined using a competitive [^{125}I]mIB4 soluble binding assay with stimulated human PMNs. The results of the binding assays indicated that the avidity of the veneered recombinant IB4 antibody was equal to that of the murine IB4 MAb (Table I). This suggests that an antibody with presumptive human allotype may be recombinantly constructed from the murine MAb by introducing numerous point mutations into its framework residues, and then expressing the V-regions fused to human kappa and gamma-4 constant domains, without a loss in avidity for the antigen. It can thus be inferred that the point mutations within the framework regions do not alter the presentation of the murine IB4 light chain and heavy chain CDRs. In comparison, we and others have found that CDR-grafting of murine sequences into human frameworks invariably leads to loss of avidity for the ligand or antigen (Reichmann et al., 1988; Queen et al., 1989; Co et al., 1991). Thus, although these latter transmutations are possible using CDR-grafting, the successful maintenance of avidity is not assured.

6. Immunogenicity

Irrespective of the skill and approach of the molecular alchemist, success will ultimately require the metamorphosis of the original murine antibody into one that is immunologically acceptable to the human recipient. Ideally, the demonstration of this outcome would be evaluated in a preclinical model (Hakimi et al., 1991). Human MAbs have the same pharmacokinetics in rhesus monkeys as they do in humans (Ehrlich et al., 1987; Jonker et al., 1991). These monkeys received repeated doses of human MAbs, which were well tolerated and rarely resulted in immune recognition. Groups of three rhesus monkeys were injected, at weekly intervals, for 5 weeks with 1 mg of MAb (either murine,

CDR-grafted, hemichimeric, or veneered) per kilogram body weight. At various times following each injection, the level of circulating MAb and the development of anti-MAb antibodies were assayed by ELISAs. The IB4 MAbs all bound their CD18 target on rhesus PMNs, and serum half lives were initially about 3 to 4 hr. This value predominantly reflects the normal rapid turnover rate of the PMN population. Differences between the humanized versions of IB4 and its murine predecessor were reflected by the third dose of antibody, when two of the three monkeys receiving the murine antibody displayed moderate anaphylactic symptoms. This response was never seen in animals treated with the other forms of the IB4 MAb during the 6 weeks of this study. Distinguishing differences between the recombinant IB4 MAbs were most evident following their fourth dose, when peak plasma levels for the CDR-grafted MAbs (hemichimeric and fully CDR-grafted) were significantly reduced relative to the veneered MAb. This trend continued for the remainder of the observation period (Fig. 7), and could be attributed to the progressively higher levels of anti-IB4 antibodies in these animals. These findings suggest that a veneering approach to humanization may result not only in recombinant antibodies that retain all of their affinity and potency, but also in antibodies that are less immunogenic than those humanized by CDR-grafting and reshaping procedures.

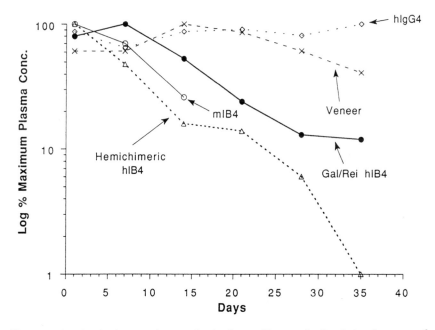

Figure 7. Alteration in rhesus peak serum levels of recombinant antibodies during the course of weekly dosing at 1 mg/kg.

7. Conclusion

Antibodies, by virtue of their exquisite specificity and high potency, have long been considered the "magic bullet" of tomorrow. They have tremendous potential in the management of immune responsiveness, the detection and treatment of cancer, and the prophylaxis and treatment of viral and bacterial infections. While their long biologic half-life suggests that MAbs would be well tolerated for chronic therapy, the specter of possible anti-antibody responses has restricted their use solely to acute indications. The recent advent of the combinatorial library approach to identifying and constructing human MAbs, as well as the development of transgenic and reconstituted mice with the capability of producing a human antibody in response to an antigen, will supersede the need to reshape rodent MAbs. Until these technologies are well proven, however, human therapy will be the provenance of the humanized antibody. Methodologies of humanization have included CDR-grafting and veneering, which both await critical evaluation in the clinic. Common to all therapeutic MAbs is the need for an economical manufacturing process. Our experience, although preliminary in scale, indicates that we may soon be able to produce 1 to 2 g/liter of recombinant antibody employing inexpensive, low-protein, serum-free medium. Two hurdles remain: the identification of efficacious MAbs, and regulatory recognition of their benefit and benignity.

References

Amit, A. G., Mariuzza, R. A., Phillips, S. E. V., and Poljak, R. J., 1986, Three-dimensional structure of an antigen–antibody complex at 2.8 Å resolution, *Science* **233**:747–753.

Anderson, K. P., Low, M.-A., L. Lie, Y. S., Keller G.-A., and Dinowitz, M., 1991, Endogenous origin of defective retrovirus like particles from a recombinant Chinese hamster ovary cell line, *Virology* **181**:305–311.

Bebbington, C. R., Renner, S., Thompson, S., King, D., Abrams, D., and Yarranton, G. T., 1992, High-level expression of a recombinant antibody forms myeloma cells using a glutamine synthetase gene as an amplifiable selectable marker, *Bio/Technology* **10**: 169–175.

Blend, M. J., 1991, Current status of tumor imaging with monoclonal antibodies, *Compre. Ther.* **17**:5–11.

Boulianne, G. L., Nobumichi, H., and Shulman, M. J., 1984, Production of functional chimaeric mouse/human antibody, *Nature* **312**:643–646.

Bruccoleri, R. E., Haber, E., and Novotny, J., 1988, Structure of antibody hypervariable loops reproduced by a conformational search algorithm, *Nature* **335**:564–568.

Chothia, C., Lesk, A. M., Tramontano, A., Levitt, M., Smith-Gill, S. J., Air, G., Sheriff, S., Padlan, E. A., Davies, D., Tulip, W. R., Colman, P. M., Spinelli, S., Alzari, P. M., and Poljak, R. J., 1989, Conformation of immunoglobulin hypervariable regions, *Nature* **342**:877–883.

Co, M. S., Deschamps, M., Whitley, R. J., and Queen, C., 1991, Humanized antibodies for antiviral therapy, *Proc Natl. Acad. Sci. USA* **88**:2869–2873.

Cockett, M. I., Bebbington, C. R., and Yarranton, G. T., 1990, High level expression of tissue inhibitor of metalloproteinases in Chinese hamster ovary cells using glutamine synthetase gene amplification, *Bio/Technology* **8**:662–667.

Daugherty, B. L., DeMartino, J. A., Law, M.-F., Kawka, D. W., Singer, I. I., and Mark, G. E., 1991, Polymerase chain reaction facilitates the cloning, CDR-grafting, and rapid expression of a murine monoclonal antibody directed against the CD18 component of leukocyte integrins, *Nucleic Acids Res.* **19**:2471–2476.

Davies, D. R., Padlan, E. A., and Sheriff, S., 1990, Antibody–antigen complexes, *Annu. Rev. Biochem.* **59**:439–473.

DeMartino, J. A., Daugherty, B. L., Law, M.-F., Cuca, G. C., Alves, K., Silberklang, M., and Mark, G. E., 1991, Rapid humanization and expression of murine monoclonal antibodies, *Antibody Immunoconjugates Radiopharm.* **4**:829–835.

Devereux, J., Haeberli, P., and Smithies, O., 1984, A comprehensive set of sequence analysis programs for the VAX, *Nucleic Acids Res.* **12**:387–395.

Eary, J. F., 1991, Fundamentals of radioimmunotherapy, *Nucl. Med. Biol.* **18**(1):105–108.

Ehrlich, P. H., Harfeldt, K. E., Justice, J. C., Moustafa, Z. A., and Ostberg, L., 1987, Rhesus monkey responses to multiple injections of human monoclonal antibodies, *Hybridoma* **6**:151–160.

George, D. G., Barker, W. C., and Hunt, L. T., 1986, The protein identification resource (PIR), *Nucleic Acids Res.* **14**:11–16.

Gomoll, A. W., Lekich, R. F., and Grove, R. I., 1991, Efficacy of a monoclonal antibody (MoAb 60.3) in reducing myocardial injury resulting from ischemia/reperfusion in the ferret, *J. Cardiovasc. Pharmacol.* **17**:873–878.

Gorelick, K., Scannon, P. J., Hannigan, J., Wedel, N., and Ackerman, S. K., 1990, Randomized placebo-controlled study of E5 monoclonal antiendotoxin antibody, in: *Therapeutic Monoclonal Antibodies* (C. A. Borrebaeck and J. W. Larrick eds), Stockton Press, New York, pp. 253–261.

Gorman, S. D., Clark, M. R., Routledge, E. G., Cobbold, S. P., and Waldmann, H., 1991, Reshaping a therapeutic CD4 antibody, *Proc. Natl. Acad. Sci. USA* **88**:4181–4185.

Güssow, D., and Seemann, G., 1991, Humanization of monoclonal antibodies, *Methods Enzymol.* **203**:99–121.

Hakimi, J., Chizzonite, R., Luke, D. R., Familletti, P. C., Bailon, P., Kondas, J. A., Pilson, R. S., Lin, P., Weber, D. V., Spence, C., Mondini, S. J., Tsien, W.-H., Levin, J. L., Gallati, V. H., Korn, L. Waldmann, T. A., Queen, C., and Benjamin, W., 1991, Reduced immunogenicity and improved pharmacokinetics of humanized anti-Tac in cynomolgus monkeys, *J. Immunol.* **147**:1352–1359.

Hale, G., Clark, M. R., Marcus, R., Winter, G., Dyer, M. J. S., Phillips, J. M., Reichmann, L., and Waldmann, H., 1988, Remission induction in non-Hodgkin lymphoma with reshaped human monoclonal antibody campath-1H, *Lancet* **2**:1394–1399.

Jones, P. T., Dear, P. H., Foote, J., Neuberger, M. S., and Winter, G., 1986, Replacing the complementarity-determining regions in a human antibody with those from a mouse, *Nature* **321**:522–525.

Jonker, M., Schellekens, P. T., Harpprecht, J., and Slingerland, W., 1991, Complications of monoclonal antibody (MAb) therapy: The importance of primate studies, *Transplant. Proc.* **23**(1):264–265.

Kabat, E. A., Wu, T. T., Reid-Miller, M., Perry, H. M., and Gottesman, K. S., 1987, *Sequences of Proteins of Immunological Interest*, U.S. Department of Health and Human Services, Bethesda.

Kettleborough, C. A., Saldanha, J., Heath, V. J., Morrison, C. J., and Bendig, M. M., 1991, Humanization of a mouse monoclonal antibody by CDR-grafting: The importance of framework residues on loop conformation, *Protein Eng.* **4**(7):773–783.

Larson, S. M., 1991, Radioimmunology: Imaging and therapy, *Cancer* **67**:1253–1260.

LoBuglio, A. F., Wheeler, R. H., Trang, J., Haynes, A., Rogers, K., Harvey, E. B., Sun, L., Ghrayeb, J., and Khazaeli, M. B., 1989, Mouse/human chimeric monoclonal antibody in man: Kinetics and immune response, *Proc. Natl. Acad. Sci. USA* **86**:4220–4224.

Lueders, K. K., 1991, Genomic organization and expression of endogenous retrovirus-like elements in cultured rodent cells, *Biologicals* **19**:1–7.

Morrison, S. L., Johnson, M. J., Herzenberg, L. A., and Oi, V. T., 1984, Chimeric human antibody molecules: Mouse antigen-binding domains with human constant region domains, *Proc. Natl. Acad. Sci. USA* **81**:6851–6855.

Nakamaye, K. L., and Eckstein, F., 1986, Inhibition of restriction endonuclease Nci I cleavage by phosphorothioate groups and its application to oligonucleotide-directed mutagenesis, *Nucleic Acids Res.* **14**:9679–9698.

Neuberger, M. S., Williams, G. T., Mitchell, E. B., Jouhal, S. S., Flanagan, J. G., and Rabbits, T. H., 1985, A hapten-specific chimaeric IgE antibody with human physiological effector function, *Nature* **314**:268–270.

Padlan, E. A., 1991, A possible procedure for reducing the immunogenicity of antibody variable domains while preserving their ligand-binding properties, *Mol. Immunol.* **28**: 489–498.

Page, M. J., and Sydenham, M. A., 1991, High level expression of the humanized monoclonal antibody campath-1H in Chinese hamster ovary cells, *Bio/Technology* **9**:64–68.

Queen, C., Schneider, W. P., Selick, H. E., Payne, P. W., Landolfi, N. F., Duncan, J. F., Avdalovic, N. M., Levitt, M., Junghans, R. P., and Waldmann, T. A., 1989, A humanized antibody that binds to the interleukin 2 receptor, *Proc. Natl. Acad. Sci. USA* **86**:10029–10033.

Reichmann, L., Clark, M., Waldmann, H., and Winter, G., 1988, Reshaping human antibodies for therapy, *Nature,* **332**:323–327.

Risler, J. L., Delorme, M. O., Delacroix, H. and Henaut, A., 1988, Amino acid substitutions in structurally related proteins, a pattern recognition approach: Determination of a new and efficient scoring matrix, *J. Mol. Biol.* **204**:1019–1029.

Schwartz, R. M., and Dayhoff, M. O., 1979, in: *Atlas of Protein Sequence and Structure* (M. O. Dayhoff, ed.), National Biomedical Research Foundation, Washington, D.C.

Seemann, G., Bosslet, K., and Sedlacek, H.-H., 1990, Recombinant monoclonal antibodies in tumor therapy, *Behring Inst. Mitt.* **87**:33–47.

Taylor, J. W., Ott, J., and Eckstein, F., 1985, The rapid generation of oligonucleotide-directed mutations at high frequency using phosphorothioate-modified DNA, *Nucleic Acids Res.* **13**:8765–8785.

Tempest, P. R., Bremner, P., Lambert, M., Taylor, G., Furze, J. M., Carr, F. J., and Harris, W. J., 1991, Reshaping a human monoclonal antibody to inhibit human respiritory syncytial virus infection in vivo, *Bio/Technology* **9**:266–271.

Tjandra, J. J., Ramadi, L., and McKenzie, I. F. C., 1990, Development of human antimurine antibody (HAMA) response in patients, *Immunol. Cell Biol.* **68**:367–376.

Verhoeyen, M., Milstein, C., and Winter, G., 1988, Reshaping human antibodies: Grafting an antilysozyme activity, *Science* **239**:1534–1536.

Waldmann, T. A., 1991, Monoclonal antibodies in diagnosis and therapy, *Science,* **252**:1657–1662.

Ziegler, E. J., Fisher, C. J., Sprung, C. L., Straube, R. C., Sadoff, J. C., Foulke, G. E., Wortel, C. H., Fink, M. P., Dellinger, R. P., Teng, N. N. H., Allen, I. E., Berger, H. J., Knatterud, G. L., LoBuglio, A. F., Smith, C. R., and the HA-1A Sepsis Study Group, 1991, Treatment of gram-negative bacteremia and septic shock with HA-1A human monoclonal antibody against endotoxin. A randomized, double-blind, placebo-controlled trial, *N. Engl. J. Med.* **324**:429–436.

Index